Energy Use Efficiency

Energy Use Efficiency

Editor

Almas Heshmati

MDPI • Basel • Beijing • Wuhan • Barcelona • Belgrade • Manchester • Tokyo • Cluj • Tianjin

Editor
Almas Heshmati
Jönköping University
Sweden

Editorial Office
MDPI
St. Alban-Anlage 66
4052 Basel, Switzerland

This is a reprint of articles from the Special Issue published online in the open access journal *Energies* (ISSN 1996-1073) (available at: https://www.mdpi.com/journal/energies/special_issues/ Energy_Use_Efficiency).

For citation purposes, cite each article independently as indicated on the article page online and as indicated below:

LastName, A.A.; LastName, B.B.; LastName, C.C. Article Title. *Journal Name* **Year**, *Volume Number*, Page Range.

ISBN 978-3-0365-0354-7 (Hbk)
ISBN 978-3-0365-0355-4 (PDF)

Contents

About the Editor . . . . **vii**

Preface to "Energy Use Efficiency" . . . . **ix**

Sebastian Cuadros, Yeny E. Rodríguez and Javier Contreras
Dynamic Data Envelopment Analysis Model Involving Undesirable Outputs in the Electricity
Power Generation Sector: The Case of Latin America and the Caribbean Countries
Reprinted from: *Energies* **2020**, *13*, 6624, doi:10.3390/en13246624 . . . . **1**

Bożena Babiarz and Władysław Szymański
Introduction to the Dynamics of Heat Transfer in Buildings
Reprinted from: *Energies* **2020**, *13*, 6469, doi:10.3390/en13236469 . . . . **21**

Bartlomiej Gawin and Bartosz Marcinkowski
Setting up Energy Efficiency Management in Companies: Preliminary Lessons Learned from
the Petroleum Industry
Reprinted from: *Energies* **2020**, *13*, 5604, doi:10.3390/en13215604 . . . . **49**

Anna Kowalska-Pyzalska, Katarzyna Byrka and Jakub Serek
How to Foster the Adoption of Electricity Smart Meters? A Longitudinal Field Study of
Residential Consumers
Reprinted from: *Energies* **2020**, *13*, 4737, doi:10.3390/en13184737 . . . . **65**

Seungho Jeon, Minyoung Roh, Almas Heshmati and Suduk Kim
An Assessment of Corporate Average Fuel Economy Standards for Passenger Cars in
South Korea
Reprinted from: *Energies* **2020**, *13*, 4533, doi:10.3390/en13174533 . . . . **85**

Minyoung Roh, Seungho Jeon, Soontae Kim, Sha Yu, Almas Heshmati and Suduk Kim
Modeling Air Pollutant Emissions in the Provincial Level Road Transportation Sector in Korea:
A Case Study of the Zero-Emission Vehicle Subsidy
Reprinted from: *Energies* **2020**, *13*, 3999, doi:10.3390/en13153999 . . . . **99**

Oleg Badunenko and Subal C. Kumbhakar
Energy Intensity and Long- and Short-Term Efficiency in US Manufacturing Industry
Reprinted from: *Energies* **2020**, *13*, 3954, doi:10.3390/en13153954 . . . . **121**

Diego Seuret-Jimenez, Tiare Robles-Bonilla and Karla G. Cedano
Measurement of Energy Access Using Fuzzy Logic
Reprinted from: *Energies* **2020**, *13*, 3266, doi:10.3390/en13123266 . . . . **143**

Selamawit G. Kebede and Almas Heshmati
Energy Use and Labor Productivity in Ethiopia: The Case of the Manufacturing Industry
Reprinted from: *Energies* **2020**, *13*, 2714, doi:10.3390/en13112714 . . . . **157**

Fernando Enzo Kenta Sato and Toshihiko Nakata
Energy Consumption Analysis for Vehicle Production through a Material Flow Approach
Reprinted from: *Energies* **2020**, *13*, 2396, doi:10.3390/en13092396 . . . . **179**

Chia-Nan Wang, Thi-Duong Nguyen and Min-Chun Yu
Energy Use Efficiency Past-to-Future Evaluation: An International Comparison
Reprinted from: *Energies* **2019**, *12*, 3804, doi:10.3390/en12193804 . . . . **197**

Samuel Lotsu, Yuichiro Yoshida, Katsufumi Fukuda and Bing He
Effectiveness of a Power Factor Correction Policy in Improving the Energy Efficiency of
Large-Scale Electricity Users in Ghana
Reprinted from: *Energies* **2019**, *12*, 2582, doi:10.3390/en12132582 . . . . **213**

Jihu Zheng, Rujie Yu, Yong Liu, Yuhong Zou and Dongchang Zhao
The Technological Progress of the Fuel Consumption Rate for Passenger Vehicles in
China: 2009–2016
Reprinted from: *Energies* **2019**, *12*, 2384, doi:10.3390/en12122384 . . . . **225**

Ying Han, Jianhua Shi, Yuanfan Yang and Yaxin Wang
Direct Rebound Effect for Electricity Consumption of Urban Residents in China Based on the
Spatial Spillover Effect
Reprinted from: *Energies* **2019**, *12*, 2069, doi:10.3390/en12112069 . . . . **239**

Xiaoyan Zheng and Almas Heshmati
An Analysis of Energy Use Efficiency in China by Applying Stochastic Frontier Panel
Data Models
Reprinted from: *Energies* **2020**, *13*, 1892, doi:10.3390/en13081892 . . . . **255**

About the Editor

Almas Heshmati is a Professor of Economics at Jönköping University, Sweden. He held similar positions at Sogang University, Korea University, Seoul National University, and the University of Kurdistan Hawler. He was a Research Fellow at the World Institute for Development Economics Research (WIDER), the United Nations University during 2001–2004. From 1998 until 2001, he held the position of Associate Professor of Economics at the Stockholm School of Economics. He was awarded his Ph.D. degree from the University of Gothenburg (1994), where he held a Senior Researcher position until 1998. His research interests include applied microeconomics, globalization, development strategy, efficiency, productivity and growth, with applications in manufacturing and services. In addition to the many scientific journal articles that he has published, he has also produced books on the EU Lisbon Process, Global Inequality, East Asian Manufacturing, the Chinese Economy, Technology Transfer, Information Technology, Water Resources, Landmines, Power Generation, Renewable Energy, Development Economics, World Values, Poverty, well-being and Economic Growth.

Preface to "Energy Use Efficiency"

Energy is one of the most important factors of production. Its efficient use is crucial for ensuring production, profitability of firms and environmental quality. Unlike normal goods—which generally have supply management procedures to maximize sales—energy is demand-managed, with the objective to minimize its use. Efficient energy use aims to reduce the amount of energy required to provide products and services. Energy use efficiency can be achieved in residential settings, offices, industrial production, transport and agriculture, as well as in public lighting and services. The use of energy can be reduced by using technology that is energy saving and by reducing energy-using activities. There are many benefits associated with reducing energy use, including reduced energy dependency and vulnerability, and improved energy security. Various incentive-based programs have been introduced to the industry and public to promote the development, installation and use of energy efficient technologies and equipment. The policy is, in general, oriented to protect the air, water and land, and to prevent climate change and the associated negative health impacts by reducing the generation and use of energy from fossil fuels and nuclear primary sources. Saving energy to reserve fossil fuels for future generations and conserving natural resources has double dividend effects in the form of cost efficiency and the realization of sustainability.

This Special Issue contains 15 papers on energy use efficiency in different countries, locations and economic sectors. The areas of analysis include: undesirable outputs in the electricity power generation sector; the dynamics of heat transfer in buildings; energy efficiency management in companies; adoption of electricity smart meters in residential settings; assessment of corporate average fuel economy standards for passenger cars; modelling the air pollutant emissions of the road transportation sector; manufacturing energy intensity and energy efficiency in the manufacturing industry; measurement of energy access among regions; energy use and labor productivity in the manufacturing industry; vehicle energy consumption analysis; international comparison of energy use efficiency; effectiveness of power factor correction policies; technological progress of the fuel consumption rate for passenger vehicles; directed rebound effect for electricity consumption of urban residents, and comparison of energy use efficiency at the province level. This Special Issue provides good coverage on energy use efficiency theories, methods and diverse applications, therefore contributing important knowledge to the literature.

Almas Heshmati
Editor

Article

Dynamic Data Envelopment Analysis Model Involving Undesirable Outputs in the Electricity Power Generation Sector: The Case of Latin America and the Caribbean Countries

Sebastian Cuadros [1,*], Yeny E. Rodríguez [1] and Javier Contreras [2]

[1] Escuela de Economía y Finanzas, Departamento de Estudios Contable y Financiero, Universidad Icesi, 760008 Cali, Colombia; yerodriguez@icesi.edu.co
[2] E.T.S. de Ingeniería Industrial, University of Castilla-La Mancha, 13071 Ciudad Real, Spain; javier.contreras@uclm.es
* Correspondence: scuadros@icesi.edu.co

Received: 30 October 2020; Accepted: 10 December 2020; Published: 15 December 2020

Abstract: Studying the evolution of the efficiency of the electricity generation sector is a relevant task for policy makers, and requires the undesirable outputs derived from the activity to be considered in the evaluation. In this work, we propose a dynamic slack-based Data Envelopment Analysis (DEA) model that incorporates the assumption of weak disposability between the generation of electricity from fossil sources and the CO_2 emissions caused by the sector to measure the technical efficiency of 24 Latin American and Caribbean countries in the period 2000–2016. The results show that, of the total number of countries studied, four are efficient overall, and four groupings of countries in relation to the levels of efficiency achieved are also identified. These results are important given that less-efficient countries can, through learning, increase their efficiency in electricity generation or emulate the future strategies proposed by the most efficient countries.

Keywords: dynamic DEA; efficiency measurement; electricity power generation; weak disposability; undesirable outputs

---

1. Introduction

Measuring the efficiency of production systems is an important task in economic science, and different studies have addressed this problem with different methodologies. In regulated sectors, such as the electricity sector, the evaluation of productive efficiency has been promoted. Particularly, in the activity of electricity power generation, this stimulus is created by the dependence that exists on the traditional sources of generation, such as fossil sources, considering their impact on the environment caused by emissions of greenhouse gases (GHG) [1].

Electricity generation in the Latin American region has been largely composed of two types of sources: hydroelectric and fossil energy. In 2017, these two sources made up almost 88% of the total generation, at 47.52% and 40.25%, respectively. There has been an important change in the use of different types of energy compared to 1990—a time when there was almost total dependence on generation by these two sources, at 95.76% of the total, and there was also a greater relative importance of hydroelectric energy, which made up 65.69% of total energy compared to 33.43% of fossil sources [2].

Currently, developed and developing countries are concerned about increasing the proportion of renewable sources within their energy matrixes, which has resulted from essential decisions to address climate change [3]. This is supported by the fact that the electricity generation sector is the most important for CO_2 emissions, followed by the transport sector and the industrial sector [4], although the use of renewable energy in Latin America and the Caribbean was lower in 2018, at 27.59%

of the total, compared with its usage proportion in the rest of the world, at 41.93% [5]. To this extent, the comparison of the efficiency of electricity generation activity is a relevant task for policy makers, particularly considering the emissions caused by generation activity.

The aim of this research is to evaluate the evolution of the technical efficiency of the electricity generation sector of 24 countries in Latin America and the Caribbean during the period 2000–2016, with a dynamic slack-based DEA model from an output-oriented perspective. The proposed model considers a desirable output, an undesirable output and three inputs, of which two aim to capture the temporal interdependence in the generation activity, which are called link variables. In our model, the desirable output is the generation of electricity, while the undesirable output is the total CO_2 emissions derived from the generation of electricity. Although Sánchez et al. [6] studied the evolution of the efficiency of electricity generation in Latin American countries between 2006 and 2013, they did not consider the possible temporal interdependence present in the activity, nor did they use the installed capacity in the different generation sources as inputs or assume a weak disposability between generation and CO_2 emissions.

The contribution of this research is twofold at the country level: (1) it is the first study to incorporate the dynamic component of the DEA methodology, capturing the possible temporary interdependencies that exist in the generation activity when seen as a production system; and (2) it is the first investigation to capture the assumption of weak disposability between fossil generation sources and CO_2 emissions, which affects the efficiency measures calculated by the DEA models.

The rest of the document is organized as follows. Section 2 presents the main studies related to the DEA methodology as applied to the electricity sector. Section 3 presents the methodology, which introduces the concepts that are used to capture the environmental and dynamic components, and the methodology of DEA assessment by a non-radial model is also given. Section 4 shows the descriptive statistics of the variables used in the evaluation; it also describes the electricity generation sector and presents the results of the calculated efficiency levels for the 24 countries based on the proposed dynamic DEA model. Section 5 presents the conclusions, discussions and limitations of the study.

2. Literature Review

Recently, Data Envelopment Analysis (DEA) has become one of the main tools for environmental assessment. This methodology was initially proposed by Charnes et al. [7], and since then it has become an important tool for measuring relative efficiency in different fields [8]. This tool can serve as a guideline for firms and policy makers. Since Faere et al. [9] introduced the concept of the undesirable output, the use of DEA has widely spread in environmental assessment, becoming the most popular application area within the DEA methodology [10]. This section presents a complete review of this methodology and its application in the power generation sector, and its different variations in terms of environmental models and dynamic models.

2.1. DEA in Power Generation

Several studies have implemented the DEA methodology to assess the efficiency of the electricity generation sector at the generation firm level and at the geographical level.

Some works at the firm level include that of Golany et al. [11], which measured the efficiency of power plants in Israel; the works of Shermeh et al. [12] and Khalili-Damghani et al. [13], which investigated Iran regional power companies; the work of Yang and Pollit [14] regarding Chinese coal-fired power plants; the work of Sueyoshi and Goto [15] regarding U.S. coal-fired power plants; and the work of Cherchye et al. [16], which explored U.S. fossil and non-fossil plants. The last four studies also included an environmental assessment, including the emission of polluting gases as an undesirable output.

At the geographical level, Chang and Yang [17] measured the efficiency of the power generation of municipalities in Taiwan, while Tao and Zhang [18] investigated 16 Chinese cities located in the

Yangtze River Delta. These studies introduced environmental analysis considering different pollutants of the air and water. Other works have focused on conducting electricity generation performance assessment, and they have taken countries as decision-making units; among the researchers in this area are Dogan and Tugcu [19], who evaluated the efficiency of the G-20 group; Whiteman [20] and Yunos and Hawdon [21], who investigated 95 and 27 countries of the world, respectively; Bi et al. [22] who considered 26 OECD member countries; Zhou et al. [23], who used information from 126 countries around the world; Li et al. [24] who performed an analysis for the G-20 group; and Sánchez et al. [6], who measured the efficiency of Latin American countries. These four last groups performed efficiency evaluations that considered the undesirable outputs and external costs of the activity.

2.2. Environmental DEA

Policy makers must consider environmental efficiency assessment at a country level in order to regulate to promote environmental protection and economic development. In this way, some studies have involved undesirable outputs in the definition of DEA. The treatment of these undesirable outputs within the DEA literature has been presented in three ways, according to Dyson et al. [25]: (1) inverting the anti-isotonic factor, (2) subtracting the value of the undesirable factor from a large number or (3) treating the undesirable output as an input. We have opted for the third strategy.

The DEA model for environmental assessment requires the incorporation of different production factors (desirable outputs, undesirable outputs and inputs), and this requires all variables to be greater than or equal to zero. Here, non-radial models satisfy this requirement; therefore, they can measure the efficiency of DMUs (decision-making units) that contain negative or zero values in any of their inputs or outputs.

Conventional energy efficiency measures that do not consider undesirable outputs are biased because firms can lose their productive efficiency due to a negative output [26]. Following Faere et al. [9], when evaluating the performance of producers, it makes sense to compensate for the supply of desirable outputs, as well as to penalize the provision of undesirable outputs. In other words, "positive" and "negative" factors should be treated asymmetrically when measuring a producer's performance. The performance measures outlined above, in fact, treat positive and negative factors asymmetrically, valuing the former and ignoring the latter. This extreme form of asymmetry characterizes much of the literature on measuring productivity and efficiency, so it is necessary to introduce concepts that allow the smoothing of this approach.

Unlike traditional DEA models, the model proposed by Faere et al. [9] assumes that the reduction of undesirable outputs is costly in terms of desirable outputs. To reduce undesirable outputs, part of the production of desirable outputs must be sacrificed. In the literature, this implies moving from the assumption that the technology of undesirable outputs is "freely (or strongly) disposable", where the variation of undesirable outputs does not represent any cost in terms of production, to the assumption of "weakly disposable" outputs, where such variation involves a cost, given the conceptual incorporation that implies that desirable and undesirable outputs are jointly produced. In this work, the desirable outputs $\left(y^d \epsilon R_+\right)$ are distinguished from the undesirable outputs ($y^u \epsilon R_+$) and the inputs are denoted by $x \epsilon R_+$.

According to Faere et al. [9], mathematically, the concept of strong disposability between desirable and undesirable outputs can be expressed as follows:

$$\left(y^u, y^d\right) \in P(x) \; \rightarrow \; \left(y^d - s\right) \in P(x), \; s \geq 0 \tag{1}$$

Given a vector of inputs (x) and a production possibility frontier $P(x)$, if a level y^d can be reached, then $y^d - s$ can also be produced for any $s \geq 0$.

On the other hand, it is common that certain bad outputs cannot be separated from the corresponding good outputs; therefore, to reduce a bad output, it is necessary to reduce the good output [27]. Within the DEA literature, this is the concept of weak disposability, and it can be denoted as follows:

$$\left(y^u, y^d\right) \in P(x) \;\to\; \left(\theta y^u,\, \theta y^d\right) \in P(x), \text{ with } 0 \le \theta \le 1. \tag{2}$$

Given a vector of inputs (x) and a production possibility frontier $P(x)$, on the one hand, a total decrease of the undesirable output $(y^u = 0)$ is not possible unless the desirable output is also zero $\left(y^d = 0\right)$; on the other hand, it can only be decreased proportionally $\left(y^u,\, y^d\right)$ when $0 \le \theta \le 1$. In this case, y^u and y^d are called non-separable undesirable outputs and non-separable desirable outputs, respectively. We consider that the weak disposability assumption in the activity generation activity is necessary considering that it is not possible to generate electricity using fossil fuels without incurring CO_2 emissions.

2.3. Dynamic DEA

Traditional DEA models do not consider the interdependencies between consecutive periods. This can be a problem in the case of electricity generation because the level of installed capacity available for a country is determined by the installed capacity in the immediately preceding period, which modifies the efficiency assessment [28]. Static DEA models assume that the inputs in period t are mixed with the technology of period t to produce the outputs of period t.

Färe and Grosskopf [29] were the first to incorporate variables that connect consecutive periods, called link flows, from carry-over equations into the DEA approach, allowing inputs to be stored by modeling "savings" in period t to be used in period $t + 1$. Later, Tone and Tsutsui [30] identified different kinds of carry-over activities and proposed a dynamic slack-based model.

3. Materials and Methods

This section first presents the description of the data and the source of information; subsequently, the definition of the variables that are part of the proposed model is presented; and finally, the strategy used to measure the efficiency of the electricity generation of the countries of Latin America and the Caribbean is shown.

3.1. Data and Sources

The data set used corresponded to annual data between 2000 and 2017 from 24 countries in Latin America and the Caribbean. The data collected originated from two sources: The U.S. Energy Information Administration (EIA) and the International Energy Agency (IEA). We collected the CO_2 emissions from the generation of electricity from the IEA, while GDP, installed capacity and generation of electricity were collected from the EIA.

3.2. Definition of the Variables

The proposed model includes a desirable output, an undesirable output, three inputs and two link variables, which are described below.

- Desirable output

As a desirable output, we used the generation of electricity, measured in TWh, distinguishing whether the generation sources were based on fossil sources—oil, gas and coal—or non-fossil sources—nuclear, geothermal, solar, wind, biomass and waves—to capture the assumption of weak disposability between CO_2 emissions and electricity generation through fossil sources. This strategy was used by Cherchye et al. [16] and Walheer [31] to isolate the emissions of three polluting gases, but the latter used electricity generation as a necessary input to produce CO_2 emissions and GDP.

- Undesirable output

To capture the dependence between the generation of electricity based on fossil sources and the CO_2 emissions that they incur, we discriminated between energy from clean generation sources and energy generated from fossil fuels. This strategy allowed us to capture the proportional variations between the non-separable desirable output—fossil-generation—and the CO_2 emissions, known in the DEA literature as the assumption of weak disposability and introduced by Faere et al. [9].

As an undesirable output, we used the CO_2 emissions generated by the electricity generation activity, measured in MTm. Due to the availability of information, we used observed data for 2016 and 2017, while we estimated the data for the period 2000–2015 from the CO_2 emissions from electricity and heat production in each country in 2016 and the electricity generated from fossil sources in the same year. For Guyana, we created an estimate for the entire period using the regression from other countries because of the lack of information regarding CO_2 emissions from electricity and heat production for this country. We think that this measure represented a good proxy for CO_2 emissions generated by the electricity sector considering the high R-square of the regression of 0.9886.

- Inputs

We incorporated three inputs: the gross domestic product (GDP) per capita, the installed capacity of non-fossil generation sources and the installed capacity of fossil generation sources. These last two variables were also used to capture the inter-temporal dependence of electricity generation, entering the model as link variables.

The GDP of each country has been used in different studies within the DEA methodology as a desirable output [22,32–35]. We believe that, within the productive process presented by each DMU, one of the main inputs is the GDP per capita, in the sense that high-income countries can benefit from greater technological innovation and make greater efforts in R&D to improve energy efficiency [36]. This decision to use GDP per capita as an input was also based on studies that have evaluated the causality between electricity generation and economic growth, finding a unidirectional relationship for economic growth and electricity generation [37–39]. This indicator was measured in billions $2015 PPP. In addition, Dyson et al. [25] recommended the use of type of variables to control the lack of homogeneity in the units tested.

The installed capacity has been used in different studies as an input for electricity generation. For example, Yunos and Hawdon [21] and Li et al. [24] used the installed capacity of fossil sources as an input without taking into account the different non-fossil sources of generation. In addition, Whiteman [20], Chen et al. [36] and Dogan and Tugcu [19] used the installed capacity of non-fossil sources in a disaggregated manner. This variable is measured in GW.

- Link variables

In this research, we considered that there is a dynamic component within the electricity generation sector that depends on the installed capacity for the different generation sources. The main reasons for this is as follows: (1) the level of installed capacity available for each country in year t is determined by the installed capacity level in the immediately preceding term, $t - 1$ [28]; (2) it can be assumed that the installed capacity in each country is a quasi-fixed input, and because of the large investment that this entails, it makes it difficult to adjust this to optimum levels every year [40]; and (3) the level of installed capacity in a year t has impacts on the generation in year $t + 1$, taking into account the fact that this input also functions as a warehouse, either for electricity, through batteries, or of potential generation—for example, the electricity power output that depends on the water flow in the penstock and the water accumulated in the reservoir [41].

3.3. Model Approach

To measure the efficiency of electricity generation in the 24 countries mentioned, we propose a dynamic slack-based DEA model and assume a constant return to scale (CRS). The model is based on the dynamic slack-based DEA model proposed by Tone and Tsutsui [30], which has been expanded to include undesirable outputs and to capture the assumption of weak disposability between the generation of electricity from fossil sources and emissions of CO_2, presented in Tone [42].

The model structure is represented in Figure 1. We observe n countries over T terms. At each term t, each country uses its respective inputs (GDP, non-fossil-fuel installed capacity and fossil-fuel installed capacity) to produce the desirable output (non-fossil and fossil electricity generation). A variation in fossil generation implies a proportional variation of the undesirable output (CO_2). The link variables connect consecutive terms $(1, \dots , t - 1, t, t + 1, \dots , T)$; in our model, the level of installed capacity available for each country in term t determines the installed capacity in the immediately succeeding term, $t + 1$, and is determined by the installed capacity in the immediately preceding term, $t - 1$.

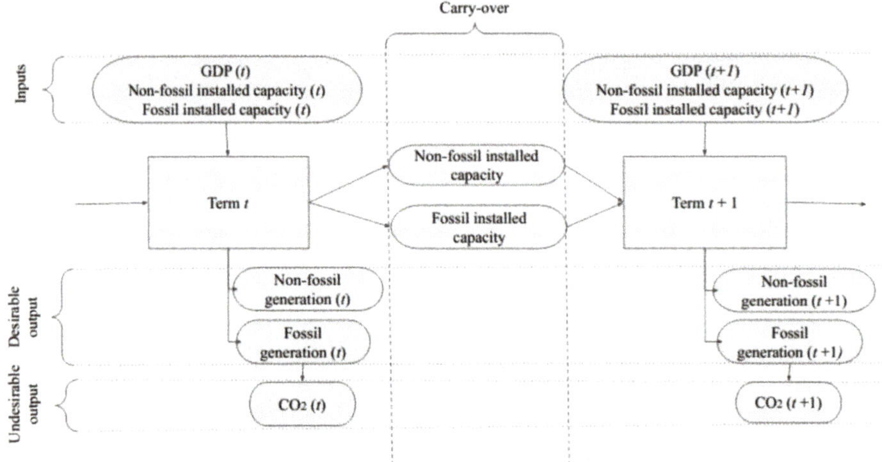

Figure 1. Model structure.

The dynamic DEA model defines a production possibility set for each term based on the observed output and input values of the DMUs in each term t.

Following Zhou and Liu [43], the maximization of the desirable output and minimization of the undesirable output can be reached with an additive DEA model with the next objective function:

$$\max SDO\_NF_{ot} + SDO\_F_{ot} + SUO\_CO_{2ot}. \tag{3}$$

However, continuing to follow Zhou and Liu [43], this model cannot produce efficiency measures directly; thus, output-oriented efficiency must be measured for each year, and the overall efficiency measure for $DMUo$ must be calculated while replacing the slacks in the following equations:

$$\tau_{ot}^* = \frac{1}{1 + \frac{1}{3}\left(\frac{SDO\_NF_{ot}}{DO\_NF_{ot}} + \frac{SDO\_F_{ot}}{DO\_F_{ot}} + \frac{SUO\_CO_{2ot}}{UO\_CO_{2ot}}\right)}; \; t = 2000, \dots, 2016 \tag{4}$$

$$\tau_o^* = \frac{1}{17} \sum_{t=2000}^{2016} \tau_{ot}^* \tag{5}$$

where a country o will be globally efficient ($\tau_o^* = 1$) if and only if $SDO\_NOF_{ot} = SDO\_F_{ot} = SUO\_CO_{2ot} = 0$; $\forall \, t = 2000, \dots, 2016$. In other words, the country will be efficient throughout the period if it is efficient in each year. It should be noted that the evaluation of efficiency for the last year is lost because temporary interdependence is introduced into the proposed model. We chose an output-oriented measure of efficiency as we aimed to evaluate, given the set of inputs, if there were deficiencies in the desirable outputs or excesses in the undesirable output.

The production possibility set for the *DMUo* (country *o*, with *o* = 1, … , 24) under a CRS is defined by Equations (1)–(9).

- Equations (1)–(3) are associated with constraints on inputs:

$$GDP_{ot} = \sum_{j=1}^{24} GDP_{jt}\lambda_j^t + S\_GDP_{ot} \tag{6}$$

The *GDP* of country "*o*" must be less than or equal to the linear combination of the *GDP* of all countries in each term *t*. The difference is the slack variable of the *GDP* of country *o* in term *t* (*S_GDP*).

$$IC\_NF_{ot} = \sum_{j=1}^{24} IC\_NF_{jt}\lambda_j^t + SIC\_NF_{ot} \tag{7}$$

The non-fossil installed capacity (*IC_NF*) of country *o* must be less than or equal to the linear combination of the non-fossil installed capacity of all countries in each term *t*. The difference is the slack variable of the non-fossil installed capacity of country *o* in term *t* (*SIC_NF*).

$$IC\_F_{ot} = \sum_{j=1}^{24} IC\_F_{jt}\lambda_j^t + SIC\_F_{ot} \tag{8}$$

The fossil installed capacity (*IC_F*) of country *o* must be less than or equal to the linear combination of the fossil installed capacity of all countries in each term *t*. The difference is the slack variable of the fossil installed capacity of country *o* in term *t* (*SIC_F*).

The equation associated with the constraint on the separable desirable output is as follows:

$$DO\_NF_{ot} = \sum_{j=1}^{24} DO\_NF_{jt}\lambda_j^t - SDO\_NF_{ot}. \tag{9}$$

The electricity generation from non-fossil sources (*DO_NF*) of country *o* must be greater than or equal to the linear combination of electricity generation from the non-fossil capacity of all countries in each term *t*. The difference is the slack variable of the electricity generation from non-fossil capacity of country *o* in term *t* (*SDO_NF*).

Equations (5) and (6) capture the assumption of weak disposability between the electricity generation from fossil sources and the emission of CO_2. A variation of the non-separable desirable output is designated by $\alpha_t DO\_F_{ot}$ and is accompanied by the same proportional variation in the non-separable undesirable output designated by $\alpha_t UO\_CO_{2ot}$. Equation (5) represents the constraint on the non-separable desirable output. Equation (6) is the constraint of the non-separable undesirable output:

$$\alpha_t DO\_F_{ot} = \sum_{j=1}^{24} DO\_F_{jt}\lambda_j^t - SDO\_F_{ot} \tag{10}$$

The electricity generation from fossil sources (*DO_F*) of country *o* must be greater than or equal to the linear combination of electricity generation from the fossil capacity of all countries in each term *t*. The difference is the slack variable of the electricity generation from the fossil capacity of country *o* in term *t* (*SDO_F*).

$$\alpha_t UO\_CO_{2ot} = \sum_{j=1}^{24} UO_{CO_{2ot}}\lambda_j^t + SUO\_CO_{2ot} \tag{11}$$

The CO_2 emissions caused by the electricity generation activity (UO_CO_2) of country o must be less than or equal to the linear combination of CO_2 of all countries in each term t. The difference is the slack variable of the CO_2 of country o in term t (SUO_CO_2).

Two carry-over equations that guarantee the continuity of the link flows between the terms t and $t + 1$ are as follows:

$$\sum_{j=1}^{24} IC\_NF_{jt}\lambda_j^t = \sum_{j=1}^{24} IC\_NF_{jt}\lambda_j^{t+1}; \; t = 2000, \ldots, 2016 \tag{12}$$

$$\sum_{j=1}^{n} IC\_F_{jt}\lambda_j^t = \sum_{j=1}^{n} IC\_F_{jt}\lambda_j^{t+1}; \; t = 2000, \ldots, 2016 \tag{13}$$

The installed capacity in non-fossil and fossil sources in each term t is determined by the respective installed capacity in term $t - 1$.

The assumption of a CRS in the production possibility set is captured by the following condition:

$$\sum_{j=1}^{24} \lambda_j^t \geq 0 \tag{14}$$

Additionally, non-negativity conditions are as follows:

$$S\_GDP_t, \; SDO\_NF_t, \; SDO\_F_t, \; SIC\_NF_t, \; SIC\_F_t, \; SUO\_CO_{2t}, \; \geq 0 \tag{15}$$

We test the CRS assumption using the following test introduced by Banker [44]:

$$F_j = \frac{\sum_{j=1}^{N}\left(\hat{\theta}_j^{CCR} - 1\right)^2}{\sum_{j=1}^{N}\left(\hat{\theta}_j^{BCC} - 1\right)^2} \tag{16}$$

where $\hat{\theta}^{CCR}$ is the calculated efficiency measure that assumes a CRS, as proposed by Charnes et al. [7], and $\hat{\theta}^{BCC}$ is the calculated efficiency measure that assumes a variable return to scale (VRS), as proposed by Banker et al. [45]. This calculated value is asymptotically F-distributed with (N, N) degrees of freedom. If not rejected, the CRS is accepted.

4. Results

This section is composed of two parts: in the first part, we show the descriptive statistics of the variables used for the 24 countries of the sample between 2000 and 2017; in the second part, we analyze the efficiency measure in two levels—at the aggregate level and at the country group level.

4.1. Descriptive Analysis of the Variables

Table 1 presents the mean and standard deviation of the set of data used at the country level, which was used to assess the relative efficiency of electricity generation.

Table 1. Mean and standard deviation at the country level.

Country (1)	Statistic (2)	Desirable Ouput		Undesirable Output	Input		
		Non-Fossil Gen. (3)	Fossil Gen. (4)	CO_2 (5)	GDP Per Capita (6)	Ins. Cap. Non-Fossil (7)	Ins. Cap. Fossil (8)
AR	Mean	41.00	70.00	44.23	18,412.81	11.78	20.53
	SD	3.09	16.80	9.79	2268.57	0.68	3.40
BO	Mean	2.30	3.85	3.64	5656.90	0.56	1.22
	SD	0.21	1.71	0.86	862.72	0.09	0.37
BR	Mean	400.33	62.29	40.93	14,394.79	90.36	18.26
	SD	62.27	34.45	22.37	1544.55	19.02	6.45
CH	Mean	27.04	31.03	21.45	19,981.79	6.37	9.48
	SD	3.83	8.83	6.83	3006.13	1.73	3.08
CO	Mean	43.89	13.62	9.84	11,522.65	9.78	4.63
	SD	7.07	5.65	3.53	1833.39	1.16	0.20
CR	Mean	8.62	0.42	1.50	13,297.63	1.94	0.54
	SD	1.24	0.34	0.55	1926.17	0.47	0.18
CU	Mean	0.75	15.78	10.95	10,079.38	0.45	5.25
	SD	0.17	1.74	0.90	2137.05	0.25	0.87
DR	Mean	1.71	12.01	8.83	11,363.87	0.57	2.69
	SD	0.58	2.37	1.48	2262.95	0.15	0.26
EC	Mean	10.42	7.51	5.98	9819.95	2.35	2.47
	SD	3.55	2.75	1.71	1232.25	0.85	0.69
ES	Mean	3.17	2.09	2.54	6742.26	0.75	0.74
	SD	0.64	0.34	0.50	542.44	0.15	0.17
GU	Mean	5.32	3.54	3.65	7097.37	1.36	1.51
	SD	1.66	0.62	0.50	472.13	0.61	0.28
HA	Mean	0.19	0.51	1.60	1736.84	0.06	0.21
	SD	0.07	0.27	0.26	59.55	0.00	0.04
HO	Mean	2.81	3.43	3.45	4118.70	0.75	0.94
	SD	1.00	0.90	0.56	378.25	0.35	0.23
JA	Mean	0.32	4.79	4.27	8742.39	0.09	1.03
	SD	0.15	1.41	0.99	239.61	0.04	0.17
MX	Mean	50.27	201.74	126.31	17,608.92	15.22	42.87
	SD	6.48	31.75	19.44	746.11	2.67	6.96
NI	Mean	1.32	2.07	2.55	4361.85	0.40	0.70
	SD	0.72	0.18	0.39	581.76	0.16	0.15
PN	Mean	4.46	2.69	2.95	16,470.64	1.20	0.87
	SD	1.44	0.68	0.46	4310.40	0.53	0.25
PR	Mean	54.27	0.00	1.24	9911.98	8.24	0.01
	SD	4.33	0.00	0.44	1439.28	0.64	0.01
PE	Mean	21.04	12.23	8.70	9956.29	3.67	5.05
	SD	3.61	6.89	3.92	2319.50	0.81	2.12
TT	Mean	0.01	7.56	5.83	29,570.22	0.01	1.92
	SD	0.01	1.64	0.84	4795.99	0.00	0.49
UR	Mean	8.40	1.28	2.04	16,976.87	2.01	1.02
	SD	2.71	1.12	0.91	3483.31	0.64	0.39
VE	Mean	74.21	32.30	22.05	16,547.00	14.33	11.30
	SD	9.66	6.16	4.95	2198.72	0.87	3.73
GU	Mean	0.00	0.82	1.90	6075.02	0.03	0.33
	SD	0.01	0.12	0.08	1059.76	0.01	0.04
SU	Mean	0.99	0.71	1.72	13,847.67	0.19	0.23
	SD	0.23	0.08	0.32	1846.36	0.00	0.04
TOTAL	Mean	31.79	20.51	14.1	11,845.57	7.19	5.58

Source: Own elaboration. Labels: AR: Argentina, BO: Bolivia, BR: Brazil, CH: Chile, CO: Colombia, CR: Costa Rica, CU: Cuba, DR: Dominican Rep., EC: Ecuador, ES: El Salvador, GU: Guatemala, HA: Haiti, HO: Honduras, JA: Jamaica, MX: Mexico, NI: Nicaragua, PN: Panama, PY: Paraguay, PE: Peru, TT: Trinidad and Tobago, UR: Uruguay, VE: Venezuela, GY: Guyana, SU: Suriname. Inst. Cap.: installed capacity.

- Electricity generation

In 2017, the 24 countries in the study had a total electricity generation of 1545.74 TWh, representing a growth of 63.62% compared to the year 2000, when the recorded generation was 944.73 TWh. Of the total generated by the 24 countries over the period 2000–2017, four countries contributed 74.27% (columns 3 and 4). These countries were Brazil 36.86%, Mexico 20.08%, Argentina 8.84% and Venezuela 8.49%.

During the period 2000–2017, the generation mostly originated from non-fossil sources, representing 60.78% of the total electricity generated. However, it is observed that there has been a wide variation in the share of electricity generation by types of sources. For example, in 2017,

Trinidad and Tobago, Guyana and Cuba had a lower share of non-fossil sources, at 0.04%, 4.04% and 4.05%, respectively, while countries such as Paraguay, Costa Rica and Uruguay had high shares, at 100%, 99.69% and 98.42%, respectively.

- CO_2 emissions

Regarding the CO_2 emissions caused by the electricity generation sector (column 5), four countries stand out as maximum polluters: Mexico, Argentina, Brazil, Venezuela and Chile, representing 75.40% of total emissions for the analyzed period. This is due to non-fossil sources being more involved in the composition of their generation matrix, or the fact that these countries have high volumes of generated electricity.

Mexico was the country with the highest level of emissions between 2000 and 2017, with an annual average of 126.31 MTm, depending on the high level of fossil sources of the total electricity generation in the period, at 80.05%. The country with the second highest level of emissions was Argentina, with an annual average of 44.2 MTm emissions, because of the participation of fossil sources in its energy matrix, which amounted to 63.06% in the period observed. The third country with the highest level of emissions was Brazil; considering that it has been strongly oriented towards electricity generation with non-fossil sources in the period, at 86.65%, the result can be explained by its high volume of generation, annually emitting an average of 40.9 MTm. Venezuela had an annual average emission level of 22.0 MTm, which is explained by the high volume of electricity generation and by the high participation of fossil sources in the studied years, at 30.33%. Finally, Chile had an annual average of 21.45 MTm of emissions of CO_2, which could be explained by its fossil-source-dominated generation of electricity, at 53.43%.

- GDP per capita

The aggregate size of the economy of the countries analyzed, captured by the GDP, increased from $6256 billion to $9633 billion from 2000 to 2017 ($2015 PPP), indicating an aggregate growth of 53.99%.

In per capita terms, large differences can be observed between countries in the studied period. The country with the highest per capita income in 2017 was Trinidad and Tobago, with 30,347 ($2015 PPP), followed by Chile with 23,782 ($2015 PPP), while the two poorest countries were Haiti and Honduras, with per capita incomes of 1767 and 4773 ($2015 PPP), respectively.

- Installed capacity

Between 2000 and 2017, the installed capacity in the region showed an expansion of 87.89%, from 222.52 GW to 418.09 GW. In addition, the weight of the installed capacity of non-fossil sources was greater than the weight of fossil sources in the period, comprising from 59.98% to 57.34% of the total capacity in the region.

Between 2000 and 2017, Brazil was also highlighted as the country with the highest average installed capacity of non-fossil sources, at 90.36 GW, and Mexico was the country with the highest installed capacity of fossil sources, at 42.87 GW.

4.2. Electricity Generation Sector and Efficiency Measure

This subsection is composed of two parts: in the first part, we analyze the global measure of efficiency based on the spatial distribution of the measure as an aggregate; in the second part, we analyze the relative efficiency individually and expose the sources of inefficiency according to the averages of the relative slacks found by the model.

4.2.1. Aggregate Spatial Analysis of the Overall Efficiency Measure

We calculate the efficiencies with the proposed model with a CRS and with a VRS to test the assumption of a CRS following the Banker test [44]. To calculate the F value, we eliminate the measured efficiency of Guyana due to the lack of information for the first five years. Our calculated F is $1.501/7.58 = 1.981$; that is smaller than 2.014, and thus, the null hypothesis of a CRS is not rejected with a p-value of 0.05.

Figure 2 presents the spatial distribution of the overall measure of efficiency, aggregated in four ranges from the information in Table 2.

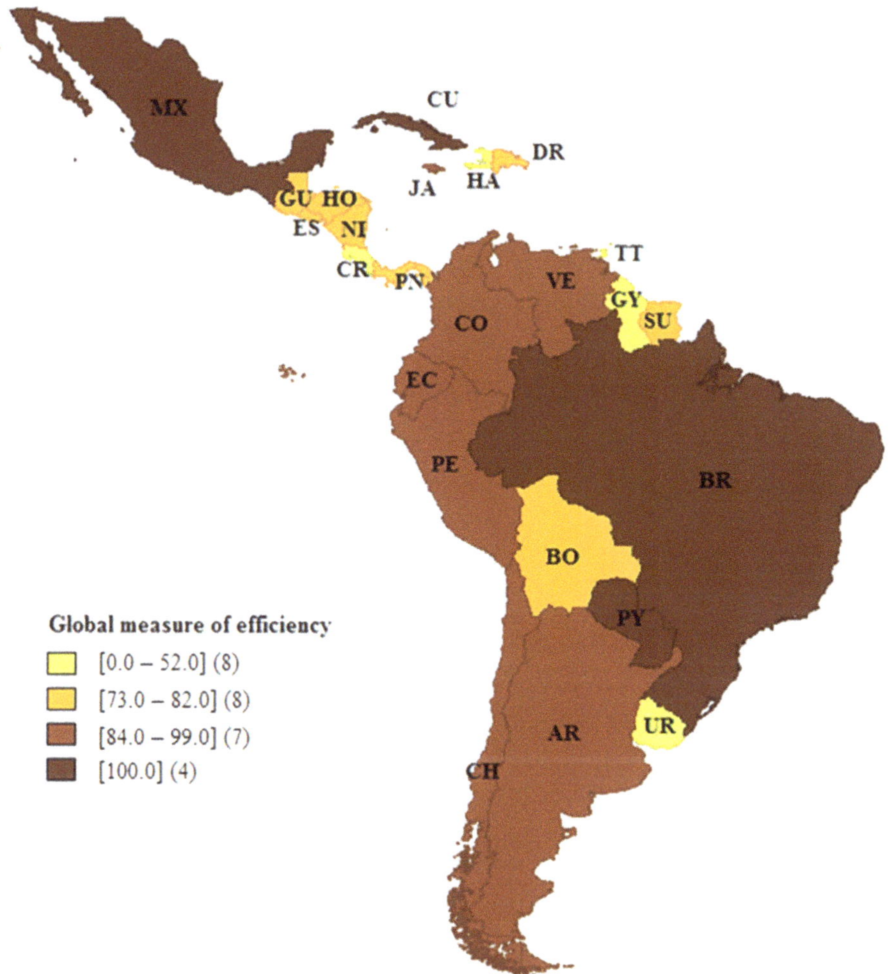

Figure 2. Global measure of efficiency of electricity generation.

Table 2. Term and global efficiency measurement.

Country	2000	2001	2002	2003	2004	2005	2006	2007	2008	2009	2010	2011	2012	2013	2014	2015	2016	Global
AR	82.57	92.87	89.28	89.22	86.94	88.41	93.86	87.20	84.21	89.02	91.00	85.32	85.82	88.18	84.85	91.76	87.14	88.10
BO	72.23	75.20	84.48	72.06	75.57	76.30	78.00	82.93	81.78	84.98	84.44	85.84	85.57	88.63	83.34	89.52	73.99	80.87
BR	100.00	100.00	100.00	100.00	100.00	100.00	100.00	100.00	100.00	100.00	100.00	100.00	100.00	100.00	100.00	100.00	100.00	100.00
CH	88.35	97.07	97.00	95.13	97.11	98.62	99.19	95.37	95.54	98.43	91.66	96.36	85.59	87.36	96.19	92.49	82.32	93.75
CO	80.96	86.02	85.70	85.13	86.97	87.60	90.71	91.16	88.84	90.06	94.03	99.03	88.25	93.35	96.09	95.91	85.76	89.74
CR	11.03	16.72	19.35	25.09	8.55	26.95	44.67	69.01	54.52	38.03	51.40	66.42	47.83	58.37	60.37	8.36	46.70	38.44
CU	100.00	100.00	100.00	100.00	100.00	100.00	100.00	100.00	100.00	100.00	100.00	100.00	100.00	100.00	100.00	100.00	100.00	100.00
DR	52.14	55.06	61.06	70.57	80.78	89.76	84.41	84.16	77.76	80.68	73.01	76.56	80.83	86.63	77.78	72.44	74.69	75.20
EC	73.25	80.01	78.74	76.03	80.42	78.83	83.61	98.19	90.91	88.59	85.26	90.61	93.52	89.85	92.51	96.47	93.74	86.50
ES	66.35	69.29	71.42	73.44	72.22	76.81	84.93	91.88	79.87	79.30	75.96	74.82	75.28	74.70	77.65	77.39	75.51	76.28
GU	76.70	76.69	76.52	88.26	81.26	86.13	79.56	84.95	77.38	79.55	79.87	75.13	87.83	73.21	86.31	84.07	66.41	79.99
HA	49.96	54.90	51.31	43.44	43.94	43.90	49.87	46.78	53.38	69.43	59.01	55.05	73.85	65.90	59.47	51.53	39.60	53.61
HO	71.12	77.64	76.76	76.72	70.19	79.56	81.45	84.52	82.35	87.98	89.53	91.79	86.53	81.27	86.53	99.52	70.47	82.00
JA	69.93	59.26	52.32	59.20	70.04	99.23	97.55	95.44	94.87	95.53	96.53	94.23	92.08	90.30	90.67	91.49	87.22	84.46
MX	100.00	100.00	100.00	100.00	100.00	100.00	100.00	100.00	100.00	100.00	100.00	100.00	100.00	100.00	100.00	100.00	100.00	100.00
NI	64.14	70.64	79.85	70.40	70.90	76.95	63.61	65.34	75.71	76.07	76.06	76.86	95.54	68.11	71.53	74.45	74.67	73.58
PN	67.80	77.50	88.06	77.44	79.37	83.06	80.85	80.63	83.49	83.00	84.84	98.90	86.08	71.61	81.69	84.12	69.33	81.05
PY	100.00	100.00	100.00	100.00	100.00	100.00	100.00	100.00	100.00	100.00	100.00	100.00	100.00	100.00	100.00	100.00	100.00	100.00
PE	70.64	100.00	77.43	79.69	88.83	91.27	100.00	100.00	100.00	100.00	100.00	100.00	100.00	100.00	100.00	100.00	100.00	94.58
TT	45.78	59.21	54.03	36.57	55.97	99.96	99.94	88.19	95.88	81.11	27.59	39.37	25.55	24.39	26.94	26.23	22.73	53.50
UR	33.81	2.13	2.96	3.67	57.36	83.77	99.39	71.25	70.72	72.93	56.50	75.12	80.25	56.45	28.64	37.69	39.51	51.30
VE	91.57	97.67	96.33	94.93	100.00	100.00	100.00	100.00	100.00	100.00	100.00	100.00	100.00	100.00	100.00	100.00	100.00	98.85
GY	-	-	-	-	-	1.27	1.71	1.76	97.88	4.00	1.86	2.00	2.24	2.05	14.90	20.07	8.91	13.22
SU	65.92	68.15	67.32	62.64	62.74	64.16	72.38	70.31	72.74	78.15	100.00	78.74	89.90	100.00	100.00	100.00	77.35	78.26

Source: Own elaboration.

According to the map, it is difficult to establish a spatial pattern that contributes to the explanation of global efficiency levels for each country. However, at least three aspects can be highlighted.

On the one hand, all of the six Central American countries—Costa Rica, El Salvador, Guatemala, Honduras, Nicaragua, and Panama—belong to the two lowest global efficiency levels. On the other hand, of the five countries in the Caribbean—Cuba, Jamaica, Dominican Republic, Haiti and Trinidad and Tobago—only Cuba is in the highest global efficiency level. Finally, of the 12 South American countries, eight are in the two highest levels—Argentina, Brazil, Chile, Colombia, Ecuador, Paraguay, Peru and Venezuela—and four are in the two lowest global efficiency levels—Bolivia, Guyana, Suriname, and Uruguay.

It is worth investigating whether there is any spatial pattern in the distribution of the global efficiency measure. Moran's Index (Moran's I) is the most commonly used measure of spatial autocorrelation to describe the degree of spatial concentration or dispersion for variables included in an analysis [46]. According to Moran (1950), Moran's *I* is calculated as follows:

$$I = \frac{N}{S} \frac{\sum_{i=1}^{N} \sum_{j=1}^{N} w_{ij}(x_i - \bar{x})(x_j - \bar{x})}{\sum_{i=1}(x_i - \bar{x})^2} \tag{17}$$

where N is the number of spatial units indexed by i and j; x is the variable of interest; $\bar{x}$ is the mean of x; and w_{ij} is a matrix of spatial weights such that (1) the diagonal elements w_{ii} are equal to zero and (2) the non-diagonal elements w_{ij} indicate the way that a region i is spatially connected with the region j. S is a scalar term that is equal to the sum of all w_{ij}.

When the Moran's *I* is positive, this implies that large values for the variable are surrounded by other large values, and when the Moran's *I* for a variable is negative, then the large values are surrounded by small values. Therefore, a positive spatial autocorrelation implies a spatial clustering for a variable, whereas a negative spatial autocorrelation suggests a spatial dispersion.

Figure 3 presents the Moran's *I* of global efficiency measures of electricity generation for 21 countries that have at least one neighbor.

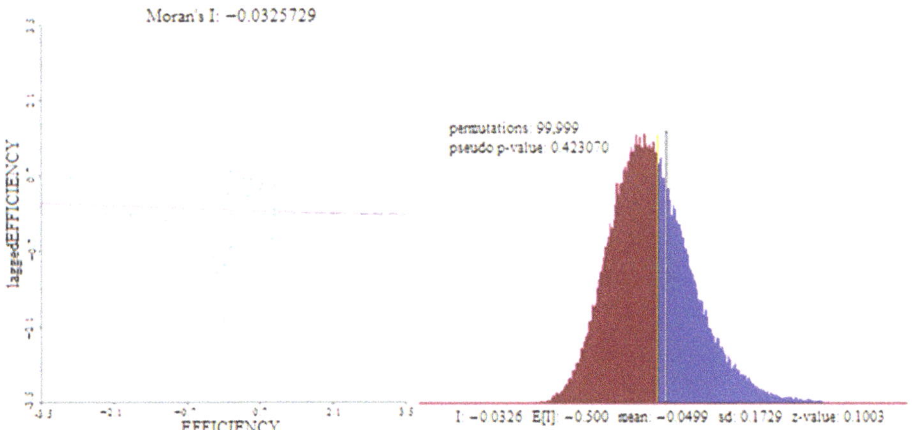

Figure 3. Moran Index of global efficiency measure of electricity generation.

According to the figure, the global efficiency measure presents a very slight negative spatial autocorrelation of −0.0326; thus, the null hypothesis of a random spatial distribution of the measure is not rejected with a *p*-value of 0.05.

4.2.2. Measure of Efficiency of Electricity Generation

Table 2 presents the evolution of the efficiency for each country and for each year of the period 2000–2016; in addition, it contains the global measure of efficiency for each country, which was calculated as the yearly average of efficiency.

On the left-hand side of Figure 4, we present the evolution of the efficiency of the 24 countries for the period 2000–2016, dividing them into the three groups. On the right-hand side, we show the participation of slacks for each country. Slacks can be interpreted as deficits in desirable outputs or excesses in undesirable output given the production possibilities set.

Our results confirm that, although there is currently a common agenda for Latin America to improve its energy efficiency, the incentives granted to increase efficiency have been heterogeneous throughout the countries in the region [47]. Usually, programs related to energy efficiency are led by public organizations [47], who tend to be more efficient in the development of multi-tasking than private firms [48]. Energy-efficiency entities are key to control and implement programs to support energy efficiency, but they are not enough by themselves to promote energy-efficiency improvements [47], and a complementary mechanism would be the use of incentives. There are different types of incentives that can be used to improve the energy efficiency of a country; among the most used in generation sector are mandatory performance standards and market-based and information-based incentives [49]. Mandatory codes and standards are regulatory instruments regarding energy efficiency. Market-based incentives are related to the development of auctions and tradable emission products, among others [49]. Finally, governance and support represent the final step for the implementation of energy-efficiency policies. This refers to the mechanisms used by governments in order to incentivize energy efficiency.

According to their level of efficiency, we have classified the countries into three groups: high efficiency level, medium–high efficiency level and low–medium efficiency level.

The first group is made up of 11 countries, four of which are not in the figure because they make up an efficient border and registered efficiency levels of 100 for all years; they are Brazil, Cuba, Mexico and Paraguay. These countries have an overall efficiency of 100, which is equivalent to a solution of zero slacks in each year, and implies that they have no deficiencies in desirable outputs or excesses in undesirable output given the set of inputs. In relation to Mexico and Paraguay, the results coincide with the work of Sánchez et al. [6], who found complete efficiency between 2006 and 2013 for these countries.

We highlight Mexico and Brazil because they have consolidated their institutional and regulatory frameworks to support energy efficiency activities [50], and have been recognized by IEA [51] for having a high coverage potential of regulatory instruments in terms of energy efficiency. Auctions focused on improving the efficiency of energy were conducted in the state of Roraima in Brazil [47], and also this country has implemented the Energy Efficiency Obligation Program [52]. In Paraguay, the National Committee for Energy Efficiency (CNEE) was created in 2011, which is responsible for the preparation and implementation of the National Plan for the Efficient Use of Energy [47]. Regarding Cuba, we consider that it is part of this ranking because the relationship between electricity generation, GDP and CO_2 emissions corresponds to an efficient behavior, confirming the results of Somoza et al. [53], who used a stochastic frontier as their methodology for analysis.

The other seven countries in the first group are Argentina, Chile, Colombia, Perú, Venezuela, Ecuador and Jamaica. In this group, a greater variability of efficiency is observed for the first years compared to the variability of the last years. For these countries, the most important source of inefficiency was non-fossil generation. In Venezuela and Argentina, their total inefficiency came from this source. Chile and Colombia presented deficiencies in the two desirable outputs, with non-fossil generation being their main source of inefficiency. Finally, Jamaica, Ecuador and Peru had deficiencies in the two desirable outputs and excesses in the undesirable output. For Jamaica and Ecuador, the main source of inefficiency was non-fossil generation followed by fossil generation, while for Peru the main source of inefficiency was CO_2 emissions followed by fossil generation.

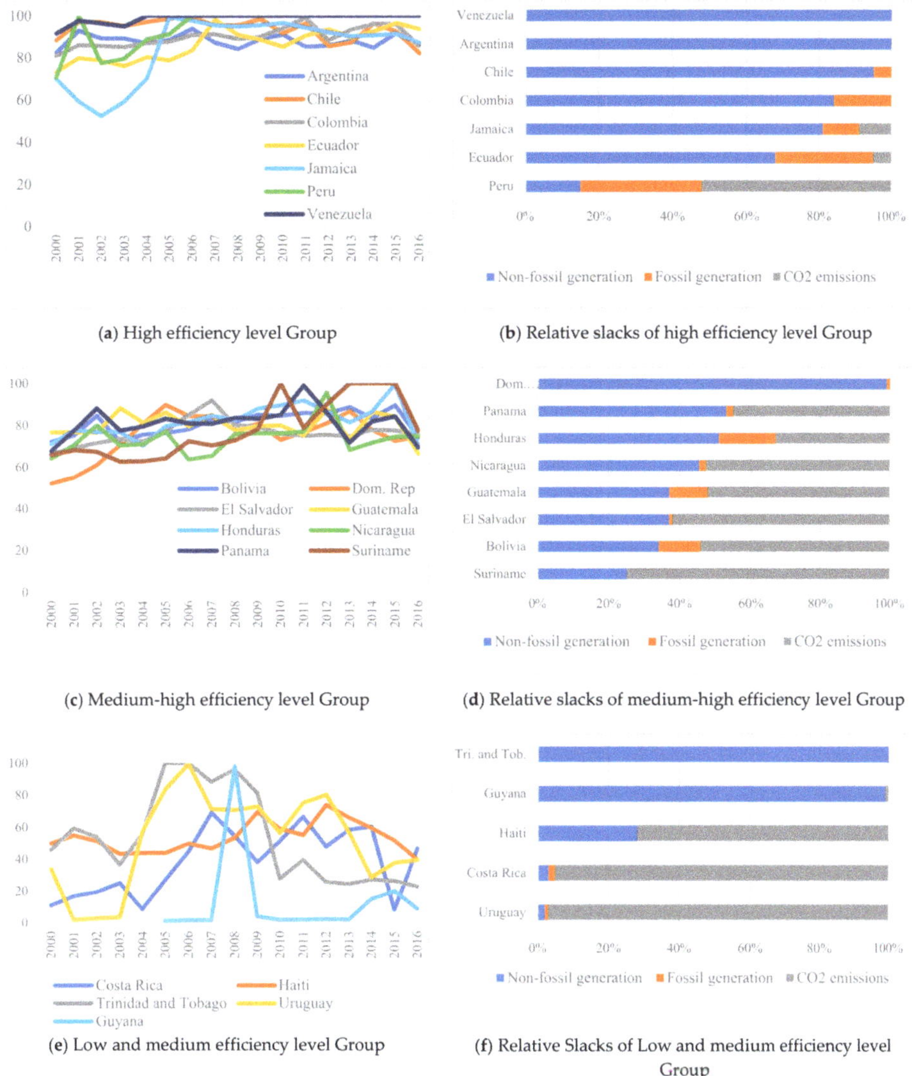

(**a**) High efficiency level Group

(**b**) Relative slacks of high efficiency level Group

(**c**) Medium-high efficiency level Group

(**d**) Relative slacks of medium-high efficiency level Group

(**e**) Low and medium efficiency level Group

(**f**) Relative Slacks of Low and medium efficiency level Group

Figure 4. Efficiency evolution and relative slacks by groups.

We note that the level of consolidation of the institutional environment of these countries is mixed. Colombia, Perú, Venezuela and Ecuador established legal and regulatory frameworks; Colombia did this in the same year as Brazil, while Peru, Venezuela and Ecuador did this long before the other countries. Chile is currently in the process of preparing or discussing a national law, while Jamaica does not include energy efficiency in its main national laws [47]. Similar to Mexico and Brazil, Chile is recognized for having defined some regulatory instruments in terms of energy efficiency [51], and also, similar to Brazil, for having an obligation scheme [52]. Colombia is the country with the highest number of uncharged entities to regulate and monitor the energy efficiency law.

In the second group, there are eight countries with medium-high efficiency levels, ranging from 73 to 82. The countries of Bolivia, Dominican Republic, El Salvador, Guatemala, Honduras, Nicaragua, Panamá and Suriname are included in this group. This group is characterized by exhibiting an increasing trend in the evolution of efficiency and degree of convergence. However, this claim should be tested. In Suriname, Bolivia, El Salvador, Guatemala and Nicaragua, the main source of inefficiency came from CO_2 emissions followed by non-fossil generation; however, of these five, only Suriname did not present slacks in fossil generation. In Honduras and Panama, the main source of inefficiency was non-fossil generation, followed by CO_2 emissions. Finally, almost all of the inefficiency of the Dominican Republic came from non-fossil generation, and it did not present excesses in CO_2 emissions.

The legal framework in terms of energy efficiency in these countries is varied. For example, in the early 2010s, Panamá developed a national law on energy efficiency, while Nicaragua did so in the mid-decade. However, Panamá is aligned with Mexican labeling standards, while the rest of the Latin Americas countries are aligned with the programs defined in the European Union or the United States [47]. The Dominican Republic, El Salvador, Guatemala and Honduras are currently developing national laws, which are either in the process of preparation or in discussion, and we highlight the fact that the Dominican Republic and Guatemala have planned to have only one uncharged entity to regulate and monitor the national law. Finally, Bolivia is the only country that has not shown any regulatory development in this matter [47].

Finally, the last group comprises five countries with medium and low global efficiency, with scores below 54. The countries are Costa Rica, Guyana, Haiti, Trinidad and Tobago and Uruguay. Regarding Costa Rica, Uruguay and Haiti, the results coincide with those obtained by Sánchez et al. [6], who found very low efficiency levels for these countries. This group presents a very high volatility in its efficiency levels, exhibiting scores above 70 and below 25, as is the case of Uruguay, Costa Rica, Guyana and Trinidad and Tobago. In this group, the most important source of inefficiency was CO_2 emissions. Trinidad and Tobago was the only country in this group in which the inefficiency measure depended on only one component: non-fossil generation. The inefficiency of Guyana and Haiti depended on two sources—non-fossil generation and CO_2 emissions—although in Guyana, non-fossil generation was the main source of inefficiency while CO_2 emissions were predominantly responsible in Haiti. Finally, the inefficiency in Costa Rica and Uruguay was caused by deficiencies in the two desirable outputs and excesses in the undesirable output, with the latter being the main source of inefficiency. Finally, the inefficiency in Costa Rica and Uruguay was caused by deficiencies in the two desirable outputs and excesses in the undesirable output, with the latter being the main source of inefficiency. Regarding Uruguay and Costa Rica, they present an average annual efficiency of around 51 and 38, respectively, although Uruguay established both legal and regulatory frameworks in the same year as Brazil and Mexico, and Costa Rica was the first country in Latin America to define a Law of Rational Use of Energy [47]. In addition, IEA [51] did not report the coverage potential of existing mandatory codes and standards in terms of energy efficiency. Haiti and Trinidad and Tobago do not include energy efficiency in any major national laws.

5. Conclusions

In this research, we have carried out an evaluation of the evolution of the technical efficiency of electricity generation for 24 countries in Latin America and the Caribbean during the period 2010–2016. We used the DEA methodology, which allowed the evaluation of the relative efficiency of different production systems for different DMUs through a dynamic model of a CRS based on slacks and incorporated the assumption of weak disposability between electricity generation from fossil sources and CO_2 emissions. Additionally, we tested the assumption of a CRS with the test proposed by Banker (1996) and concluded that the hypothesis of a CRS was not rejected. The proposed model allowed us to establish inefficiencies in the generation methods of 20 of the 24 countries studied.

When both efficient countries and sources of inefficiency are identified, the results found in the research provide relevant information for the 20 inefficient countries, because, through learning, they can adopt best practices in the productive process of generation, with those countries that make better use of their productive capacity as reference points.

The methodology used has some advantages and disadvantages that are worth noting. The advantages mainly correspond to three aspects: (i) the method does not require an explicit mathematical specification for the production or cost function, (ii) it can handle multiple inputs and outputs simultaneously and (iii) the source of the inefficiency can be identified, quantified and analyzed for each DMU.

Regarding the disadvantages, five aspects are particularly important: (i) the results are sensitive to the selection of inputs and outputs, (ii) as a non-parametric technique, the best specification cannot be corroborated, (iii) the number of efficient DMUs increases with the number of inputs and outputs, (iv) the measurement of efficiency is sensitive to outliers, and (v) the dynamic DEA assumes implicitly that there is no technological change over time. Regarding the first disadvantage, in this research, we did not have access to information associated with the labor used in the generation of electricity in each country, which, without a doubt, is an important productive factor of the activity. Therefore, for future studies, it would be interesting to introduce this variable, as previously incorporated in the study of Bi et al. [22].

Another important point is that this study focuses solely on the measurement of the technical efficiency of electricity generation, leaving aside the evaluation of the efficiency of allocation. Because of this, we did not consider the electricity rates in each country. The countries found in this study to be the most efficient do not have lower rates per unit of electricity than those that are less efficient (in terms of technical efficiency). Besides, the total losses of electricity in the transmission and distribution systems are not considered; therefore, the study does not include an evaluation of the efficiency of the electricity systems.

Finally, the results suggest that the most efficient countries have developed an institutional and legal context for energy efficiency, accompanied by other market incentives, as well as information mechanisms to improve energy efficiency. While less-efficient countries have developed the legal context recently or do not plan to do so yet, these types of countries should implement the strategies of Brazil or Mexico, which border these countries.

Author Contributions: Conceptualization, S.C. and Y.E.R.; methodology, S.C.; software, S.C.; validation, S.C. and Y.E.R.; formal analysis, S.C. and Y.E.R.; investigation, S.C. and Y.E.R.; resources, S.C. and Y.E.R.; data curation, S.C.; writing—original draft preparation, S.C., Y.E.R. and J.C.; writing—review and editing, Y.E.R. and J.C.; visualization, S.C. and Y.E.R.; supervision, Y.E.R. and J.C.; project administration, Y.E.R. All authors have read and agreed to the published version of the manuscript.

Funding: This research received no external funding.

Conflicts of Interest: The authors declare no conflict of interest.

Nomenclature

DEA	Data envelopment analysis
DMU	Decision-making units
GHG	Greenhouse gases
GW	Gigawatt
TWh	Terawatt-hours
MTm	Million metric tons
DO_NF	Desirable output: non-fossil generation
DO_F	Desirable output: fossil generation
GDP	Gross Domestic Product
IC_NF	Installed capacity: non-fossil sources
IC_F	Installed capacity: fossil sources
SDO_NF	Slack associated with desirable output: non-fossil generation
SDO_F	Slack associated with desirable output: fossil generation
SUO_CO_2	Slack associated with undesirable output: CO_2 emissions
S_GDP	Slack associated with GDP per capita
SIC_NF	Slack associated with installed capacity: non-fossil sources
SIC_F	Slack associated with installed capacity: fossil sources

References

1. Mahmoodi, M. The Relationship between Economic Growth, Renewable Energy, and CO_2 Emissions: Evidence from Panel Data Approach. *Int. J. Energy Econ. Policy* **2017**, *7*, 96–102.
2. EIA. International–Electricity. Available online: https://www.eia.gov/international/data/world/electricity/electricity-generation? (accessed on 17 March 2020).
3. REN21. Renewables 2017 Global Status Report. Available online: https://www.ren21.net/wp-content/uploads/2019/05/GSR2017_Full-Report_English.pdf (accessed on 15 January 2020).
4. Eberle, A.; Heath, G. Projected Growth in Small-Scale, Fossil-Fueled Distributed Generation: Potential Implications for the U.S. Greenhouse Gas Inventory. United States. 2018. Available online: https://www.osti.gov/biblio/1431247-projected-growth-small-scale-fossil-fueled-distributed-generation-potential-implications-greenhouse-gas-inventory (accessed on 15 January 2020).
5. IEA. 2018. Available online: https://webstore.iea.org/co2-emissions-from-fuel-combustion-2018-highlights (accessed on 15 January 2020).
6. Sánchez, L.; Vásquez, C.; Viloria, A. The data envelopment analysis to determine efficiency of Latin American countries for greenhouse gases control in electric power generation. *Int. J. Energy Econ. Policy* **2018**, *8*, 197–208. Available online: https://www.econjournals.com/index.php/ijeep/article/view/6360 (accessed on 31 November 2019).
7. Charnes, A.; Cooper, W.W.; Rhodes, E. Measuring the efficiency of decision making units. *Eur. J. Oper. Res.* **1978**, *2*, 429–444. [CrossRef]
8. Mariz, F.B.; De Almeida, M.R.; Aloise, D. A review of Dynamic Data Envelopment Analysis: State of the art and applications. *Int. Trans. Oper. Res.* **2017**, *25*, 469–505. [CrossRef]
9. Faere, R.; Grosskopf, S.; Lovell, C.A.K.; Pasurka, C. Multilateral Productivity Comparisons When Some Outputs are Undesirable: A Nonparametric Approach. *Rev. Econ. Stat.* **1989**, *71*, 90. [CrossRef]
10. Sueyoshi, T.; Yuan, Y.; Goto, M. A literature study for DEA applied to energy and environment. *Energy Econ.* **2017**, *62*, 104–124. [CrossRef]
11. Golany, B.; Roll, Y.; Rybak, D. Measuring efficiency of power plants in Israel by data envelopment analysis. *IEEE Trans. Eng. Manag.* **1994**, *41*, 291–301. [CrossRef]
12. Shermeh, H.E.; Najafi, S.; Alavidoost, M. A novel fuzzy network SBM model for data envelopment analysis: A case study in Iran regional power companies. *Energy* **2016**, *112*, 686–697. [CrossRef]
13. Khalili-Damghani, K.; Tavana, M.; Haji-Saami, E. A data envelopment analysis model with interval data and undesirable output for combined cycle power plant performance assessment. *Expert Syst. Appl.* **2015**, *42*, 760–773. [CrossRef]
14. Yang, H.; Pollitt, M.G. The necessity of distinguishing weak and strong disposability among undesirable outputs in DEA: Environmental performance of Chinese coal-fired power plants. *Energy Policy* **2010**, *38*, 4440–4444. [CrossRef]
15. Sueyoshi, T.; Goto, M. DEA environmental assessment of coal fired power plants: Methodological comparison between radial and non-radial models. *Energy Econ.* **2012**, *34*, 1854–1863. [CrossRef]
16. Cherchye, L.; De Rock, B.; Walheer, B. Multi-output efficiency with good and bad outputs. *Eur. J. Oper. Res.* **2015**, *240*, 872–881. [CrossRef]
17. Chang, D.S.; Yang, F.-C. Assessing the power generation, pollution control, and overall efficiencies of municipal solid waste incinerators in Taiwan. *Energy Policy* **2011**, *39*, 651–663. [CrossRef]
18. Tao, Y.; Zhang, S. Environmental efficiency of electric power industry in the Yangtze River Delta. *Math. Comput. Model.* **2013**, *58*, 927–935. [CrossRef]
19. Dogan, N.O.; Tugcu, C.T. Energy Efficiency in Electricity Production: A Data Envelopment Analysis (DEA) Approach for the G-20 Countries. *Int. J. Energy Econ. Policy* **2015**, *5*, 246–252. Available online: https://ideas.repec.org/a/eco/journ2/2015-01-19.html (accessed on 31 November 2019).
20. Whiteman, J.L. The technical efficiency of developing country electricity systems. In Proceedings of the 1995 International Conference on Energy Management and Power Delivery EMPD '95, Singapore, 21–23 November 1995; Volume 2, pp. 504–509.
21. Yunos, J.M.; Hawdon, D. The efficiency of the National Electricity Board in Malaysia: An intercountry comparison using DEA. *Energy Econ.* **1997**, *19*, 255–269. [CrossRef]

22. Bi, G.; Liang, N.; Wu, J.; Liang, L. Radial and non-radial DEA models undesirable outputs: An application to OECD countries. *Int. J. Sustain. Soc.* **2010**, *2*, 133. [CrossRef]

23. Zhou, P.; Ang, B.; Wang, H. Energy and CO_2 emission performance in electricity generation: A non-radial directional distance function approach. *Eur. J. Oper. Res.* **2012**, *221*, 625–635. [CrossRef]

24. Li, J.; Geng, X.; Li, J. A Comparison of Electricity Generation System Sustainability among G20 Countries. *Sustainability* **2016**, *8*, 1276. [CrossRef]

25. Dyson, R.; Allen, R.; Camanho, A.; Podinovski, V.; Sarrico, C.; Shale, E. Pitfalls and protocols in DEA. *Eur. J. Oper. Res.* **2001**, *132*, 245–259. [CrossRef]

26. Barros, J. Performance measurement of thermoelectric generating plants with undesirable outputs and random parameters. *Int. J. Electr. Power Energy Syst.* **2013**, *46*, 228–233. [CrossRef]

27. Tone, K. Dealing with Undesirable Outputs in DEA: A Slacks-based Measure (SBM), 2003 Approach. Available online: https://www.researchgate.net/profile/Kaoru_Tone/publication/284047010_Dealing_with_undesirable_outputs_in_DEA_a_Slacks-Based_Measure_SBM_approach/links/56ebdf2b08aee4707a384415/Dealing-with-undesirable-outputs-in-DEA-a-Slacks-Based-Measure-SBM-approach.pdf (accessed on 4 December 2020).

28. Koltai, T.; Lozano, S.; Uzonyi-Kecskés, J.; Moreno, P. Evaluation of the results of a production simulation game using a dynamic DEA approach. *Comput. Ind. Eng.* **2017**, *105*, 1–11. [CrossRef]

29. Färe, R.; Grosskopf, S. Introduction. In *Intertemporal Production Frontiers: With Dynamic DEA*; Springer Science and Business Media LLC: Heidelberg/Berlin, Germany, 1996; pp. 1–8.

30. Tone, K.; Tsutsui, M. Dynamic DEA: A slacks-based measure approach. *Omega* **2010**, *38*, 145–156. [CrossRef]

31. Walheer, B. Economic growth and greenhouse gases in Europe: A non-radial multi-sector nonparametric production-frontier analysis. *Energy Econ.* **2018**, *74*, 51–62. [CrossRef]

32. Amowine, N.; Ma, Z.; Li, M.; Zhou, Z.; Asunka, B.A.; Amowine, J. Energy Efficiency Improvement Assessment in Africa: An Integrated Dynamic DEA Approach. *Energies* **2019**, *12*, 3915. [CrossRef]

33. Guo, X.; Lu, C.-C.; Lee, J.-H.; Chiu, Y.-H. Applying the dynamic DEA model to evaluate the energy efficiency of OECD countries and China. *Energy* **2017**, *134*, 392–399. [CrossRef]

34. Şimşek, N. Energy Efficiency with Undesirable Output at the Economy-Wide Level: Cross Country Comparison in OECD Sample. *Am. J. Energy Res.* **2014**, *2*, 9–17. [CrossRef]

35. Tsai, W.-H.; Lee, H.-L.; Yang, C.-H.; Huang, C.-C. Input-Output Analysis for Sustainability by Using DEA Method: A Comparison Study between European and Asian Countries. *Sustainability* **2016**, *8*, 1230. [CrossRef]

36. Chen, T.-Y.; Yeh, T.-L.; Lee, Y.-T. Comparison of Power Plants Efficiency among 73 Countries. *J. Energy* **2013**, *2013*, 1–8. [CrossRef]

37. Ghosh, S. Electricity supply, employment and real GDP in India: Evidence from cointegration and Granger-causality tests. *Energy Policy* **2009**, *37*, 2926–2929. [CrossRef]

38. Lean, H.H.; Smyth, R. Multivariate Granger causality between electricity generation, exports, prices and GDP in Malaysia. *Energy* **2010**, *35*, 3640–3648. [CrossRef]

39. Yoo, S.-H.; Kim, Y. Electricity generation and economic growth in Indonesia. *Energy* **2006**, *31*, 2890–2899. [CrossRef]

40. Nemoto, J.; Goto, M. Dynamic data envelopment analysis: Modeling intertemporal behavior of a firm in the presence of productive inefficiencies. *Econ. Lett.* **1999**, *64*, 51–56. [CrossRef]

41. Jiekang, W.; Zhuangzhi, G.; Fan, W. Short-term multi-objective optimization scheduling for cascaded hydroelectric plants with dynamic generation flow limit based on EMA and DEA. *Int. J. Electr. Power Energy Syst.* **2014**, *57*, 189–197. [CrossRef]

42. Tone, K. Dealing with undesirable outputs in DEA: A Slacks-Based Measure (SBM) approach. In Proceedings of the North American Productivity Workshop 2004, Toronto, ON, Canada, 23–25 June 2004; pp. 44–45.

43. Zhou, Z.; Liu, W. DEA Models with Undesirable Inputs, Intermediates, and Outputs. In *Multiple Criteria Decision Analysis*; Springer Science and Business Media LLC: Heidelberg/Berlin, Germany, 2015; pp. 415–446.

44. Banker, R.D. Hypothesis tests using data envelopment analysis. *J. Prod. Anal.* **1996**, *7*, 139–159. [CrossRef]

45. Banker, R.D.; Charnes, A.; Cooper, W.W. Some Models for Estimating Technical and Scale Inefficiencies in Data Envelopment Analysis. *Manag. Sci.* **1984**, *30*, 1078–1092. [CrossRef]

46. Fischer, M.M.; Wang, J. Introduction. *Spat. Econom.* **2011**, 1–11. [CrossRef]

47. Ravillard, P.; Carvajal, F.; Soto, D.D.L.; Chueca, J.E.; Antonio, K.; Ji, Y.; Hallack, M.C.M. Towards Greater Energy Efficiency in Latin America and the Caribbean: Progress and Policies, Publications. Available online: https://publications.iadb.org/publications/english/document/Towards_Greater_Energy_Efficiency_in_Latin_America_and_the_Caribbean_Progress_and_Policies.pdf (accessed on 4 December 2020).

48. Williamson, E. The Economic Institutions of Capitalism. *J. Econ. Issues* **1987**, *21*, 528–530. [CrossRef]

49. Wiese, C.; Larsen, A.; Pade, L.-L. Interaction effects of energy efficiency policies: A review. *Energy Effic.* **2018**, *11*, 2137–2156. [CrossRef]

50. Gerarden, T.D.; Newell, R.G.; Stavins, R.N. Assessing the Energy-Efficiency Gap. *J. Econ. Lit.* **2017**, *55*, 1486–1525. [CrossRef]

51. International Energy Agency. Market-based Instruments for Energy Efficiency: Policy Choice and Design. 2017. Available online: https://www.raponline.org/knowledge-center/market-based-instruments-energy-efficiency-policy-choice-design/ (accessed on 4 December 2020).

52. Lees, E.; Bayer, E. Toolkit for Energy Efficiency Obligations; Regulatory Assistance Project. Brussels, Belgium. 2016. Available online: https://www.raponline.org/knowledge-center/toolkit-for-energy-efficiency-obligations/ (accessed on 4 December 2020).

53. Somoza, J.; Bano, J.; Llorca, M. La medición de la eficiencia energética y su contribución en la mitigación de las emisiones de CO^2 para 26 países de América Latina y el Caribe. *Econ. Desarro.* **2014**, *152*, 87–106. Available online: http://scielo.sld.cu/scielo.php?pid=S0252-85842014000200006&script=sci_abstract&tlng=es (accessed on 4 December 2020).

Publisher's Note: MDPI stays neutral with regard to jurisdictional claims in published maps and institutional affiliations.

Article

Introduction to the Dynamics of Heat Transfer in Buildings

Bożena Babiarz [1,2,*] and Władysław Szymański [2]

[1] Department of Heat Engineering and Air Conditioning, The Faculty of Civil and Environmental Engineering and Architecture, Rzeszow University of Technology, 12 Powstańców Warszawy Av., 35-959 Rzeszow, Poland
[2] Polish Association of Sanitary Engineers and Technicians, 35-959 Rzeszów, Poland; wszym9@wp.pl
* Correspondence: bbabiarz@prz.edu.pl; Tel.: +48-017-865-1445

Received: 13 October 2020; Accepted: 5 December 2020; Published: 7 December 2020

Abstract: Changing climatic conditions cause the variability of the parameters of the building's surroundings, which in turn causes both the gains and losses of heat to change over time. There is variability in both daily and annual cycles. Meeting the requirements of thermal comfort in rooms requires maintaining the required parameters, including constant temperature. Heat gains and losses must be balanced, and this balance is ensured through appropriate heating systems. At the same time, the above means that the demand for heating buildings is not constant but depends on external weather conditions and the energy efficiency of the building. This, in turn, affects the thermal inertia, causing changes in the partition temperature to occur slower than the changes in air temperature. Therefore, the amplitude of the heating power changes is not proportional to the amplitude of the outside air temperature change. The paper presents an example of the analysis of thermal dynamics in buildings. Various aspects of heat transfer in the building were investigated taking into account the transient conditions. The variability of temperature over time at different depths of the partition was analysed, showing the results graphically. The periodic variability of the outside air temperature and the intensity of solar radiation were described by the Fourier series. Moreover, the article shows the influence of the thermal insulation thickness of the external wall on the annual amplitude of temperature changes and on the duration of the heating season, which is important from the point of view of optimization.

Keywords: dynamics; heat transfer in buildings; heat losses; buildings; thermal power; heating

---

1. Introduction

The issue of heat transfer dynamics is closely related to the subject of the energy efficiency of buildings, which is important at the design and construction stages, as well as during the operation of buildings or their parts. This is visible in many legal regulations and policies aimed at improving the energy efficiency of buildings. This is due to the fulfillment of the provisions of Art. 7 of Directive 2002/91/EC [1] and Art. 20 of Directive 2010/31/EU on the energy performance of buildings [2], according to which European Union Member States must take measures to provide all participants in the construction process with a wide range of information on different methods and practices for improving the energy performance of buildings. Moreover, Art. 12 of Directive 2012/27/EU on energy efficiency [3] obliges EU Member States to take appropriate measures to promote and enable the efficient use of energy by consumers. The provisions of the above directives have been implemented in the Polish legal system through Art. 11 sec. 1 of the Act of 15 April 2011, amended on 20 May 2016 on energy efficiency, [4] and Art. 40 of the Act of 29 August 2014 on the energy performance of buildings [5]. According to Directive (EU) 2018/844 of the European Parliament and of the Council of 30 May 2018 [6], amending Directive 2010/31/EU on the energy performance of buildings and Directive 2012/27/EU on energy efficiency, clear and ambitious targets for the renovation of the existing building stock have a great significance. Therefore, efforts to improve the energy performance of buildings would actively contribute to enhance the energy independence of the Union and would also have enormous potential to create jobs in the Union. In this context, Member States should take into account the need to clearly link their long-term renovation strategies to relevant initiatives to support skills development and training in the construction and energy efficiency sectors. These provisions oblige the minister responsible for construction, spatial planning, and development and housing to conduct information, education, and training activities regarding available energy efficiency improvement measures, as well as to conduct a campaign of information to improve the energy performance of buildings. The purposefulness and methodology of determining the energy performance of a building result from the regulations [7,8]. Activities in the field of improving the energy efficiency of buildings, which are the subject of many scientific publications, such as [9], are part of shaping the climate and energy policy, ensuring, inter alia, the reduction of greenhouse gas emissions and constitute one of the most important challenges resulting from membership in the European Union. The Union is committed to efforts to develop a sustainable, competitive, safe and low-carbon energy system while maintaining the security of heat and energy supplies. In the context of infrastructure responsible for ensuring the security of heat supply, an important issue is its assessment in the aspect of supply security, taking into account economic and environmental conditions, presented in the paper [10]. The results of energy saving calculations may be of interest for the investors, engineers, and policy makers who intend to minimize the difference between the planned and real energy savings analyzed in the paper [11]. Aspects of energy savings and energy supply management in buildings have been analyzed in many publications, e.g., papers [12,13], with elements of heat supply safety simulation presented in the paper [14]. These issues play an important role in social, technical, and political terms. These aspects are related also with heat losses in the buildings and district heating systems, which was also underlined in the work [15]. The analysis of sensitivity of energy distribution for residential buildings is presented in the paper [16]. The energy reduction effects of the thermal labyrinth system were analyzed in the paper [17].

In changing climatic conditions, phenomena occurring in buildings are influenced by a number of parameters, such as: air temperature and humidity, wind direction and speed, cloudiness, azimuth and height of the sun, solar radiation, or even the management of the surroundings. The analysis of external climate parameters, such as temperature, air humidity and wind conditions, for the needs of outdoor thermal comfort have been included in paper [18].

The heat balance in buildings results from the analysis of heat losses and gains. It is made for a building, assuming appropriate parameters in order to select appropriate devices to meet the requirements of thermal comfort. Example of analysis of indoor air parameters contain the work [19,20]. Thermal comfort optimization in microgrids equipped with renewable energy sources and energy storage units was analyzed in the paper [21].

Issues of heat transfer in heat exchangers was emphasized in the work [22]. There is a correlation between some parameters of the isolation of buildings and the wind free stream velocity and wind-to-surface angle. In the work [23], it has been shown that the convective heat transfer coefficient value strongly depends on the wind velocity. The influence of the thermal insulation thicknesses of external walls on heating cost from the ecological and economic assessment is analyzed in [24].

The way to achieve high energy efficiency of buildings along with the required quality of the internal environment are advanced technologies in both control [25] and construction where, for example, phase change materials can be used [26]. Examples of building energy management analyzes using increased thermal capacitance and thermal storage management are shown in work [27]. Tools for increasing energy efficiency in the examples [28] and in the integration of HVAC systems are presented in the works [29,30].

In order to analyze the modelling and simulation of heat transfer in buildings, the theory and application of this type of tool are collected and characterized by Clarke [31]. The methods to analyzing building energy and control systems are often used, such as using the equation-based, object-oriented

Modelica in the paper [32]. Methods based on the coupling of three different types of simulation models, namely spectral optical model, computational fluid dynamics model, and building energy simulation, are presented in [33]. Physical phenomena, notably optical, thermodynamic, and fluid dynamic processes, have been analyzed for commercial buildings with double-skin façades. The modelling of heat transfer taking advantage of heat energy accumulation in building walls is the goal of the work [34]. The paper is focused on the future optimization of a control strategy. The issue of simulating heat transfer through point thermal bridges is the subject of the paper [35].

On the basis of the literature analysis of the subject of this work, it can be stated that there is a lack of ordering and development of methods for analyzing heat transfer dynamics using changeable external conditions. Existing works in this field mainly deal with the issues of energy control and control of the HVAC system's parameters, while there are no studies taking into account the changeability of atmospheric conditions and their impact on the dynamics of heat transfer. We realize that this is important in a changing climate, where the heating season and the summer season stand out. The issue presented in the article may be helpful in the analysis of the thermal inertia of a building in order to optimize the operation of HVAC systems.

The literature review presented in the article confirms that the approach to dynamics of heat transfer in buildings, proposed in the manuscript, is innovative with regard to the analysis of the impact of the variability of external conditions on energy efficiency and it was not previously applied in this way. This issue was the subject of this work.

2. Methodology

The room can be treated as a closed object, limited by building partitions, located in the space, treated as the surroundings. Due to the lack of a thermodynamic equilibrium between them, there are energy interactions between the room and the surroundings. To explain it simply, there is a heat transfer, considered as heat losses or gains.

Heat gains can originate from:

- heating, Q_H,
- the sun through non-transparent partitions, Q_{SE},
- the sun through transparent partitions, Q_{os},
- permeating from adjacent rooms, Q_{iw},
- people, Q_l,
- devices, Q_u,
- lighting, Q_e

Heat losses can be caused:

- by penetrating through the external partitions Q_{si},
- by penetrating into adjacent Q_{sw} rooms,
- for Q_V ventilation.

The thermal balance includes heat fluxes presented schematically in Figure 1.

Heat gains from people Q_l, from devices Q_u, and from lighting Q_e can be taken together as internal heat gains Q_w.

$$Q_w = Q_l + Q_u + Q_e \tag{1}$$

Due to variable environmental parameters, both heat gains and losses change over time. For a room to be kept at a constant temperature, the heat gains and heat losses must be balanced, and this balance is ensured through appropriate heating systems. At the same time, the above means that the demand for heat for heating facilities is not constant but depends on external weather conditions. The weather conditions that determine heat exchange are the temperature of the outside air and solar radiation.

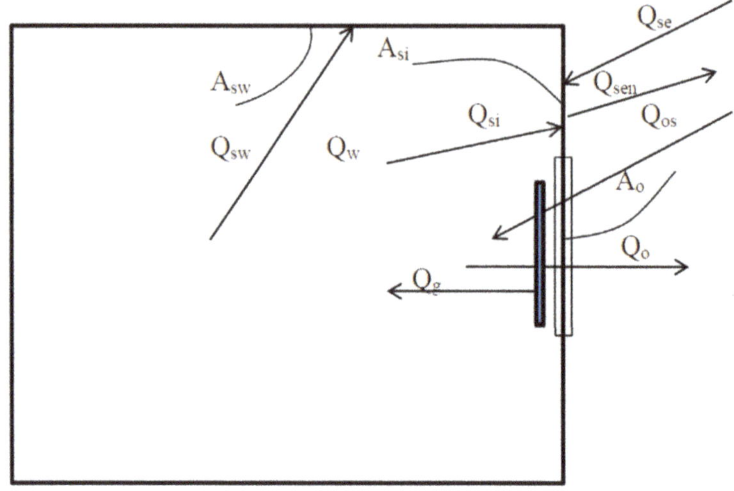

Figure 1. Balance sheet diagram.

Due to the thermal inertia, changes in partition temperature are slower than changes in air temperature. Therefore, the amplitude of changes in heating power Q_g is not proportional to the amplitude of outside air temperature change T_e.

The equation for the heat balance of the room is as follows:

$$Q_g + Q_{os} + Q_w = Q_o + Q_{si} + Q_V + Q_{sw} \tag{2}$$

The individual components of Equation (1) have been explained below.
Heat transferred from indoor air to the wall:

$$Q_{si} = \alpha_i \cdot A_{si} \cdot (T_i - T_{wi}) \tag{3}$$

Heat transferred from indoor air to interior walls:

$$Q_{sw} = \alpha_i \cdot A_{sw} \cdot (T_i - T_{sw}) \tag{4}$$

The heat of solar radiation penetrating the window [8]:

$$Q_{os} = z \cdot w_s \cdot I \cdot A_o \tag{5}$$

Heat from internal sources:

$$Q_w = q_A \cdot A_f \tag{6}$$

Heat to prepare the ventilation air:

$$Q_V = V \cdot \rho \cdot c_p \cdot (T_i - T_e) \tag{7}$$

or

$$Q_w = V_{Ve} \cdot A_f \tag{8}$$

Heat loss through the window:

$$Q_o = U_o \cdot A_o \cdot (T_i - T_e) \tag{9}$$

The temperature of the internal surface of an outer wall is a result of the influence of the inner environment and the heat conduction in this wall.

The heat conduction in the wall is caused by the temperature distribution, which is unsteady due to changing external climatic conditions.

The temperature of the outer surface of the outer wall is influenced by the transfer of heat to the outside air, the absorption of solar radiation, and the emission of radiation to the sky.

Heat transferred from the outer wall surface to the outside air:

$$Q_{se} = \alpha_e \cdot A_{se} \cdot (T_{we} - T_e) \tag{10}$$

Radiant heat losses from the outer wall to the skyfall:

$$Q_{sen} = \sigma \cdot \varepsilon_{sn} \cdot A_{se} \cdot \left(T_n^4 - T_{we}^4\right) \tag{11}$$

Solar radiation of heat absorbed by the outer surface of the outer wall [8]:

$$Q_{re} = \varepsilon_s \cdot A_{se} \cdot I \tag{12}$$

Signs:

α_i—coefficient of heat transfer from the wall surface to the internal air,
α_e—heat transfer coefficient from the external wall surface to the outside air,
A_{si}—the surface of the outer wall inside the room,
A_{sw}— the surface of the internal walls of the room,
A_{se}—the outer surface of the outer wall,
A_f—reference surface (floors),
A_o—window area,
V—ventilation air stream,
q_A—indicator of internal heat sources,
V_{Ve}—ventilation rate,
z—shading coefficient,
w_s—radiation transmittance coefficient,
$\sigma_s = 5.67 \times 10^{-8}$ [W/m^{2}K^4]—radiation constant,
ε_{sn}—radiation absorption coefficient of the outer surface of the outer wall,
ε_s—radiation emission coefficient of the outer surface of the outer wall,
I—solar radiation intensity,
T_i—internal air temperature,
T_e—outside air temperature,
T_{wi}—temperature of the inner surface of the outer wall,
T_{sw}—surface temperature of internal walls,
T_{we}—external wall surface temperature,
T_n—skyfall temperature.

Equation (2) shows the required heat output to heat the room during the heating season.

$$Q_g = (Q_o + Q_{si} + Q_V + Q_{sw}) - (Q_{os} + Q_w) \tag{13}$$

The use of a heating device with the required thermal power and automatic temperature control results in maintaining the room temperature in accordance with the regulations [7].

In most cases, this temperature is taken as a constant value (T_i = const).

Outside the heating season, the heating devices are turned off (Q_g = 0), and the internal air temperature is determined based on the thermal balance.

After using the thermal balance equations and transformations, the internal air temperature is described by the equation:

$$T_i = \frac{z \cdot w_s \cdot I \cdot A_o + q_A \cdot A_f + \left(U_o \cdot A_o + V \cdot \rho \cdot c_p\right) \cdot T_e + \alpha_i \cdot A_{si}\, T_{wi} + \alpha_i \cdot A_{sw}\, T_{sw}}{\left(U_o \cdot A_o + V \cdot \rho \cdot c_p\right) + \alpha_i \cdot A_{si} + \alpha_i \cdot A_{sw}} \tag{14}$$

Due to the variable temperature of the external air, outside the heating season also the temperature of the inside air is variable. The temperature of the inner surface of the walls is also variable.

This temperature is the result of heat transfer through the external wall between the room and the outside air.

With the simplifying assumption that the outside air temperature changes periodically according to the cosine function, the solution to the problem of heat conduction in a semi-infinite medium is described by the equations, according to [36,37]:

It was assumed that the air temperature changes according to the equation:

$$T_f = T_{fo} \cdot \cos(\omega \cdot t) \tag{15}$$

in which

$$\omega = \frac{2 \cdot \pi}{t_o} \tag{16}$$

$$\nu = \frac{\omega}{2 \cdot \pi} \tag{17}$$

where:

T_{fo}—amplitude of air temperature changes,

t—time,

ω—period of temperature changes,

t_o—change period time,

ν—frequency of changes,

The temperature T at depth x, below the surface, is described by the following dependencies:

$$T = C_2 \cdot T_{fo} \cdot e^{-\sqrt{\frac{\omega}{2 \cdot a}} \cdot x} \cdot \cos\left(\omega \cdot t + \sqrt{\frac{\omega}{2 \cdot a}} \cdot x + C_2\right) \tag{18}$$

$$C_1 = \frac{1}{\sqrt{1 + 2 \cdot \frac{\lambda}{\alpha} \cdot \sqrt{\frac{\omega}{2 \cdot a}} + 2 \cdot \left(\frac{\lambda}{\alpha} \cdot \sqrt{\frac{\omega}{2 \cdot a}}\right)^2}} \tag{19}$$

$$C_2 = -\mathrm{arctg} \frac{1}{1 + \frac{\alpha}{\lambda} \cdot \sqrt{\frac{2 \cdot a}{\omega}}} \tag{20}$$

where:

λ—thermal conductivity,

a—thermal diffusivity of the area,

α—coefficient of heat transfer from air to the surface.

The constant C_1 determines the degree of air temperature reduction resulting from the transfer of heat, while the constant C_2 means the delay in wave propagation due to the transfer of heat to the surface.

3. Dynamics of Heat Transfer through an External Wall—Case Study

The external wall of the room is subjected to variable outside air temperature. Moreover, the external surface is influenced by the variable intensity of solar radiation. Climatic conditions cause periodic temperature variability in daily and annual cycles. Climate changes also cause changes in the temperature of external surfaces such as external building envelopes. Surface temperature changes are transferred deep into the material according to the principles of heat transfer. With a sufficient wall thickness, it can be treated as a semi-infinite medium for which the heat conduction problem has been analytically solved.

For example data:

$T_{fo} = 20\,°C,$
$a = 0.485 \times 10^{-6}\ m^2/s,$
$t_o = 24\ h,$
$\alpha = 12\ W/(m^2{\cdot}K),$
$\lambda = 0.82\ W/(m^2{\cdot}K).$

The tendencies of temperature changes in the semi-infinite medium is shown in Figure 2. A negative coordinate x indicates temperature changes in the adjacent fluid.

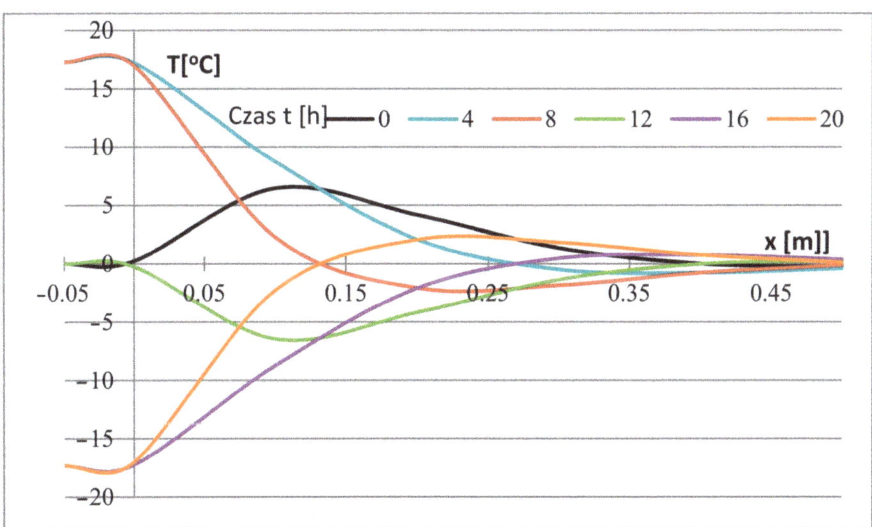

Figure 2. Temperature changes along the depth in a semi-infinite medium [38].

The graphically presented temperature changes in the medium and equations [36,37] allow for the following conclusions:

- There are temperature oscillations in the material with the same period in each plane, but with a phase shift in relation to the surface.
- The amplitude of temperature changes on the surface is smaller than the amplitude of changes in air temperature (C1 < 1).
- There is a phase shift in temperature changes (C2 < 0).

- The amplitude of temperature changes in the material quickly decreases with depth.
- The lower the frequency of temperature changes, the greater the amplitudes at the same depth.

The variability of temperature over time at different depths is shown in Figure 3. The presented considerations apply to a semi-infinite medium. However, they may be the basis for the analysis of temperature distribution in media with a finite, sufficiently large thickness. An example of such a medium are the external walls of a building, which is influenced by external air of variable temperature. There is a daily and annual periodicity of changes. Daily changes do not significantly affect the temperature distribution inside the wall under the surface, especially on the inner wall surface (from the room side). Annual changes, on the other hand, affect heat losses and, consequently, seasonal heat demand for heating.

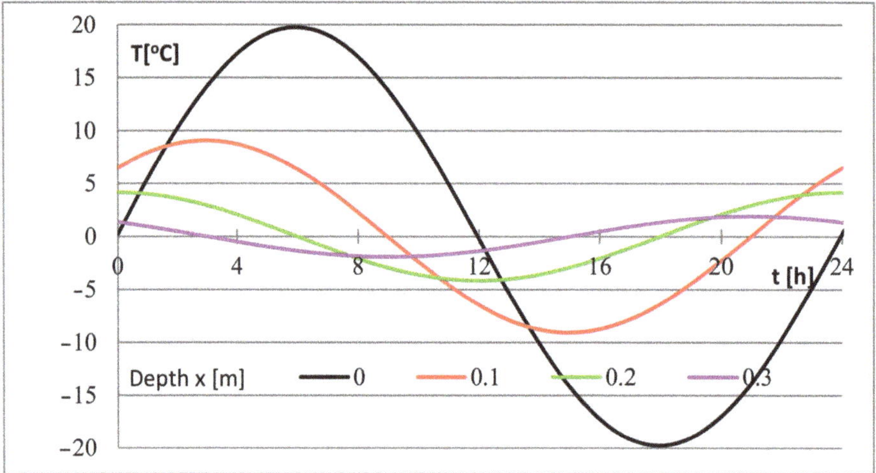

Figure 3. An example of temperature changes with time at depth x in a semi-infinite medium [38].

In relation to the semi-infinite medium model, the following differences should be taken into account:

- Changes in outside air temperature are periodic, but a description of the changes with a simple cosine function would be a simplification.
- In addition to the transfer of heat between the outer surface and the air, there is a heat exchange by radiation (solar radiation).
- The outer wall is usually multi-layered and therefore heterogeneous.

The periodic variability of outdoor air temperature and solar radiation intensity can be described by the Fourier series of the form [39]:

$$y = \frac{1}{2}a_0 + \sum_{i=1}^{k} i \cdot a_i \cdot \cos(\omega \cdot t) + \sum_{i=1}^{k} i \cdot b_i \cdot \sin(\omega \cdot t) \tag{21}$$

The coefficients of the Fourier series should be determined on the basis of real data obtained, e.g., from measurements.

The temperature of the outside air and the intensity of solar radiation are climatic parameters and they are measured at meteorological stations. The results of long-term measurements are available on the website of the Ministry of Infrastructure [40].

For the selected weather station, we can read the average monthly and hourly average outside temperature values determined from multi-year measurements.

Using these values, data can be approximated by a Fourier series. For the Rzeszow-Jasionka meteorological station, the annual changes in the values of average daily outside air temperatures are shown in Figure 4, where the line resulting from the approximation of the data by the Fourier series ($T_{e\,(apr)}$) is also plotted.

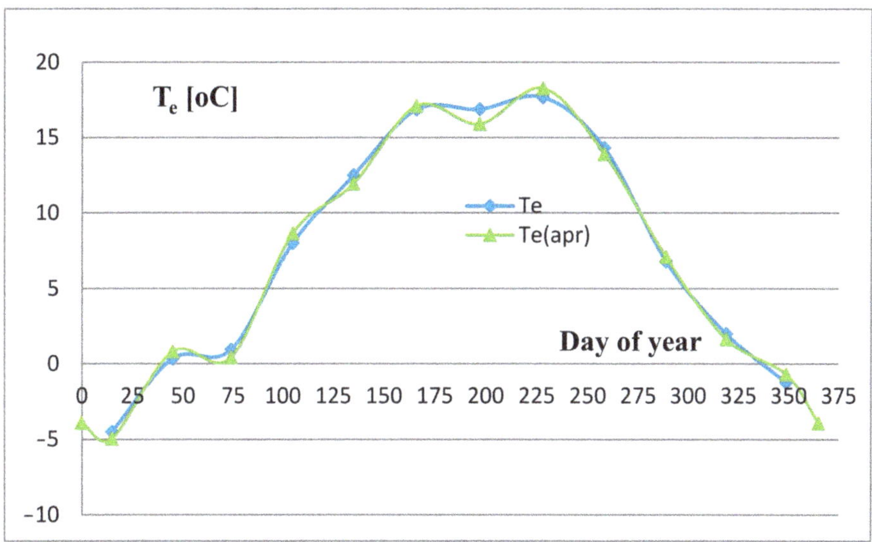

Figure 4. Annual variability of outside air temperature for Rzeszow city.

The parameters of the Fourier series for approximation calculations of the outside air temperature are presented in Table 1.

Table 1. Parameters of the Fourier series for approximation calculations of the outside air temperature.

a_0	a_1	a_2	a_3	a_4	a_5	a_6
7.5416667	−10.20434	−0.475828	0.4847788	−0.658554	−0.804472	0.1787785

b_0	b_1	b_2	b_3	b_4	b_5	b_6
0	−2.751565	0.3030664	0.1518707	−0.543962	−0.141402	−1.064303

As can be seen from the diagram, there is a high agreement of the measurement data with the results obtained from the approximation equation. Significant differences occur in the summer, outside the heating season.

A similar procedure can be applied to record the variability of solar radiation, but due to the consideration of the external wall, only the radiation to the vertical plane will be important.

The Fourier series parameters for the approximation calculations of the radiation intensity per vertical surface (Wh/m^2d) are summarized in Table 2.

Exemplary results for the Rzeszow-Jasionka actinometrical station are shown in Figure 5.

The comparison of the measurement results and those calculated from the Fourier series equation for the direction of the southern and northern radiation is presented in Figure 5.

Marking N in Figure 5 depicts the radiation changes from the North. Labeling S denotes imaging of changes in radiation from the southern side. Marking N-a represents the results of approximation of radiation changes from the North using the Fourier series. Then, by analogy, S-a refers to the results of the approximation of radiation changes from the south side.

Table 2. Fourier series parameters for approximation calculations of the radiation intensity.

Azimuth	a_0	a_1	a_2	a_3	a_4	a_5	a_6
S-E	2148	−1086.93	−290.441	−0.1582	−57.45	61.83312	88.37249
S	2184	−808.599	−411.05	−75.1664	−56.1412	45.27268	88.73687
S-W	2052	−989.688	−246.853	−64.7049	−62.5093	29.71963	82.69103
W	1694	−1152.13	−57.8825	−50.2884	−37.8367	19.73632	68.45242
N-W	1310	−972.617	32.30067	−60.987	−11.8132	9.586983	51.39972
N	1124	−760.472	15.24531	−75.8967	1.364588	11.80952	45.10691
N-E	1358	−1040.44	28.53617	−17.6399	−36.229	17.43642	54.88028
E	1804	−1271.09	−98.6696	13.37886	−43.3159	49.50059	75.13932
S-E	b_1	b_2	b_3	b_4	b_5	b_6	b_1
S	247.2218	89.42463	44.11991	−26.3514	−60.4461	13.67889	247.2218
S-W	220.6692	208.4086	60.60816	−27.2299	−62.535	−10.4756	220.6692
W	312.2481	211.5765	38.31833	−18.5326	−63.0764	6.145611	312.2481
N-W	358.7338	142.2012	0.810727	−13.0379	−51.4592	27.8375	358.7338
N	294.664	75.47576	−16.0399	−8.85253	−30.6874	37.86873	294.664
N-E	262.2807	59.18627	21.32114	−22.0431	−30.7906	42.58792	262.2807
E	317.4635	19.13467	25.85404	−18.7495	−33.7939	45.45755	317.4635

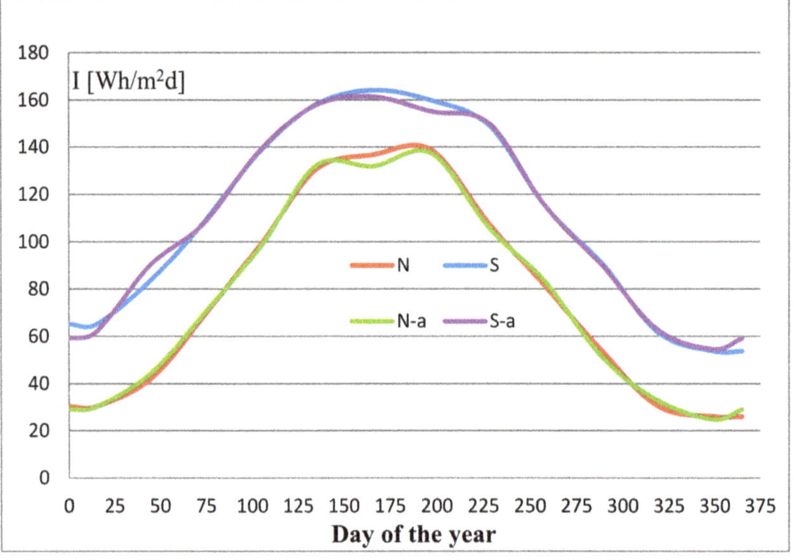

Figure 5. Comparison of the annual variability of the intensity of solar radiation measured and from the approximation equation for Rzeszow city.

Remarks on the compliance of the approximation with the data are the same as in the case of the approximation of the outside air temperature.

Increasing the accuracy of the approximation is possible by increasing the amount of Fourier series components. For the assumed purpose of the analysis of the temperature distribution in the external wall, the assumed accuracy is sufficient.

4. Heat Losses through the External Wall—Case Study

The external wall separates the outside environment and the interior of the room. Such a wall is subjected to an externally variable air temperature as well as solar radiation. Climatic conditions cause periodic temperature variability as well as the variability of radiation intensity, and these changes occur in daily and annual periods. The resulting wall temperature distribution makes it difficult to provide a strict analytical solution, especially in the case of a multi-layer partition.

Simplified methods can be used to solve special cases, e.g., the finite difference method (MRS).

The analysis of heat loss through the external partition is made below, with the following assumptions.

On the side of the inner wall (in the room), the air temperature remains constant. This is the case during the heating season with automatic temperature control (thermostatic valves).

The heat transfer from the inside air to the surface takes place by transfer, with a transfer factor taking into account the radiation.

On the outer side of the outer wall, heat is transferred to the outside air by taking over.

On the outer side of the outer wall, heat transfer also takes place through radiation to the outer space (the sky).

The outside air temperature changes throughout the year according to the climatic conditions.

Solar radiation falls on the outer surface of the wall with periodic, annual variability.

The values of external temperature and radiation intensity were taken as daily averages.

The image of the partition in question, with significant values marked, is shown in Figure 6.

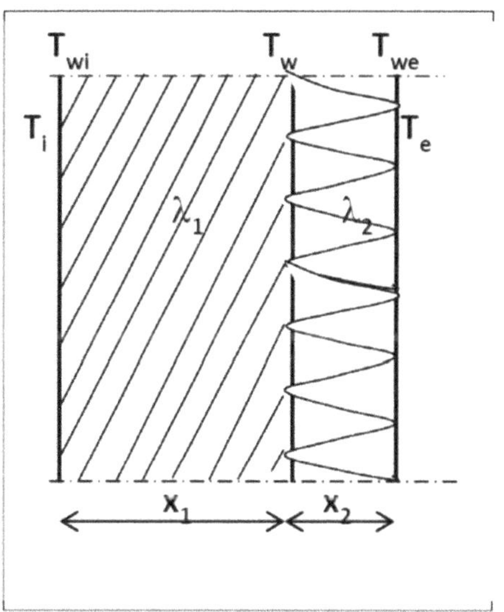

Figure 6. Cross-section through the outer wall.

The variable temperature distribution in the wall was determined by the finite difference method based on heat transfer equations, taking into account the thermal balance of the room.

Due to the variable temperature of the outside air, the temperature of the inside surface of the outside walls is also variable.

This temperature is the result of heat transfer through the external wall between the room and the outside.

When applying the finite difference method (FDM) to solve the transient heat conduction in the outer wall, formulas based on the calculation scheme are obtained (Figure 7).

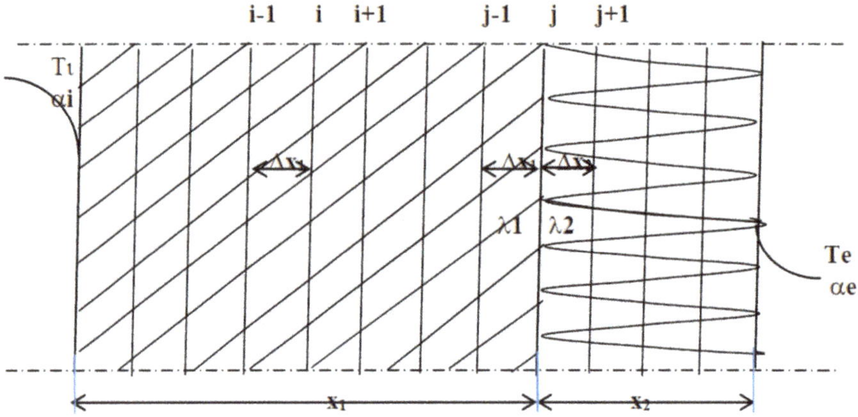

Figure 7. Calculation scheme.

The formulas resulting from the discretization of areas were used for the calculations. Inner surface temperature of outer wall:

$$T_{wi} = \frac{T_i + \frac{\lambda_i}{\alpha_i . \Delta x_1} \cdot T_1}{1 + \frac{\lambda_i}{\alpha_i . \Delta x_1}} \tag{22}$$

Surface temperature at the boundary of the layers:

$$T_w = \frac{\frac{\lambda_1}{\Delta x_1} \cdot T_{j-1} + \frac{\lambda_2}{\Delta x_2} \cdot T_{j+1}}{\frac{\lambda_1}{\Delta x_1} + \frac{\lambda_2}{\Delta x_2}} \tag{23}$$

The temperature of the outer surface of the outer wall was determined from the balance sheet (Figure 8):

$$Q_p + Q_{re} = Q_{se} + Q_{sen} \tag{24}$$

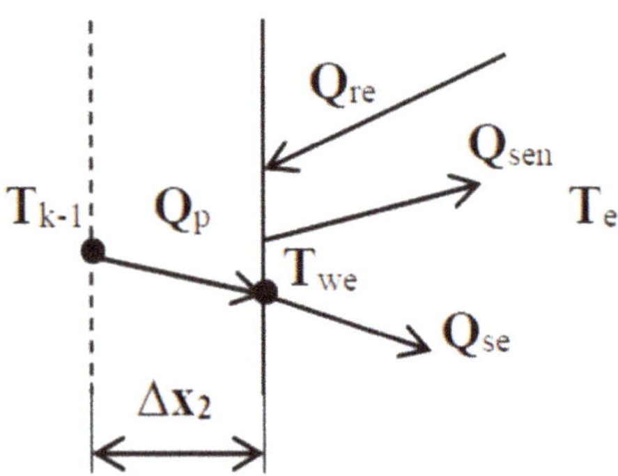

Figure 8. Balance diagram of the external surface.

Taking into account the Equations (10)–(12) and the heat conduction equation under the surface of the partition (24), it can be concluded that:

$$Q_p = \frac{\lambda_2}{\Delta x_2} \cdot A_{se} \cdot (T_{k-1} - T_{we}) \tag{25}$$

After substitution and transformations, the temperature of the outer surface of the outer wall is described by the relationship:

$$T_{we} = \frac{\varepsilon_s \cdot I + \frac{\lambda_2}{\Delta x_2} \cdot T_{k-1} + \alpha_e \cdot T_e - \varepsilon_n \cdot \left(T_{we}^4 - T_n^4\right)}{\alpha_e + \frac{\lambda_2}{\Delta x_2}} \tag{26}$$

Due to the presence of T_{we}, the equation is solved by the iteration method.

Signs:

T_n—skyfall temperature

$\sigma_s = 5.67 \times 10^{-8}$ [W/m^{2}K^4]—radiation constant

ε_s—surface radiation absorption coefficient

ε_{sn}—surface radiation emission coefficient

α_e—heat transfer coefficient from the external wall surface to the outside air

λ_2—thermal conductivity coefficient of the insulating layer

Δx_2—step of discretization of the insulating layer

The above equations are completed with boundary conditions.

- The heating season was assumed to occur during the period when the outside air temperature is lower than 12 °C ($T_e < 12$ °C).
- If, during the heating season, the heat losses are greater than the heat gains, the heating control system maintains a constant internal temperature ($T_i = T_{io}$).
- If in the heating season the heat losses are lower than the heat gains, the control system switches the heating off and the internal temperature results from the thermal balance.
- Outside the heating season, the internal temperature results from the thermal balance.

5. Results and Discussion

5.1. Assumptions and Output Parameters

Using the previously given equations, calculations of the parameters of the room and the external partition were made for variable external conditions [41].

The calculations were made for several variants of rooms with general assumptions:

- A living room.
- The heat transfer coefficient on the inside and outside is constant in accordance with the standard [42].
- Natural ventilation with the intensity resulting from meeting the requirements of the standard [43].
- The thermal conductivity coefficient of the wall construction material and insulation is constant.
- Internal heat gains are constant.

Data for calculations, with values (according to [43]):

$q_A = 6.8$ W/m^2—single-family buildings,

$q_A = 7.0$ W/m^2—multi-family buildings,

$V_{Ve} = 0.31 \times 10^{-3}$ m^3/(s$'$m^2)—single-family buildings,

$V_{Ve} = 0.32 \times 10^{-3}$ m^3/(s$'$m^2)—multi-family buildings.

Relationship between V_{Ve} coefficient and the number of exchanges n:

$$n = \frac{3.6 \cdot V_{Ve}}{h} \tag{27}$$

h—room height.

Thermal conductivity:

λ_1 = 0.82 W/mK (brick wall),
λ_2 = 0.032 W/mK (polystyrene insulation).

Room dimensions:

a = 3.0 m—the width of the room,
b = 4.0 m—room depth,
h = 3.0 m—room height,
A_o = 1.96 m^2—window area.

Meteorological data:
Sample calculations were made using the meteorological data for the Rzeszów-Jasionka measuring station.

The table of hourly changes in outside temperature and changes in the intensity of solar radiation was used, provided by the website of the Ministry of Infrastructure [40].

The example assumes the location of the outer wall towards the north.

It is assumed that the final result of the calculations will be the determination of the variability of the heat demand for space heating over the year, taking into account the variability of climatic conditions and the inertia of building partitions.

The basis for determining the heat demand is Equation (2). The components of the equation were determined by Equations (3)–(10) using the assumptions made above.

The calculations were made for a living room with an external wall made of brick, 50-cm thick, for the cases: without insulation and with insulation with a layer of polystyrene 5-, 10-, 15-, 20-, and 25-cm thick. The results of the calculations are presented in the following charts.

5.2. Calculation of Indoor Air Temperature

In the heating season, the heat losses exceed the heat gains in the room, which are supplemented by the heating system. The regulation of this system allows for the maintenance of the constant temperature T_i = 20 °C, assumed in the example. In summer, the temperature of the indoor air is the result of the heat balance of gains and losses. Temperature fluctuations increase as the thickness of the insulation increases.

The calculation results of the average daily temperature of indoor air obtained using the Equation (14) are presented graphically in Figure 9.

During the heating season, the regulation ensures a constant temperature. In summer, there is a temperature change, and these fluctuations increase with increasing insulation thickness. The labels in Figures 9 and 10 were signed by adding units of insulation thickness as "Ti-0 cm" and analogously "Twi-0 cm".

From the graph in Figure 9 you can read the time during which heating is required, i.e., the duration of the heating season. This time depends on the insulation of the outer wall (insulation thickness).

5.3. Calculation of the Surface Temperature of the Inner Outer Wall

The temperature of the inner surface of the outer wall is the result of heat conduction through the wall and taking heat from the inner air. It clearly depends on the wall insulation. It was calculated according to the Equation (22) and the results are presented in Figure 10. This temperature and its fluctuations throughout the year clearly depend on the wall insulation.

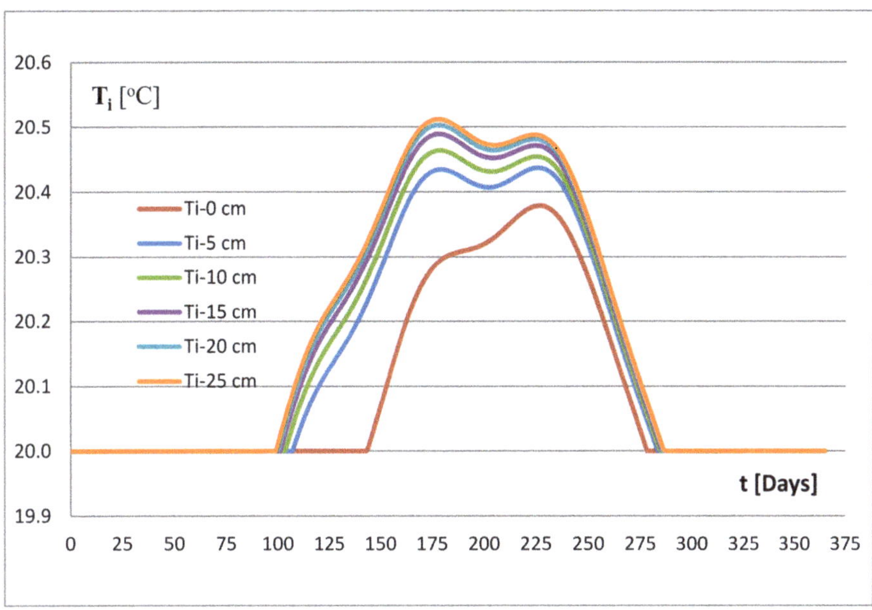

Figure 9. Indoor air temperature.

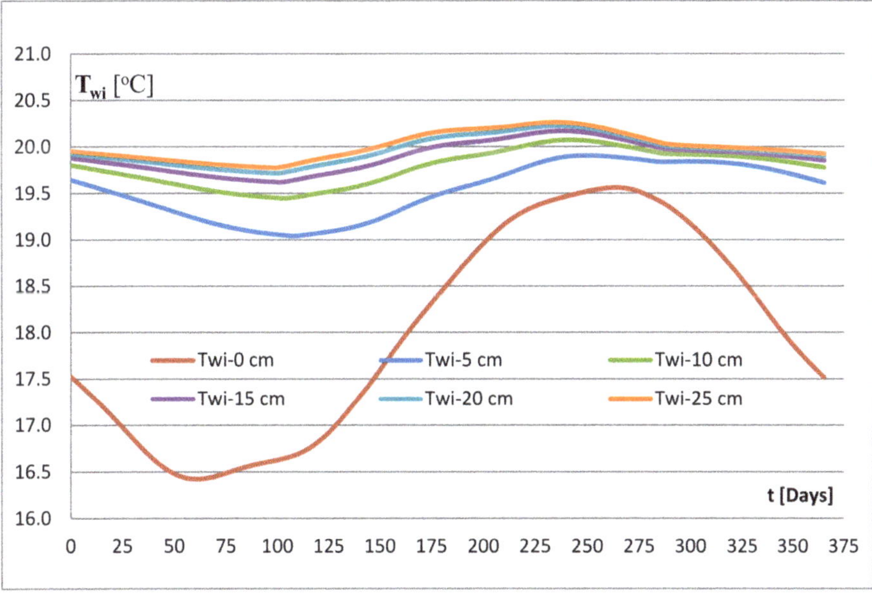

Figure 10. The temperature of the inner surface of the outer wall.

The annual amplitude of changes in this temperature decreases with increasing insulation thickness. In the absence of insulation, the lowest temperatures appear around the beginning of March, and for an insulated wallm around mid-April.

Fluctuations in the average daily temperature ΔTi (Figure 11) with efficient regulation in the heating season are very small during the year (<0.51 °C) and depend to a small extent on the wall insulation. On the other hand, the temperature fluctuations of the inner surface of the outer wall T_{wi} depend on the insulation. For a wall without insulation, the maximum difference in a year is 3.14 °C. The difference in temperature values when changing the insulation thickness from 5 to 25 cm, equal to 0.49–0.86, is small and it has little effect on thermal comfort.

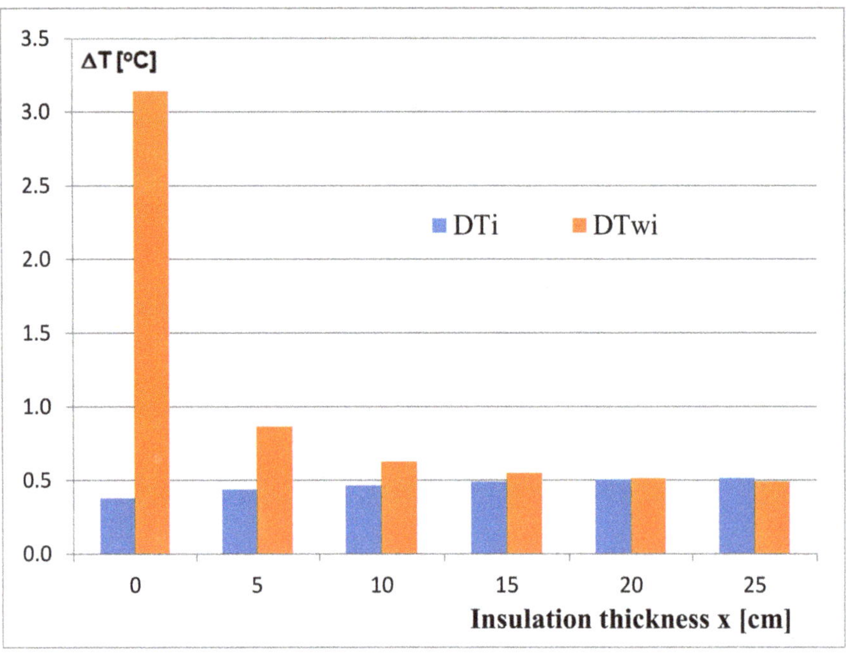

Figure 11. Maximum temperature differences on the inner surface of the outer wall.

5.4. Analysis of Thermal Power Variability for Heating Purposes

Using the proposed methodology, calculations of thermal power for heating were performed for the conditions assumed in Section 2. The diagram given in Figure 12 shows the dependence of this power on the insulation thickness.

The chart of changes in heat demand for heating (Figure 11) clearly shows the division into the heating season and the summer season. The demand in the heating season depends (which is obvious) on the insulation of the outer wall. The length of the heating season t_g also depends on the insulation performance.

The maximum heat output, which is the basis for the selection of heating devices, is also different for each case. The comparison of these powers is shown in Figure 13. The diagram also includes heating powers Q_{maxn} calculated on the basis of the applicable standards. Q_{max} denotes the maximum thermal power for a given case of the partition structure, calculated according to the considered methodology. Q_{maxn} means the maximum thermal power for a given case of the partition structure, calculated on the basis of the applicable standard PN EN 12831 titled "Heating installations in buildings. Design heat load calculation method".

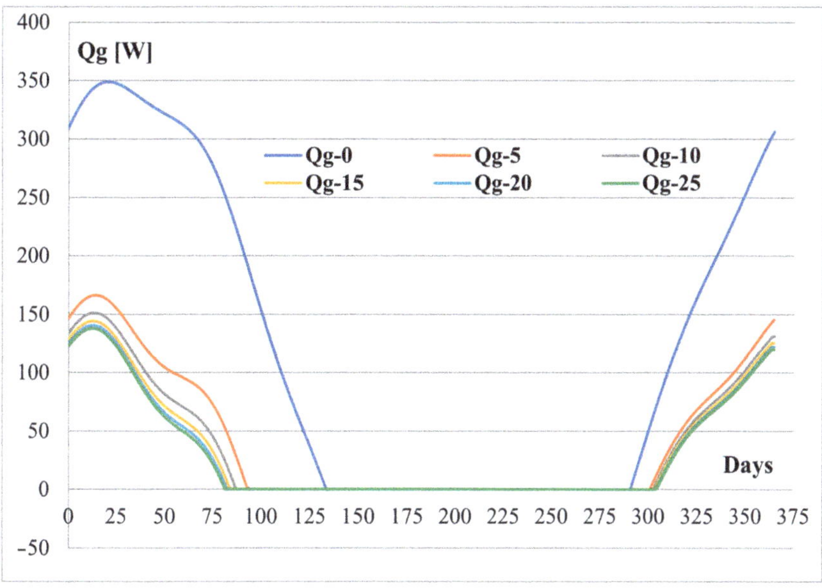

Figure 12. Variability of thermal power for heating over time.

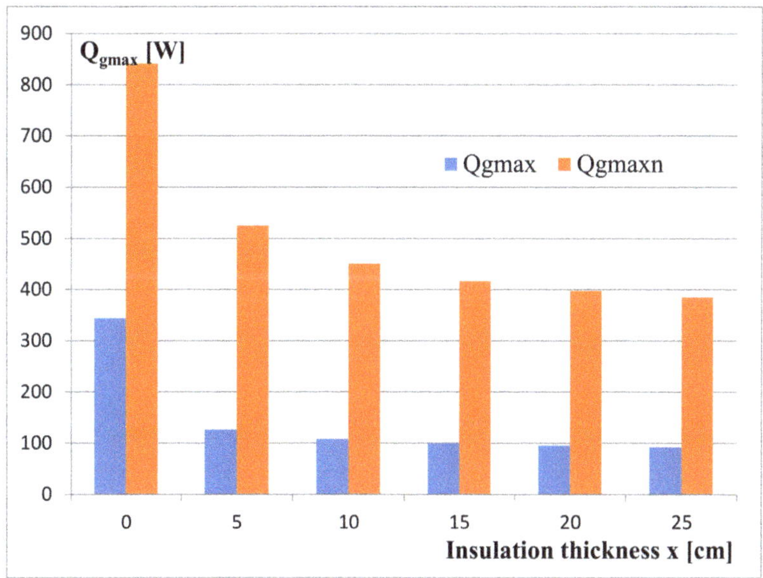

Figure 13. The dependence of the maximum heat output of heating on the thickness of the insulation.

Among the calculated thermal powers presented in Figure 12, the maximum powers Q_{gmax} necessary to ensure thermal comfort were selected. Such powers Q_{gmaxn} were also calculated for the considered cases in accordance with the standard. The values are presented in the Table 3 and in the diagram in Figure 13.

Table 3. Results of thermal power calculations for various insulation thickness.

Parameter	Insulation Thickness x [cm]					
x [cm]	0	5	10	15	20	25
Q_{gmax}	344.02	126.41	108.16	100.02	95.56	92.60
Q_{gmaxn}	840.98	525.07	449.96	416.30	397.21	384.90
Q_{gmax}/Q_{gmaxn}	0.409	0.241	0.240	0.240	0.241	0.241

The following conclusions can be drawn from the presented values:

- Taking into account the thermal inertia of the partitions makes it possible to perform a significant reduction in the demand for thermal power, and thus the use of smaller devices.
- For the considered partition, the calculated thermal power determined by the adopted method is only about 40% of the value calculated according to the standard.
- The reduction in thermal power depends on the thermal inertia of the partitions.
- Due to the low thermal capacity, the insulation thickness has a little effect on the maximum heat output.

5.5. The Dependence of the Time of the Heating Season on the Thickness of the Insulation

The dependence of the length of the heating season on the insulation thickness is shown in Figure 14.

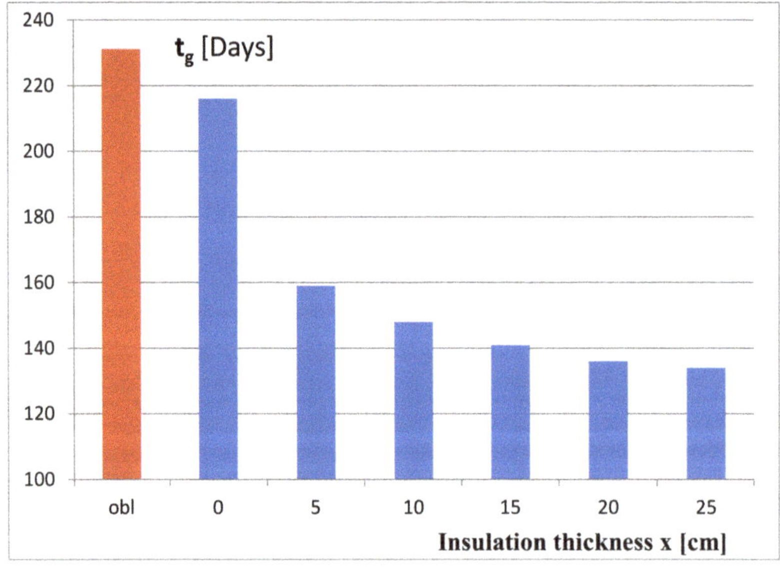

Figure 14. Variability of the duration of the heating season as a function of insulation thickness.

The diagram in Figure 14 clearly shows the dependence of the heating season on the "insulation" of external walls. Increasing the thickness of the insulation shortens the heating season, which contributes to reducing the annual heat consumption for heating purposes. For comparison, the graph also shows (red) the time determined on the basis of the annual variability of the outside air temperature. This time is approximately in line with the calculation method adopted for the wall without external insulation (x = 0). This confirms the correctness of the adopted calculation methodology and the correctness of the performed calculations.

5.6. Validation of the Model by Using Fragmented Temperature Measurements

In order to assess the compliance of the adopted calculation methodology with the actual conditions, fragmentary temperature measurements were performed. Measurements were made in a real facility in a single-family residential building, on an external brick wall with the following parameters:

- Wall thickness: 50 cm,
- Thickness of the outer polystyrene insulation: 10 cm
- Thermal conductivity of the wall: $\lambda = 0.82$ W/(m·K)
- Thermal conductivity of the insulation: $\lambda = 0.032$ W/(m·K)

Following temperatures measurements were taken:

- T_i—internal air temperature
- T_{wi}—temperature of the inner surface of the outer wall
- T_e—outside air temperature

The purpose of the measurements is to show the effect of thermal inertia of the partitions on the changes in internal temperature. The measure of the influence of inertia is the amplitude of changes in the internal temperature and the temperature of the internal surface of the external partition as a function of the amplitude of changes in the external air temperature.

Measurements were taken continuously and the values were averaged over an hourly period for analysis.

This impact was initially analyzed on the basis of measurements conducted from 20–29 February 2020 and from 18–20 July 2020. The results presented in the table below were obtained for these cases. Results of calculations and measurements have been presented in Table 4.

Table 4. Results of calculations and measurements.

Period		From 20 to 29 February 2020			From 18 to 20 July 2020		
Parameter		T_i [°C]	T_{wi} [°C]	T_e [°C]	T_i [°C]	T_{wi} [°C]	T_e [°C]
	Tmin	20.00	19.25	−4.40	20.00	20.23	13.60
Calculation	Tmax	20.20	19.33	7.80	21.17	20.77	23.70
	ΔT	0.20	0.08	12.20	1.17	0.54	10.10
	Tmin	20.10	18.90	−0.30	23.50	22.40	15.20
Measurement	Tmax	21.40	19.90	11.20	24.90	23.80	28.30
	ΔT	1.30	1.00	11.50	1.40	1.40	13.10

Exemplary results for several hours in the heating season (February) are shown in Figure 15 as a function of successive hours throughout the year. The graph also shows the corresponding temperatures for these hours obtained from calculations (T_{io}, T_{wio}, T_{eo}).

On the horizontal axes of the graphs (Figure 15 and following), the time is defined as successive hours of the year. Such a system was adopted for the purpose of comparing the results of measurements and calculations, because the calculations used climatic data for the following hours of the year. The hours for calculations and measurements were kept consistent.

Supplementary markings:

- T_i—measured internal air temperature
- T_e—measured outside air temperature
- T_{wi}—measured temperature of the inner surface of the outer wall
- T_{io}—internal air temperature adopted for calculations
- T_{eo}—outside air temperature taken for calculations
- T_{wio}—calculated temperature of the inner surface of the outer wall

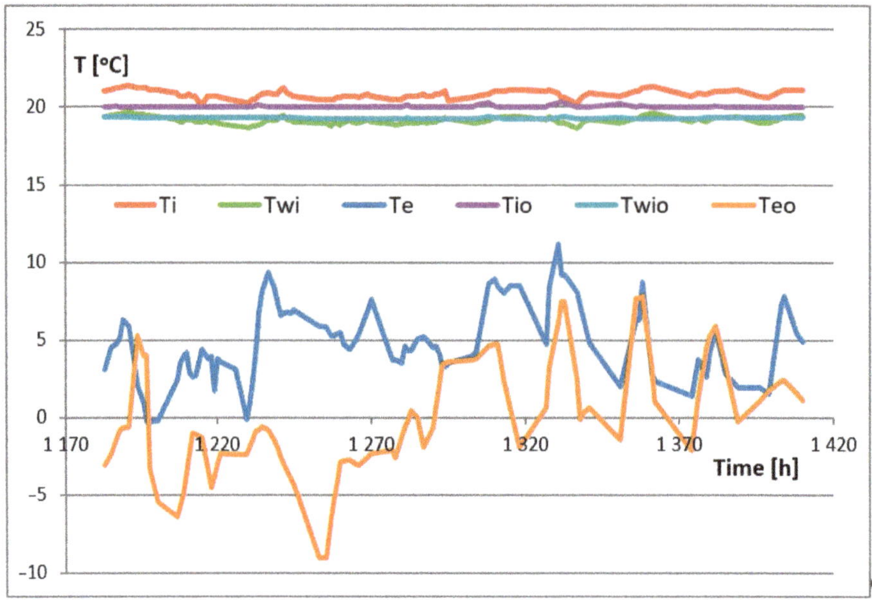

Figure 15. Variability of the measurement and calculation temperatures during the time from 20 to 29 February 2020.

During the heating season with variable outside air temperature, there are small changes in the temperature of the inner surface of the outer wall. The temperature value is comparable with the value calculated according to the adopted method.

In the summer, we can observe a similar tendency. A detailed comparison will be the purpose of further research. We initially present sample measurement results for summer season. We presented obtained results of researches in the Figure 16.

In order to verify the correctness of the adopted methodology, the amplitudes of differences in temperature between the internal air and the internal surface of the external wall were determined and presented in the Table 5.

During the measurement period, there were significant fluctuations in the outside temperature and due to this there were also changes in the internal temperature, despite the regulation of the heating with a thermostatic valve. For this reason, there was a change in the temperature difference between the inside air and the wall surface. The course of amplitude changes during the period from 20 to 29 February 2020 is shown in Figure 17.

Table 5. The amplitudes of temperature differences between the internal air and the internal surface of the external wall.

Period	From 20 to 29 February 2020		From 18 to 20 July 2020	
Parameter	Measurement $(T_i - T_{wi})$ [°C]	Calculation $(T_{io} - T_{wio})$ [°C]	Measurement $(T_i - T_{wi})$ [°C]	Calculation $(T_{io} - T_{wio})$ [°C]
Max	1.95	1.00	1.50	0.35
Min	1.05	0.63	0.20	−0.34
Average	1.69	0.74	9.70	0.01

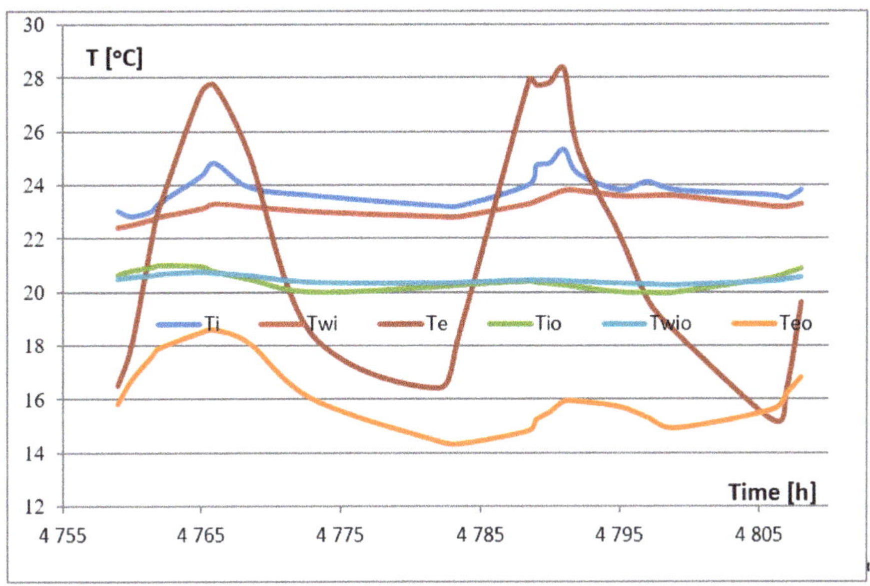

Figure 16. Variability of the measurement and calculation temperatures during the time from 18 to 20 July 2020.

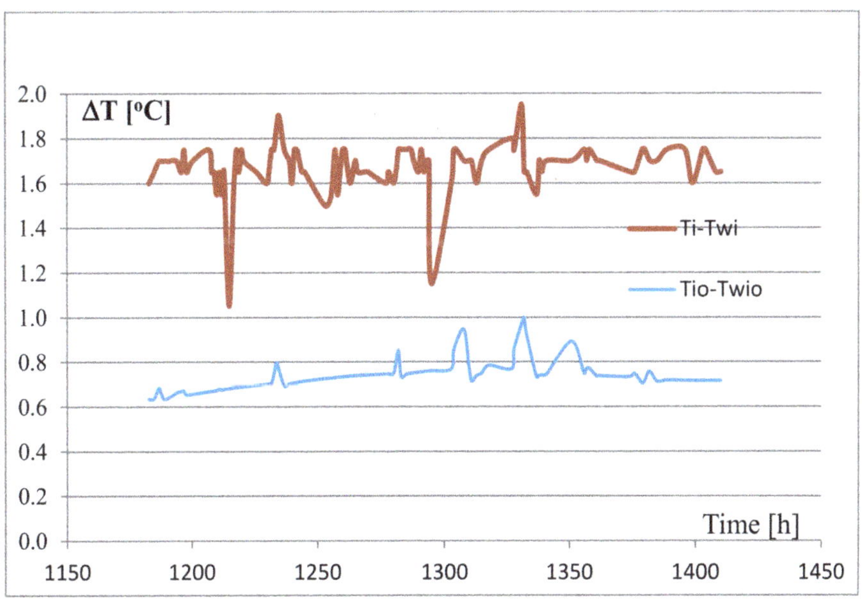

Figure 17. Difference in temperature of internal air and internal wall surface during the time from 20 to 29 February 2020.

Analogically, the course of amplitude changes during the time from 18 to 20 July 2020 is presented in Figure 18.

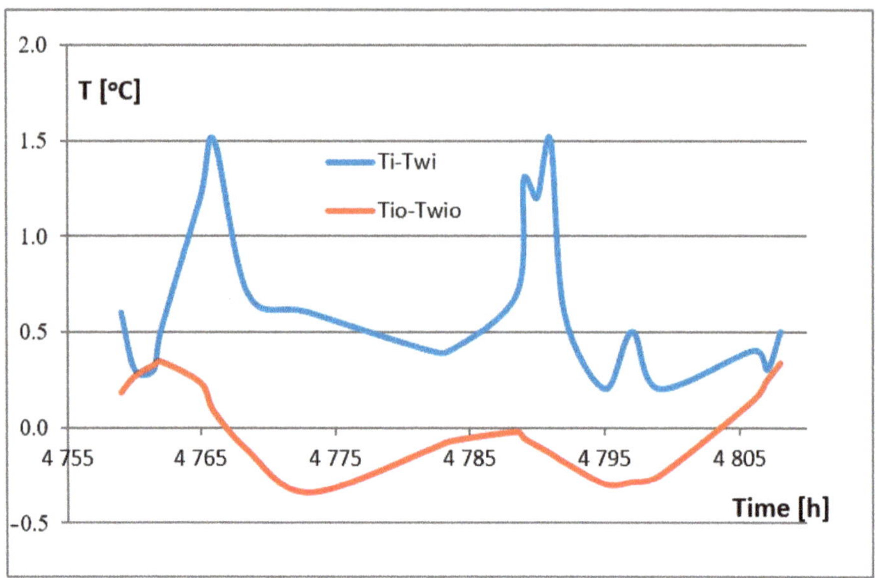

Figure 18. Difference in temperature of internal air and internal wall surface during the time from 18 to 20 July 2020.

In order to verify the correctness of the adopted methodology, additional, more precise temperature measurements were made for the period from 9 to 17 November 2020. The measurement results are shown in Figures 19 and 20.

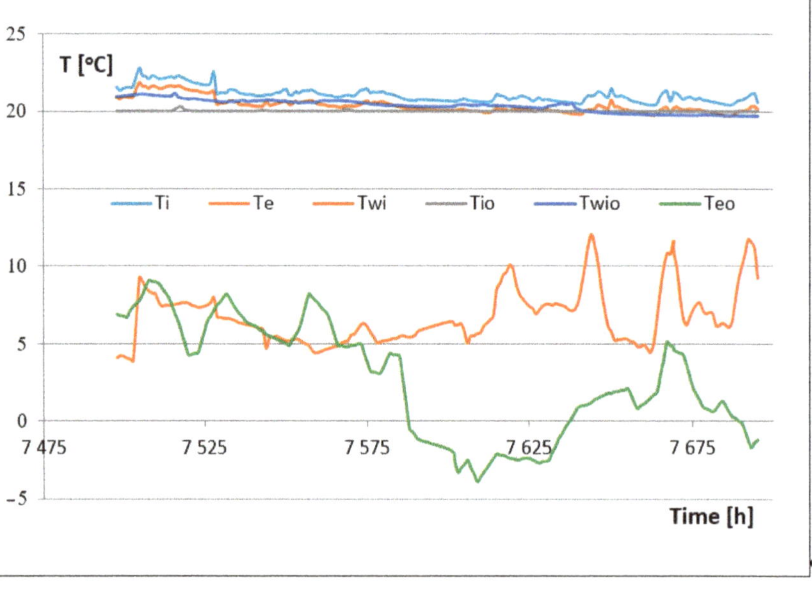

Figure 19. Variability of the temperature during the time from 9 to 17 November 2020.

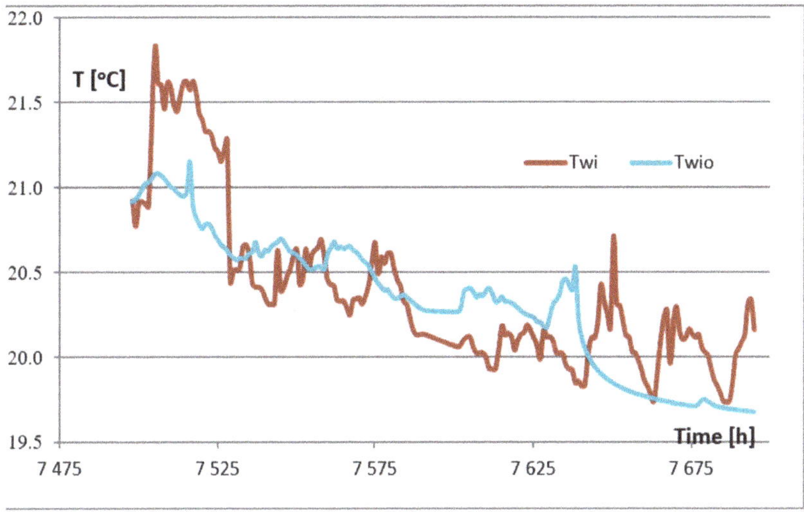

Figure 20. Variability of internal surface temperature for the period from 9 to 17 November 2020.

The measurement results were compared with the values used in the calculations in the adopted model and are presented in Table 6.

Table 6. Results of calculations and measurements for the period from 9 to 17 November 2020.

	Measurement			Calculation		
Parameter	T_i [°C]	T_{wi} [°C]	T_e [°C]	T_{io} [°C]	T_{wio} [°C]	T_{eo} [°C]
Tmin	20.42	19.73	3.87	20.00	19.68	−3.90
Tmax	22.78	21.81	12.03	20.30	21.16	9.10
ΔT	2.36	2.08	9.08	0.30	1.48	13.00

The influence of the thermal inertia of partitions on changes in internal temperature was analyzed. The results of measurements and calculations are shown in Figure 20.

Changes in the temperature of the outside air cause slight changes in the temperature of the inside surface of the outside wall. The value of the wall surface temperature is comparable with the value calculated in accordance with the adopted method, which is shown in Figure 20.

In order to better visualize the compliance of the measurement results with the calculations, relative deviations calculated on the basis of the differences in the measurement and calculation results were determined. The deviations were applied to the temperature of the inner surface of the outer wall, because this temperature determines the transfer of heat for all three analyzed measurement series.

$$\delta = 100 \cdot \frac{T_{wi} - T_{wio}}{T_{wi}} \ [\%] \tag{28}$$

when:

δ—relative deviation of the measured and calculated internal surface temperature.

The changes of the relative deviations of the values measured for the period from 9 to 17 November 2020 is shown in Figure 21. The relative deviations for all three measurement series are presented in Table 7.

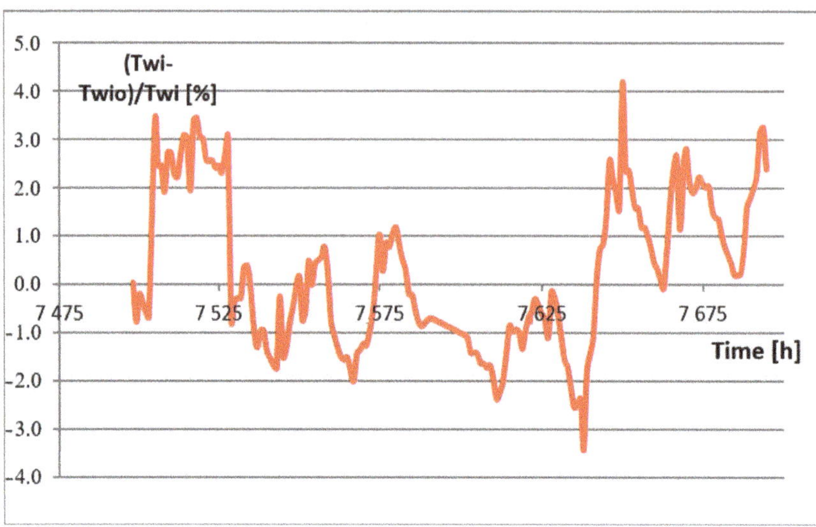

Figure 21. Relative measurement deviations for the period from 9 to 17 November 2020.

Table 7. Relative deviations of the measured and calculated internal surface temperature.

Period	From 20 to 29 February 2020	From 18 to 20 July 2020	From 9 to 11 November 2020
Max [%]	17.3	14.39	4.19
Min [%]	−3.47	8.61	−3.43
Average [%]	−0.85	12.34	0.45

A small value of the average deviation of the measurement and calculation results during the heating season indicates the correctness of the adopted analysis method. The big difference in July is caused by very large changes in outside air temperatures, deviating from typical meteorological years. It should be emphasized that such a result is obtained from measurements of a real object (not in a laboratory) under very variable climatic conditions. Moreover, the temperature of the indoor air in the summer with no air conditioning is directly related to the temperature of the outdoor air and is not regulated.

Furthermore, complete measurements are planned to fully assess the compliance of the calculation model.

The presented data show the following conclusions:

- In the winter period (February), the measured amplitude of changes in the outside temperature of 11.50 °C corresponds to the amplitude of changes in the internal temperature of 1.30 °C and the amplitude of changes in the temperature of the inside surface of the external wall, which equals 1.00 °C.
- Similar relations exist for the period from 18 to 20 July 2020.
- In any case, the average amplitude of changes in the temperature of the internal surface of the external wall is smaller than the amplitude of changes in the temperature of internal air.
- The same conclusion follows from the calculation results. Quantitative differences are caused by the random selection of the measurement period and the uniqueness of the climate parameters.
- Reducing the amplitude has a positive effect on maintaining thermal comfort.

6. Summary and Conclusions

The current regulations define the rules for calculating the heat demand for heating based on the difference between the maximum and minimum outside air temperature for the climate zone. The calculations do not take into account the thermal inertia of building partitions.

The main aim of the article is to determine the influence of the thermal inertia of building partitions on the building's heating system, taking into account changing climatic conditions.

The process of heat transfer through the partitions is described in the literature, and the novelty in this respect is the use of boundary conditions appropriate for the nature of the partition, i.e., variable outside air temperature and solar radiation intensity. These conditions are taken into account in the literature—also in standards—, but in a simplified way through indicators covering longer periods (usually monthly). This variability is not considered when constructing an ordered diagram of heat loads, which affects the determination of the heating season time.

The thermal power of heating devices is assumed on the basis of calculations made for the steady state with extreme parameters, without taking into account the thermal inertia. In the presented study, the influence of the thermal inertia of the external building partition on the heat exchange between the building's interior and its surroundings was analyzed. Different thicknesses of the insulation of the outer partition were considered. The ongoing dynamic processes were analyzed computationally using the finite difference method (FDM).

Based on the obtained calculation and measurement results, the following conclusions can be drawn:

- Maximum power for heating, determining the selection of heating devices, is lower than the values determined according to the applicable calculation rules. The difference between the thermal power calculated on the basis of the applicable PN EN 12831 standard and the maximum power for a given case of the partition structure, calculated in accordance with the considered methodology presented in the article, decreases with increasing insulation thickness and equals from 41% to 24%.
- The reduction in heat output depends on the thickness of the insulation, but to a much lesser extent than the increase in thickness. The economic correlation between these values requires further analysis.
- The duration of the heating season is also dependent on the insulation of external partitions and it is definitely shorter than that determined on the basis of changes in external temperature.
- With the efficient regulation of the internal temperature in the heating season, the insulation of the external partition allows for very small temperature fluctuations of the internal surface of the partitions, which has a positive effect on the thermal comfort of the rooms.
- The results of the research on temperature variability over time presented in the article confirm the fact that the temperature fluctuations of the surface of the partitions, which reflect the heat transfer from the internal air to the partition surface, and the heat conduction inside the partition and heat transfer from the partition to the outside air are much less variable than the fluctuations of outdoor and indoor temperatures. With the amplitude of changes in the outside air temperature in summer amounting to 11.5 °C, the amplitude of changes in the internal temperature of the surface of the analyzed external wall is 1.00 °C, while the amplitude of changes in the internal temperature is 1.30 °C. In winter, these fluctuations demonstrate similar relationships with greater differences.
- The performed research confirms the compliance of the adopted model of heat transfer dynamics testing. Average temperature deviations measured and calculated in the winter season range from −0.85% to +0.45%. A significant deviation in the summer season is due to the lack of temperature regulation.

The development of issues related to the dynamics of heat transfer, taking into account the variability of weather conditions and thermal inertia in buildings, will be the subject of future research conducted by the authors of this article.

Author Contributions: Conceptualization, W.S. and B.B., methodology, W.S. and B.B; writing—original draft preparation, B.B. and W.S.; writing—review and editing, B.B. and W.S. All authors have read and agreed to the published version of the manuscript.

Funding: This research received no external funding.

Conflicts of Interest: The authors declare no conflict of interest.

References and Notes

1. Directive 2002/91/EC of the European Parliament and of the Council of 16 December 2002 on the Energy Performance of Buildings. Official Journal of the European Union, L 1/65.
2. Directive 2010/31/EU of the European Parliament and of the Council of 19 May 2010 on the Energy Performance of Buildings (Recast). Official Journal of the European Union, L 153/13.
3. Directive 2012/27/EU of the European Parliament and of the Council of 25 October 2012 on Energy Efficiency, Amending Directives 2009/125/EC and 2010/30/EU and Repealing Directives 2004/8/EC and 2006/32/EC. Official Journal of the European Union, L 315/1.
4. The Act of 20 May 2016 on energy efficiency (Journal of Laws of 2016, item 831), as amended.
5. The Act of 29 August 2014 on the energy performance of buildings (Journal of Laws of 2014, item 1200), as amended.
6. Directive (EU) 2018/844 of the European Parliament and of the Council of 30 May 2018 Amending Directive 2010/31/EU on the Energy Performance of Buildings and Directive 2012/27/EU on Energy Efficiency. Official Journal of the European, L 156/75.
7. Regulation of the Minister of Infrastructure of 12 April 2002 on Technical Conditions to be Met by Buildings and Their Location (Journal of Laws No. 75, item 690), as amended.
8. Regulation of the Minister of Infrastructure of 27 February 2015 on the Methodology for Determining the Energy Performance of a Building or Part of a Building and Energy Performance Certificates (Journal of Laws of 2015, item 376).
9. Satchwell, A.J.; Cappers, P.A.; Deason, J.; Forrester, S.P.; Frick, N.M.; Gerke, B.F.; Piette, M.A. A Conceptual Framework to Describe Energy Efficiency and Demand Response Interactions. *Energies* **2020**, *13*, 4336. [CrossRef]
10. Babiarz, B. Aspects of heat supply security management using elements of decision theory. *Energies* **2018**, *11*, 2764. [CrossRef]
11. Cholewa, T.; Życzyńska, A. Experimental evaluation of calculated energy savings in schools. *J. Therm. Anal. Calorim.* **2020**, 1–8. [CrossRef]
12. Babiarz, B. Heat supply system reliability management. Safety and reliability: Methodology and Applications. In Proceedings of the European Safety and Reliability Conference, ESREL 2014, Wroclaw, Poland, 14–18 September 2014; Taylor & Francis Group: London, UK, 2015; pp. 1501–1506.
13. Cholewa, T.; Balaras, C.A.; Nižetić, S.; Siuta-Olcha, A. On calculated and actual energy savings from thermal building renovations—Long term field evaluation of multifamily buildings. *Energy Build.* **2020**, *223*, 110145. [CrossRef]
14. Babiarz, B.; Kut, P. District heating simulation in the aspect of heat supply safety. In Proceedings of the VI International Conference of Science and Technology INFRAEKO 2018 Modern Cities. Infrastructure and Environment, Cracov, Poland, 7–8 June 2018. [CrossRef]
15. Chicherin, S.; Mašatin, V.; Siirde, A.; Volkova, A. Method for Assessing Heat Loss in A District Heating Network with A Focus on the State of Insulation and Actual Demand for Useful Energy. *Energies* **2020**, *13*, 4505. [CrossRef]
16. Basinska, M.; Koczyk, H.; Szczechowiak, E. Sensitivity analysis in determining the optimum energy for residential buildings in Polish conditions. *Energy Build.* **2015**, *107*, 307–318. [CrossRef]
17. Rim, M.; Sung, U.J.; Kim, T. Application of Thermal Labyrinth System to Reduce Heating and Cooling Energy Consumption. *Energies* **2018**, *11*, 2762. [CrossRef]
18. Dec, E.; Babiarz, B.; Sekret, R. Analysis of temperature, air humidity and wind conditions for the needs of outdoor thermal comfort. In Proceedings of the 10th Conference on Interdisciplinary Problems in Environmental Protection and Engineering EKO-DOK 2018 E3S Web of Conferences 44, Dolny Śląsk, Poland, 8–10 April 2019. [CrossRef]

19. Krawczyk, D.A.; Wądołowska, B. Analysis of indoor air parameters in an education building. *Energy Procedia* **2018**, *147*, 96–103. [CrossRef]
20. Oldewurtel, F.; Sturzenegger, D.; Morari, M. Importance of occupancy information for building climate control. *Appl. Energy* **2013**, *101*, 521–532. [CrossRef]
21. Korkas, C.D.; Baldi, S.; Michailidis, I.; Kosmatopoulos, E.B. Occupancy-based demand response and thermal comfort optimization in microgrids with renewable energy sources and energy storage. *Appl. Energy* **2016**, *163*, 93–104. [CrossRef]
22. Xue, Y.; Ge, Z.; Du, X.; Yang, L. On the Heat Transfer Enhancement of Plate Fin Heat Exchanger. *Energies* **2018**, *11*, 1398. [CrossRef]
23. Emmel, M.G.; Abadie, M.O.; Mendes, N. New external convective heat transfer coefficient correlations for isolated low-rise buildings. *Energy Build.* **2007**, *39*, 335–342. [CrossRef]
24. Dylewski, R. Optimal Thermal Insulation Thicknesses of External Walls Based on Economic and Ecological Heating Cost. *Energies* **2019**, *12*, 3415. [CrossRef]
25. Krasoń, J.; Miąsik, P.; Lichołai, L.; Dębska, B.; Starakiewicz, A. Analysis of the Thermal Characteristics of a Composite Ceramic Product Filled with Phase Change Material. *Buildings* **2019**, *9*, 217. [CrossRef]
26. Kim, D.; Cox, S.J.; Cho, H.; Yoon, J. Comparative investigation on building energy performance of double skin façade (DSF) with interior or exterior slat blinds. *J. Build. Eng.* **2018**, *20*. [CrossRef]
27. Wilson, M.; Luck, R.; Mago, P.J.; Cho, H. Building Energy Management Using Increased Thermal Capacitance and Thermal Storage Management. *Buildings* **2018**, *8*, 86. [CrossRef]
28. Zheng, X.; Heshmati, A. An Analysis of Energy Use Efficiency in China by Applying Stochastic Frontier Panel Data Models. *Energies* **2020**, *13*, 1892. [CrossRef]
29. Seo, B.; Yoon, Y.B.; Yu, B.H.; Cho, S.; Lee, K.H. Comparative Analysis of Cooling Energy Performance Between Water-Cooled VRF and Conventional AHU Systems in a Commercial Building. *Appl. Therm. Eng.* **2020**, *170*, 114992. [CrossRef]
30. Kim, C.-H.; Lee, S.-E.; Lee, K.H.; Kim, K.-S. Detailed Comparison of the Operational Characteristics of Energy-conserving HVAC Systems during the Cooling Season. *Energies* **2019**, *12*, 4160. [CrossRef]
31. Wetter, M. Modelica-based modelling and simulation to support research and development in building energy and control systems. *J. Build. Perform. Simul.* **2009**, *2*. [CrossRef]
32. Manz, H.; Frank, T.h. Thermal simulation of buildings with double-skin façades. *Energy Build.* **2005**, *37*. [CrossRef]
33. Clarke, J.A. *Energy Simulation in Building Design*; Butterworth-Heinemann: Oxford, UK, 2001.
34. Gerlich, V. Modelling of Heat Transfer in Buildings. In Proceedings of the 25th Conference on Modelling and Simulation, Krakow, Poland, 7–10 June 2011. [CrossRef]
35. Simões, N.; Prata, J.; Tadeu, A. 3D Dynamic Simulation of Heat Conduction through a Building Corner Using a BEM Model in the Frequency Domain. *Energies* **2019**, *12*, 4595. [CrossRef]
36. Madejski, J. *Theory of Heat Transfer*; Szczecin University of Technology: Szczecin, Poland, 1998. (In Polish)
37. Staniszewski, B. Heat transfer. In *Theoretical Basics*; PWN: Warsaw, Poland, 1979.
38. Szymański, W. Heat transfer. Fundamentals, Federation of Scientific and Technical Associations Council in Rzeszów, Rzeszów 2019 (In Polish).
39. Bronsztejn, I.N.; Siemiendiajew, K.A. Maths. In *Encyclopedic Guide*; PWN: Warsaw, Poland, 1968.
40. Typical Meteorological Years and Statistical Climatic Data for Energy Calculations of Buildings. Available online: www.mib.gov.pl (accessed on 15 September 2020).
41. PN-EN ISO 52016-1: 2017 Energy performance of buildings—Energy needs for heating and cooling, internal temperatures and sensible and latent heat loads—Part 1: Calculation procedures (ISO 52016-1:2017).

42. PN-EN-ISO 6946: Building components and building elements. Thermal resistance and heat transfer coefficients. Calculation Method.
43. PN-83/B-03430 Ventilation in residential buildings, collective residence and public utility buildings. Requirements.

Publisher's Note: MDPI stays neutral with regard to jurisdictional claims in published maps and institutional affiliations.

Article

Setting up Energy Efficiency Management in Companies: Preliminary Lessons Learned from the Petroleum Industry

Bartlomiej Gawin and Bartosz Marcinkowski *

Department of Business Informatics, Faculty of Management, University of Gdansk, Piaskowa 9, 81-864 Sopot, Poland; bartlomiej.gawin@ug.edu.pl
* Correspondence: bartosz.marcinkowski@ug.edu.pl

Received: 19 September 2020; Accepted: 24 October 2020; Published: 27 October 2020

Abstract: In the era of expensive energy carriers and care for the climate, companies are keen to take action towards bolstering energy efficiency. Businesses often lack data on actual energy consumption to date, are rarely equipped with adequate analytical tools, and do not have the know-how regarding the transition itself. Developing energy efficiency management (EEM) for a given enterprise requires many steps, which ultimately unleash analytical potential and seamlessly integrate the EEM framework with the business model of a given company. This study scrutinizes and formalizes a reference process of pilot EEM implementation that involves external business partners in a multi-facility organization. The process is tailored to the specificity of the company's operations as well as its technical and management capabilities regarding energy efficiency. The proposed approach, phased in time and involving multiple stakeholders, should be especially useful for practitioners running EEM-related projects characterized by uncertain and changing requirements.

Keywords: energy use efficiency; energy efficiency management; design science research; pilot implementation

---

1. Introduction

Historically, economic growth proved to be linked to levels of energy consumption. Tugcu, Ozturk, and Aslan argued that this is the case for both renewable and non-renewable energy consumption [1]. At the same time, rising energy consumption results in excessive carbon dioxide (CO_2) emission as the most prevalent long-lasting greenhouse gas [2] and the most important greenhouse gas with regards to human activity. In effect, climate change may be positioned among crucial challenges of modern times. Energy efficiency (EE) is highlighted as the largest emissions growth restraint [3] and one of the widely acknowledged measures to meet the goal of keeping the increase in average global temperature well below 2 °C above pre-industrial levels according to the Paris Agreement [4]. Thus, societies put considerable hopes on it to mitigate the negative environmental impact and achieve sustainable economic development in the long run [5].

EE may be considered a distinctive feature of the products concerned [6], as well as a habitat preservation method that does not incur excessive costs, and at the same time, it provides companies with the opportunities for positive publicity [2]. The decision to take action to improve EE might be internal and resulting from purely economic reasons. That said, the pressure to shift towards more energy-efficient products, services, solutions, and behavioral patterns in terms of energy savings achieved is also likely to be enacted by governing bodies and institutions through energy efficiency policies (EEPs) [7]. Meeting the objectives of an EEP has a direct impact on the activity of companies operating within particular industries. All the actions taken by managers and employees of a given company to implement the rules and measures enforced by an EEP requires a properly developed business model [8]. On top of that, the results of these actions need to be monitored in order to verify whether the intended effects are achieved or not, and corrective actions are undertaken when necessary [9]. By tailoring business models, adapting business processes, rehauling, and deploying new IT solutions that support day-to-day EE-oriented management practices, the companies effectively establish energy efficiency management (EEM). The motivation behind this research was to extend the EE body of knowledge by understanding how consideration of EEPs and management practices necessary to implement them affects business models of real-world organizations.

Involving companies in EE programs is likely to exert influence on most components of their business models, regardless of the transformation strategies in place [10]. As implementing EEPs may come with substantial risks regarding the cash flow and bottom line of a company [11], such transition requires a good understanding of the constraints in place and demands on specific companies, as well as the core processes. Having a model of EEM tailored to the organization requires designing and implementing a process that features exploratory work and experiments. In our experience, all stages towards EEM—from the initial idea, through to evaluation, ensuring cost-effectiveness, actual financing, and implementation—should be considered. Therefore, the goal of this paper is to design and formalize the reference process of a pilot EEM implementation that involves external business partners in a multi-facility organization. The empirical study is fueled by three research questions:

RQ1: What key process steps and sources of information ought to be involved in the pilot implementation of EEM in a multi-facility organization?

RQ2: What level of telemetry and data processing systems involvement is expected to maintain EEM?

RQ3: How the structure of a contract between the organization that sets up EEM and an external EE solution provider should look like?

After the introduction, related research is discussed and the method is presented in Sections 2 and 3, respectively. Subsequently, the artifact as understood within the design science research is introduced, particularized, introduced to a real-life business organization, and validated. Section 5 discusses some best practices regarding post-pilot EE-focused activities between parties involved in setting up EEM based on the feedback collected during the study. Section 6 introduces the implications and limitations of the study, which are followed by conclusions.

2. Research Background

2.1. EE and its Determinants

Energy efficiency may be regarded as an integrative strategy for delaying climate change, taking care of energy security, and exerting a positive impact on economic development. EE implies using less energy to perform the same task; that is, eliminating energy waste [12]. In quantitative terms, EE constitutes a ratio between service outputs (result) and the energy input required to provide it [13]. Factors such as habits, attitude, awareness of EE measures, and perceptions of involved individuals have an impact on the EE [14]. On top of that, Chai and Baudelaire linked EE to organizational aspects and measurements [15]. Minimizing waste and reducing time or transport distances between succeeding production processes can improve EE, which is important from both environmental and business points of view since increased energy prices and costs related to emitting greenhouse gases affect the company's competitiveness [16].

Unfortunately, many organizations fail to take up the implementation of efficient EE measures due to financial determinants, insufficient information, and limited in-house skills [17]. That is why the concept of an energy efficiency gap, i.e., the discrepancy between actual and optimal energy use [18], has been introduced. This gap shows a paradox where the adoption of energy-efficient solutions is withheld despite anticipating a positive return on investment [19]. Fresner et al. argued that greater recognition of EE requires identification of what sort of direct and indirect benefits could be gained from adopting energy-efficient technologies [17].

Several factors contribute to the propagation of EE. Those include, but are not limited to:

- strictly market-related factors, such as awareness of actual energy costs, anticipating high market prices of energy in the future and subsequent attempts to constrain energy-related company costs, or availability of favorable loans for EE financing [20];
- advances within the organization and management, including the adoption of environmental management systems [21], benchmarking against competitors within a given industry, or enhancing supply chain management within a company [22];
- technological progress [23];
- environmental regulation at both national and regional levels [24], including increasing energy tariffs [25], drawing voluntary programs and agreements between industry and governing bodies that feature negotiated targets and timetables, as well as threatening to introduce future taxes/regulation [26].

The necessity to reduce costs were acknowledged as the main driver to EE, whereas corporate social responsibility, regulatory compliance, and available opportunities to implement EE, were found to have no significant effects on EE results [17]. These considerations were confirmed among others by Thollander et al., who noted that information-related determinants, such as the public sector being a role model, municipal membership in an EE program, or pressure from non-government organizations, had the least impact on the behavior of decision-makers [27].

2.2. Energy Efficiency Policies

EEPs might be regarded as abstract solutions to bolstering EE in a given legislative context. The development of an effective EEP is not a one-off activity. It is, in fact, a continuous, dynamic process that should establish conditions and rules for energy consumers and direct that change toward environmental and economic benefits. The policy-making cycle combines the design, implementation, and setting up of multiple criteria for evaluating policy instruments' impacts in a closed, repetitive loop. However, EEPs are not universal in nature or freely transferable between markets. To develop a policy tailored to the sector of the economy, one must understand the EE maturity level, and then develop a customized policy as well as to adapt it to the specifics of a business branch. Bukarica and Tomšić introduced the notion of the energy efficiency market as a concept for establishing EEPs; aside from appliance manufacturers, energy auditors, smart meter software designers, cogeneration developers, and their respective backgrounds, the market also covers sponsors, owners, authorities, and institutions that provide financing and assistance in the implementation of EE projects [6].

Many countries brought up initiatives targeted at promoting low-carbon development and improving EE in every sector of the economy that primarily features regulations and taxation [28]. On the other hand, Avgerinou, Bertoldi, and Castellazzi stress that the European Union and other major economies introduced policies and measures that are not punitive in nature [29]. Financial support policies constitute an important tool in that regard. In the Chinese market, two types of EE credits were implemented [30], enabling running EE projects with institutional support from financial institutions to counteract discontinuing investments due to the capital scarcity. As households and commercial building upgrades and retrofits are concerned, several programs may act as a benchmark for future ventures. From the USA market alone, the American Recovery and Reinvestment Act (ARRA) directed as much as $58 billion towards EE; programs under ARRA targeted, inter alia, insulation of low-income homes as well lighting/appliance upgrades with more energy-efficient solutions [31].

Some instruments covered by policies are put in place in order to prompt interest in EE projects, while others aim to advance such projects from the early stages of development towards real-life implementation [6]. At the company level, EEPs must be integrated into organizational economic planning, technical and management conditions, and business development processes. A properly developed EEP ought to increase the maturity of management, the organizational structure of the company, as well as the competencies behind it; the joint effort of practitioners and researchers contributes to advancing the understanding of the relationship between improving EE in business organizations, the change processes responsible for that, and the drivers that affect these processes [19].

We argue that effective application of austerity policies proves to be a big challenge, however, their preparation, implementation, and continual improvement is even harder.

2.3. Energy Efficiency Management

Energy efficiency management applies to many business areas: data centers [29], manufacturing [17,19], wastewater treatment plants [32], industry [19,28], and facility management [33] to name just a few. EEM practices, tools, and models have been promoted as promising means of reducing energy consumption or improving energy efficiency [2]. Such practices can have a substantial impact on the profitability of not only energy-intensive companies but also those with low energy costs since the reduced energy expenditures directly lead to increased profits [16]. Research conducted by Backlund et al., highlighted that having both long-term energy strategies in place and employing committed energy managers with high skill sets proved to be important factors behind spurring EE in industrial companies [34]. While there is clearly a vast potential for improved efficiency in technology, Schulze et al. stressed that available sources addressing the implementation of various efficiency measures are highly biased towards this perspective and require further best-practices for achieving enhanced EE by introducing new routines and implementing customized processes within energy management [35].

Companies and individuals might not be aware of their actual energy-related expenses. Consequently, consumers and corporate decision-makers often do not possess sufficient information regarding the net benefits of investment in technologies that have higher EE levels [14] and credible information is crucial. Households that have information on their energy bill or energy consumption are not only more likely to invest in energy-efficient light sources and appliances, but there is also strong evidence that households who regularly perform low-cost energy conservation measures are also more likely to spend money to bolster EE [36]. Telemetry systems that have dedicated hardware components (i.e., sensors, meters) and IT solutions integrated with them are used to monitor and control energy consumption. Software components of such telemetry environments are often referred to as energy management systems. The data acquired may be then processed using business intelligence (BI) analytical systems, which enable seeking root causes of high electric power consumption as well as monitoring the effectiveness of activities in the area of EE.

In various organizations, EE projects ultimately aim to establish an EE management model that enables the reduction of both electric power consumption and CO_2 emissions. Fernando and Hor analyzed a number of studies to come up with a list of activities that typically comprise the energy management process [2]:

- introducing review techniques that involve professionals who represent highly diversified business disciplines;
- scrutinizing historical data;
- performing energy audits;
- preparing feasibility analyses of energy improvement plans prepared by a business organization and possible implementation of those plans;
- conducting training in energy efficiency.

While both researchers and practitioners generally agree that business organizations may take advantage of multiple EE options, Fresner et al. point out that the more sophisticated of these options are often simply ignored [17]. Harris, Anderson, and Shafron showed that already in 2000, energy audits were likely to be among the first steps that any company might undertake when bolstering its EE, as the implementation rates of such audits were found to be high [37]. Although larger firms were overrepresented in their analysis, EE audits proved cost-effective and declared worthwhile by as many as 93% of the companies surveyed. Therefore, energy audits remain to be one of the leading instruments for introducing EEPs to overcome barriers to EE and to promote it. An energy audit may be considered a helpful tool for identifying opportunities and ascribing value to energy consumption to justify spending resources on EE projects [2]. Viable alternatives to energy audits feature a range of external energy services that cover contractual arrangements and funding mechanisms behind improving energy efficiency in a measurable way [27] and shifting towards a sustainable energy supply. Such services include energy performance contracting, third-party financing as well as contract energy management. A long-term EE strategy ought to be set up to take advantage of available options. According to Cai et al., an energy-saving and emission reduction (ESER) strategy promotes the sustainability of the manufacturing industry in green transition [28]. In order to address some ESER shortcomings (primarily the short completion timeframe required, lack of process standardization, and tariff-related issues), to improve the EE as well as to reduce waste emissions effectively, they introduce an enhanced concept based on lean principles, i.e., lean energy-saving and emission reduction.

3. Methods

To achieve the research goal, the design science research (DSR) method was used. DSR enabled evaluating artifacts from both a user-related and technical perspective [38]. Similarly to action research, the DSR addresses practical challenges while contributing to both practice and theory [39], thus gaining increasing recognition among information systems researchers in the process [40]. It was also successfully used to make designs that provide superior utility in the context of business process management [41]. As the current research delivered a formalization of a pilot EEM implementation process that involved external business partners in a multi-facility organization, the area of application of the method might be considered unconventional. It was, however, fully in line with the systematic literature review delivered by Offermann et al., who identified such artifact types as novel system designs, methods, languages/notations, algorithms, guidelines, requirements, patterns, or metrics [42].

Hevner et al. set forth seven guidelines that enhance the scientific rigor of the DSR approach [43]:

G1: an innovative, purposeful artifact ought to be created;
G2: the artifact must yield utility for a specified problem domain;
G3: the artifact is to be submitted for attentive evaluation;
G4: the artifact needs to address a heretofore unsolved challenge or feature a more effective solution to a well-known problem;
G5: the artifact itself must be strictly defined, internally consistent, and represented in a formal way;
G6: developing the artifact should involve a search process whereby a problem space is constructed, and a mechanism posed or enacted to find an effective solution;
G7: the outcomes of the DSR must be tellingly reported to practitioners responsible for putting them into practice, to the scientific community perfecting them, as well as to decision-makers, whose organizations shall take advantage of the artifact.

In order to (1) reliably assess the scope of data collected during the pilot EEM implementation process along with (2) determine the technological capability regarding automatic data collection; (3) narrow down the list of activities to those that show the greatest potential in terms of bolstering the EE, and; (4) correctly plan its duration, a party with prior experience in offering EE services was required. Therefore, while developing the artifact, the researchers worked hand-in-hand with the staff of SDC Ltd., a specialized company running engineering activities and related technical consultancy with know-how on deploying telemetry in office buildings and retail facilities. Previous ventures that SDC had engaged in had differed greatly in scale and were not preceded by a standardized preparatory phase that enabled the assessment of the potential energy savings policies. Notwithstanding, their post-implementation reports, project schedules, and work order notes constituted primary data sources while developing the methodology.

The measurement solution used by the company is proprietary. A single control device was installed in the electrical switchboard of each customer's facilities. The device periodically transmitted a JSON frame with the electricity consumption snapshot, along with a list of additional parameters. The development of the artifact required testing the technological capabilities regarding tightening the interval of transmitting a set of data from the sensors experimentally, as well as analyzing the validity of taking additional parameters into account. It was concluded that the scalability of the analytics required admitting CO_2 concentration, humidity, and internal temperature. Datasets were captured by an application implemented in the proven PHP/MySQL tandem. Some of its data processing functionalities were written in Scala. The application processed data and enabled the visualization of electricity consumption. It was possible to generate simple reports. More robust data analyses were performed by a BI solution based on the Qlik Sense engine. The solution took advantage of a data warehouse and enabled multi-dimensional data analyses by combining data from other sources. Those included but were not limited to, physical parameters of the facility, external weather data, and media-related invoices.

4. Results

4.1. Conceptualization of the Artifact

The artifact delivered by this study is shown at a high level of abstraction in Figure 1. The abstract representation of the process was formalized using BPMN, as expected per the G5 guideline, and detailed in Table 1. The artifact introduced key steps, recommended sources of information regarding each facility, analytical determinants, EEM application areas that featured the most likely cost reduction opportunities as well as the projected timespan of the pilot EEM implementation. On top of that, the minimum number of facilities that ought to be equipped with telemetry was specified, and reference supporting IT solutions were singled out. Thus, the methodology directly addressed RQ1–2.

The pilot itself was designed for deployments in networked retail organizations. Under typical conditions, most dispersed facilities of such organizations do not have the means to diversify energy carriers and are strongly dependent on electricity prices. Regardless of the above, heating, ventilation, air conditioning, and refrigeration (HVACR) hardware had a significant share in their energy balance. Finally, energy consumption, at least partly, resulted from the necessity to operate some devices in a continuous mode (fuel pumps, air curtains). The pilot was generally expected to last 4 months since the telemetry had been deployed in all designated locations. The time required for installation and parametrization of measurement systems was dependent on the type of system used—in our case, the customer was provided with a complete solution to reduce the technological risk of the project. A reference value of two months was required to populate the telemetry database and to collect complementary EE-targeted data from the customer side. After these two months, austerity priorities were set and a test of the implementation of the measures resulting from these priorities was carried out. Further monitoring, adjustment, and a summary of pilot works were expected to last up to 2 months.

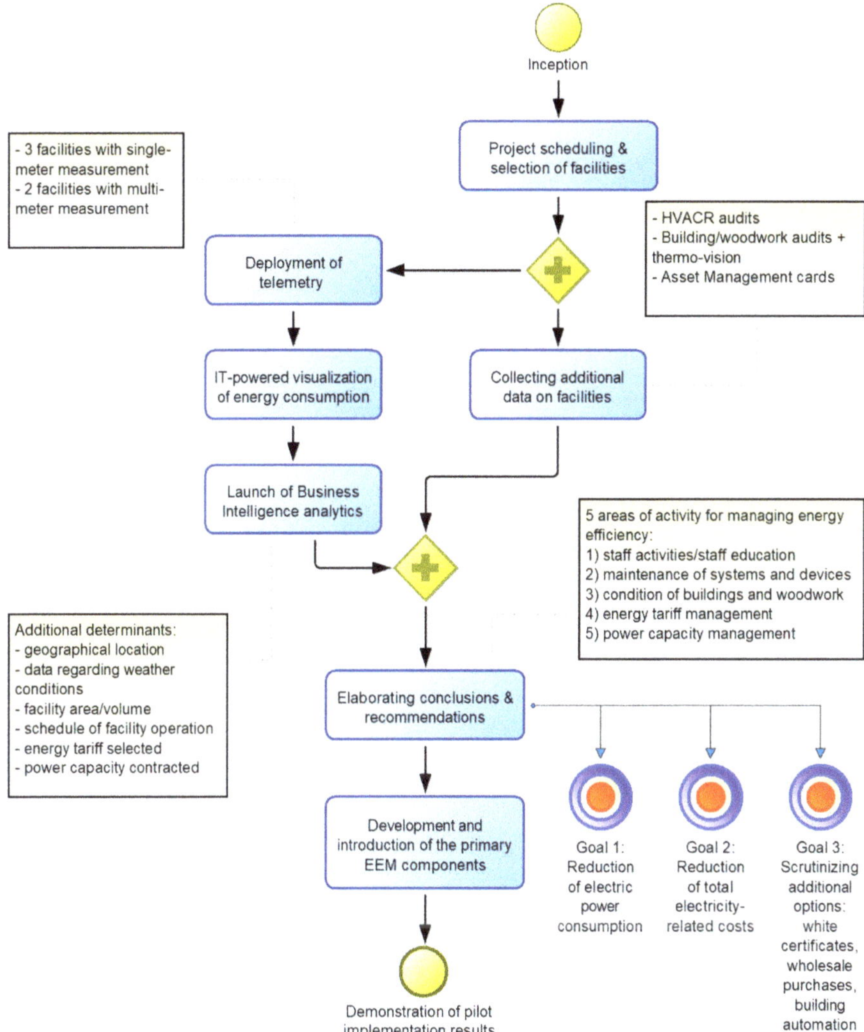

Figure 1. Reference process of a pilot energy efficiency management (EEM) implementation within a multi-facility organization.

Table 1. Description of pilot EEM implementation reference process steps.

Step	Description
Inception	The prerequisites for the pilot works include the readiness of both the provider of a telemetry system for measuring power consumption and the customer to jointly develop and test the foundations of the EEM on a small pool of facilities.
Project scheduling & selection of facilities	Five facilities shall be designated by the customer to create a pool of entities covered by pilot works. Measurement of the power consumption exclusively on the main power input is assumed regarding two facilities. Another two shall be provided with a multi-meter environment. Such an approach enables performing test measurements for both configurations and identifying their advantages and disadvantages. On top of selecting the facilities, a timetable and rules for telemetry installations shall be worked out.

Further steps to be carried out simultaneously			
Step	**Description**	**Step**	**Description**

Table 1. *Cont.*

Step	Description		
Deployment of telemetry	Putting the measurement environment in place will not require ceasing the operations of any facility. Short breaks are expected on some retail circuits (HVACR, light sources). The estimated duration of a single installation varies from 2 to 6 h per facility.		
IT-powered visualization of energy consumption	Consumption-related data are to be visualized using a dedicated IT solution. Access to reports shall be provided via a web browser.	*Collecting additional data on facilities*	In line with telemetry readings, additional data on facilities shall be collected and processed. The sources of knowledge include: • HVACR audits • building/woodwork audits • thermo-vision inspections • events generated as a part of the facility management service • customer-side internal data regarding asset management
Launch of Business Intelligence analytics	Power consumption measurements are to be benchmarked against several determinants: • the geographical location of a given facility • weather conditions for this particular location • facility area or volume • schedule of facility operation • energy tariff selected • power capacity contracted		

Follow-through upon completing the aforementioned activities	
Step	**Description**
Elaborating conclusions & recommendations	A document addressing cost reduction opportunities for the customer shall be delivered. In particular, options to be explored in the following areas of EEM are to be unveiled: • staff activities/staff education • maintenance of systems and devices • condition of buildings and woodwork • energy tariff management • power capacity management
Development and introduction of the primary EEM components	Priority items that may be immediately streamlined in terms of EE are to be pointed out in this phase. Such potential decisions include, but are not limited to, switching tariffs, rehauling or fine-tuning devices, introducing simple automation of devices, forcing certain behaviors of the staff, etc.
Demonstration of pilot implementation results	A physical meeting of the interested parties that summarizes the works accomplished throughout the implementation.

4.2. Implementation and Validation of the Artifact

DSR guidelines [43] directly stated that upon creation, the artifact needs to be evaluated (G3) and tellingly reported to interested parties (G7). In our case, both of those scientific rigor-oriented steps were effectuated jointly by the research teams and business partners as a part of the project aimed at enhancing EE within the network of facilities owned/franchised by the Polish subsidiary of a global player representing the petroleum industry. Royal Dutch Shell PLC, headquartered in the Netherlands, is included on a Global 2000 List of the World's Largest Public Companies, with $394 billion total assets and over $311 billion of revenue [44]. Its retail arm by the end of 2019 covered approx. 45 thousand facilities that operated based on different business models in close to 80 countries and handled over 30 million customer visits on a daily basis [45].

The pilot works implementation process unveiled two deviations from the original design of the artifact, i.e., the number of areas for EEM covered by pilot works and the length of the pilot itself. While the latter might be classified as minor, the former deserved special consideration. Pilot works were launched by selecting a pool of facilities (in this case: petrol stations) from among a total of over 400 facilities located in Poland. The methodology provided for designating two groups of facilities—with the first group being measured only at the main power input, and the second configured with multiple meters. Electrical circuits measured by the more sophisticated configuration included interior lighting sources of the station, air conditioning, and exterior lighting sources of a station (roofing, driveways, illuminated advertising, etc.). All the determinants necessary to initiate analytics (locations, weather conditions, volumes, schedules of operation, relevant contractual details between the customer and all energy suppliers involved) were provided, integrated, and double-checked.

Installing respective meters paved the way for collecting data on electric power consumption. Upon considering both the needs of the participating organization and the technical constraints, it was decided to send aggregated data packages from individual facilities every 15 min. Electric power consumption was plotted as hourly bars and supported by the BI system in place. It related power consumption to the aforementioned determinants. Simultaneously, the collection of additional data to support these activities was launched. In line with the construction of the artifact, energy audits were conducted as a part of pilot works, and technical inspections of electrical devices, thermo-vision, and building/woodwork audits constituted the primary sources of additional data. Values and informational content, upon being subjected to analysis, influenced the EEM of each petrol station involved.

Of the five areas of activity for EEM that the initial concept envisaged, specific actions to be completed were listed within areas 1, 4, and 5 (see Figure 1). Areas 2 and 3 were excluded from the pilot works. Activities in the areas undertaken for implementation were carried out until the end of the pilot's timeframe. The work was concluded with several recommendations, including guidelines for staff behavior at stations, benchmarking the existing electric power tariffs against optimal ones, selection of new tariffs for each facility, and proposing reductions in contracted power levels. It was empirically confirmed that the actions within these three areas resulted in a net reduction in power costs at the designated petrol stations without compromising business continuity.

Ultimately, the pilot EEM implementation lasted from 1 January 2019 to the end of May. Project inception required 2 weeks and involved preparing measurement equipment, assembling installation teams, training on safety at work at petrol stations as well as scheduling telemetry installation. The actual deployment of telemetry in all five facilities totaled one month. Beginning in mid-February, every telemetry system started generating measurement data and forwarding it to the data processing IT solutions. While measurements were kept live until the end of the pilot works, at the end of February, early feedback and recommendations were determined. Between the beginning of March and the end of May, recommendations were introduced in all facilities, and the results were observed. Upon the pilot project completion, concluding workshops were held.

As energy services tend to be settled under performance-based contracts [27], the pilot served as a data source for elaborating pessimistic, realistic, and optimistic scenarios for increasing EE during the course of the actual contract. The pilot showed a statistically significant potential to reduce the nominal energy consumption. The specificity of petrol stations' operations (24/7 mode, excluding the short period where the customer-side systems settle daily transactions) contributed to power consumption not being reduced by late evenings and during night times. On the contrary, consumption is statistically lowest during (late) mornings and mid-day. This is when few light sources are used, and the Central European climate implies less intensive use of heating systems. Regardless of the above, the selection of energy tariffs, which were rational when signing contracts with energy suppliers, became less favorable over time. Tariff systems tended to become increasingly complicated, and the unbundled energy distribution market in Poland makes identifying optimal solutions on one's own, time-consuming. In this case, having an up-to-date tariff database that is for the needs of the entire customer portfolio significantly facilitates optimization processes. Whereas tariffs individually tailored to the profile of each facility where possible, show that time periods in which energy prices are higher than the base tariff (see Figure 2, the median case with a linear tariff), are more than compensated for over the rest of the day.

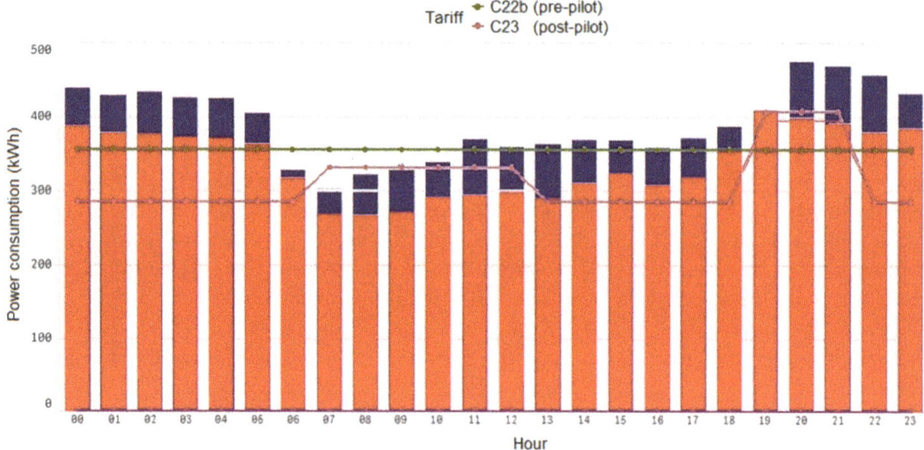

Figure 2. Differences in energy consumption and tariff adjustment.

Given the time necessary for reflection, the participating organization committed to up-scale their EE-oriented efforts to the entire network of Polish facilities by signing a 3-year contract at the end of June 2019. The pilot's pessimistic scenario for increasing EE was adopted as the benchmark for the contract. In its first year, which also covered the 5-month-long telemetry installation phase, the provider of the energy services took upon oneself achieving cumulative savings of 2.25% on one-third of the base amount. Throughout the second year of the contract, a significantly higher level of EE improvement was set: no less than 10% on two-thirds of the base amount. Such a scale of commitment was in line with the long-term estimates formulated in the post-pilot reports and the adjustments for unsupported areas of activity (see Table 2). The calculation took into account both the saturation of the telemetry system with data and the development and implementation of specific policies, as well as getting to know the facilities, personnel, and internal processes of the petrol company. Pilot-based projection indicated the need for a conservative approach in the final year (cumulative savings of 14% of the total base amount) due to reaching expected limits regarding opportunities for reducing electricity consumption without affecting the continuity of business.

Table 2. Post-pilot estimates regarding the scale of savings in the long run.

Area of Activity	Estimated Reduction in Power Consumption [% of Current kWh]	Estimated Cost Reduction [% of Pre-Implementation Invoice Amounts]
Staff activities/staff education	7%	7%
Maintenance of systems and devices	(7%)	(7%)
Condition of buildings and woodwork	(5%)	(6%)
Energy tariff management	n/a	8%
Power capacity management	n/a	3%
Total		18% (31%)

5. Discussion

The study demonstrated that EEM is, on one hand, a long-term process that requires continuous monitoring and improvement as well as adapting to constant changes both in the business environment and within the organization. While the necessity to take advantage of enhanced procedures and adapt business processes were explored before [35], lessons learned from this study are somewhat polemic with previous conclusions that companies are generally uninterested in ventures lasting beyond 3 years [17]. The recommended best practice in this respect is to stage the target contract so that periodical milestones/review points and success fees were considered (Table 3). In this very case, the progress is being monitored at weekly and monthly intervals. On top of that, once a year an official presentation of the savings recorded and details of the methods to achieve them is given to a wider audience of customer-side professionals, as indicated in the contract.

On the other hand, such a contract proves parameterizable to a large extent. The study showed that EEM implementation features selecting a set of options that do not exhaust all the possibilities but are favorable in a given horizon and budget/investment capabilities. In this case, three EE-improvement areas (human behavior, tariffs, capacity management) indeed required an investment in a telemetry system as well as developing and implementing an EEP. However, compared to the other preselected areas, it involved no additional costs related to upgrading/replacing electrical systems or conducting a general overhaul of a facility or its components (such as woodwork). Such fundamental ventures depend on the firmness of financial commitment from the investor and the lack thereof put the capability of meeting the conditions for success fee at risk. Hence, the recommended best practice in this regard is to scale down the list of potential EE improvement areas within the first 30% of the contract.

First-year performance indicators exceeded both the contract reference values and the average savings of approx. 5.0% reported by Fresner et al. based on the analysis of 280 EE-targeted projects across seven different countries [17]. It should be pointed out that the adoption of the pilot's pessimistic scenario did not increase the accuracy of forecasting the actual performance of the contract, which might be assumed based on the study by Fowlie et al. [46]. Similarly, actual energy savings in Zivin and Novan's analysis were off by 21% [47]. Nevertheless, the energy service implemented following the application of the methodology was beyond the scope of the energy audits covered by those analyses. Moreover, small/medium-sized enterprises have a natural tendency to avoid the more intrusive measures that a corporate body can easily afford, and whether the COVID-19 pandemic has had a noticeable impact on the results remains to be seen. Adopting realistic scenarios as benchmarks for similar contracts in our opinion requires pinpointing methods for calculating savings generated, which will take into account factors such as volatility of electric power prices, weather anomalies (atypically warm winters or hot summers), as well as fluctuations in equipment lists of relevant petrol stations. The latter involves, in particular, extra electrical hardware such as HVACR devices or (super) chargers for electric cars.

There is a need to highlight the strong feedback of the practitioners that the success of the project and assuring the efficacy of developed measures requires the involvement of employees from various levels, as well as adapting some business processes. This is in line with Johansson and Thollander's contribution, who listed top-management support as a leading success factor regarding EEM practices [16].

Table 3. Key activities within the EEM framework.

Year	Phase of Contract	Description
1	*The inception of the EE improvement contract*	Based on the feedback of the pilot project implementation, the contract between the interested parties shall be drafted and then carried into effect. The contract itself aims to reduce the electric power consumption/costs owing through the design, implementation and continuously enhanced EEM measures on the customer side. Accomplishing the following tasks:
	Introduction of the EEM measures	• installation of the dedicated telemetry system in all facilities covered by the contract (refining the list of facilities and deployment timetable is required; measurement environments in individual facilities shall be integrated with relevant service provider's IT solutions at this stage) • taking advantage of EE improvement opportunities within telemetry-ready facilities • identifying further options and areas for scaling up the austerity policy
	Progress towards the goal (1st-year milestone)	Summarizing the first year of the contract. Verifying the achievement of the adopted partial goals and indicators. Laying down detailed goals and indicators for the following year.
2	*Refinement and monitoring of the EEM measures*	Based on the experience gained and taking advantage of the database being populated with annual telemetry data, EE improvement shall go on. This involves in particular: • increased maturity of tariffs/contracted power management • promoting, implementing, and enforcing austerity policies within organizational structures of the customer • the popularization of best practices regarding reducing power consumption costs • assigning a high priority to the project within the organization and promoting the results achieved • stimulating the customer-side involvement in EE throughout all management levels • planning all future investments taking the EEM into account
	Progress towards the goal (2nd-year milestone)	Summarizing the second year of the contract. Verifying the achievement of the adopted partial goals and indicators. Laying down detailed goals and indicators for the final year.
3	*Development and diffusion of the EEM measures*	Cleaning the project backlog of implementation-related activities, continuous improvement, and verifying the efficacy of measures in place. Adjusting all internal business processes on the customer side in line with the EE improvement best practices and available toolset.
	Progress towards the goal (3rd-year milestone)	Summarizing the third year of the contract. Verifying the achievement of the adopted partial goals and indicators. Laying down detailed goals and indicators for the foreseeable future.
N/A	*Continuous use and development of EEM*	Further workflows related to developing techniques, tools, and best practices in EEM. Maintaining results achieved to date.

6. Implications and Limitations

6.1. Implications for Theory and Practice

The current study's results suggest several implications for both researchers and practitioners in the field of energy efficiency management. First of all, EE-enhancement ventures can all too often be brought down to a direct transposition of success stories advertised across other industries by the boards of implementing companies or capitalizing on to-date experiences of operational departments of energy service providers in the shortest timeframe possible. As multinational companies stretch their dispersed networks of facilities across locations with diametrically different local media tariffs, weather conditions, construction-related laws and practices, facility work schedules, or even cultural habits, such approach often fails to deliver [17,47]. The current study's findings strongly imply that a multiple-stakeholder perspective must be employed while implementing EEM. Various stakeholders' needs, expectations, and constraints must be carefully estimated, managed, and reconciled. Managing such a diverse set of considerations requires time, which illustrates the second major implication stemming from our study: time-related complexity. In this respect, our findings suggest that in order to manage EEM-related endeavors effectively, an approach phased in time ought to be adopted. Such an approach should allow practitioners to reasonably estimate and mitigate risks associated with the full-scope EEM implementation projects and make an informed decision about the project launching. Empirical data obtained thanks to the implementation of the artifact can also be a valuable argument while obtaining external financing, as both escalating the telemetry onto the entire facility network and overhauling individual facilities go in line with discernible investments. The phased approach to managing IT adoption projects is especially advised in changeable economic settings that result in a volatile nature of project requirements [48].

We extended the list of activities that usually cover the energy management process [2] by pinpointing the areas within which actions should be taken to manage energy efficiency. While fully acknowledging prior conclusions that energy audits allow us to determine what affects electricity losses [17], we demonstrated that pilot actions enable establishing why these losses occur—which in turn paves the way for a long-term organizational EE strategy. The framework proposed and discussed in the study may be applied in organizations of similar characteristics to those researched in the current paper. In particular, adopting an implementation approach phased in time and the use of suggested IT solutions such as sensors, data gathering, and analytical tools should help managers to better capture and manage multi-faceted considerations experienced by EEM projects. As a result, a fuller insight into the energy-related considerations might impact strategic decisions such as those related to a company's business model or business process reengineering.

6.2. Limitations and Potential Future Research Directions

The primary limitations of the current study are associated with the research setting, which features business units (petrol stations) of a global company operating on the Polish petroleum market. First of all, although facilities of this type might appear similar to other businesses—such as retail or grocery stores—when energy-related considerations are taken into account, it should be born in mind that a petrol station has its intricacies which might impact the generalization of findings to other industries. Secondly, the country of investigation within the current study, Poland, is an example of a transition economy and, as compared to the most industrialized economies, reveals a number of specific considerations such as the lack of a strategic ICT role, insufficient customer orientation, and the critical role of people-related issues [49]. Therefore, the generalization of the current study's results to other economic settings should be done with caution.

It would appear that a cross-industry and cross-country investigation into EEM is an important direction of future research. A promising focus for potential studies is mapping energy consumption carriers per industry and per specific field that are either rigid or susceptible to intensive optimization without entertaining risk factors. Researchers and practitioners alike are also interested in how extensively the increased EE of retail facilities translates into the total carbon footprint of the products or services provided.

7. Conclusions

Within the current study, the DSR's artifact—being a formalized version of a reference EEM implementation process for multi-facility companies—was conceptualized, implemented, evaluated, and escalated from pilot works towards a country-wide project. The target EE improvement contract were spread over 400 petrol stations. Our research explored key process steps, sources of information (RQ1), reference telemetry, and data processing solutions (RQ2) as well as guidelines for framing EE improvement contracts between implementing organizations and contractors (RQ3). The study enabled us to identify areas for improving EE. It was confirmed that the involvement of customer-side staff in the implementation, verification, and continuous improvement of EEM measures was crucial. Close cooperation between the petroleum company that hosted the pilot works and the supplier of telemetric and analytical tools that fully committed itself to support EE improvement activities also contributed to the success of the project.

Nowadays, an effective EEM without mature IT is virtually impossible. A multitude of complex tariffs used by multiple power suppliers, analytical challenges, patterns, and projections that quickly lose relevance, as well as the relative diversity of implementation environments, all highlights the value of IT for EEM. Detailed knowledge regarding the volume of electric power consumption and its distribution over time ought to be adequately captured, up-to-date, and easy to access. Such serviceable knowledge enables matching appropriate energy tariffs perfectly and ordering power volumes that are technically and economically justified. Capable tools combined with know-how and efficiency of operations lead to achieving significant savings.

Author Contributions: Conceptualization, B.G.; methodology, B.M.; validation, B.G.; formal analysis, B.M. and B.G.; investigation, B.M.; writing—original draft preparation, B.M.; writing—review and editing, B.G.; visualization, B.M.; supervision, B.M.; project administration, B.G.; funding acquisition, B.G. All authors have read and agreed to the published version of the manuscript.

Funding: This study was funded by the National Centre for Research and Development (PL)—competition for micro, small and medium enterprises that have received the Seal of Excellence in SME Instrument competitions, phase II (Horizon 2020). Grant number POIR.01.01.01-00-0003/19. The co-author of the article, B.G., is the author of the application, which obtained funding for research and development works.

Acknowledgments: We would hereby like to thank Piotr Soja for his constructive criticism regarding the initial version of the manuscript.

Conflicts of Interest: The authors declare no conflict of interest. The funders had no role in the design of the study; in the collection, analyses, or interpretation of data; in the writing of the manuscript, or in the decision to publish the results.

References

1. Tugcu, C.T.; Ozturk, I.; Aslan, A. Renewable and non-renewable energy consumption and economic growth relationship revisited: Evidence from G7 countries. *Energy Econ.* **2012**, *34*, 1942–1950. [CrossRef]
2. Fernando, Y.; Hor, W.L. Impacts of energy management practices on energy efficiency and carbon emissions reduction: A survey of malaysian manufacturing firms. *Resour. Conserv. Recycl.* **2017**, *126*, 62–73. [CrossRef]
3. International Energy Agency. Global Energy & CO_2 Status Report. 2019. Available online: https://iea.org/reports/global-energy-co2-status-report-2019 (accessed on 12 October 2020).
4. United Nations. Report of the Conference of the Parties on its Twenty-First Session. Addendum Part Two: Action Taken by the Conference of the Parties at its Twenty-First Session. Available online: https://unfccc.int/resource/docs/2015/cop21/eng/10a01.pdf (accessed on 12 October 2020).
5. Saboori, B.; Sulaiman, J.; Mohd, S. Economic growth and CO_2 emissions in malaysia: A cointegration analysis of the environmental kuznets curve. *Energy Policy* **2012**, *51*, 184–191. [CrossRef]
6. Bukarica, V.; Tomšić, Ž. Energy efficiency policy evaluation by moving from techno-economic towards whole society perspective on energy efficiency market. *Renew. Sustain. Energy Rev.* **2017**, *70*, 968–975. [CrossRef]

7. Cooremans, C.; Schönenberger, A. Energy management: A key driver of energy-efficiency investment? *J. Clean. Prod.* **2019**, *230*, 264–275. [CrossRef]
8. May, G.; Kiritsis, D. Business model for energy efficiency in manufacturing. *Procedia Cirp.* **2017**, *61*, 410–415. [CrossRef]
9. Marcinkowski, B.; Gawin, B. Data-driven business model development—Insights from the facility management industry. *J. Facil. Manag.* **2020**, *earlycite*. [CrossRef]
10. Gauthier, C.; Gilomen, B. Business models for sustainability: Energy efficiency in urban districts. *Organ. Environ.* **2016**, *29*, 124–144. [CrossRef]
11. Hilorme, T.; Zamazii, O.; Judina, O.; Korolenko, R.; Melnikova, Y. Formation of risk mitigating strategies for the implementation of projects of energy saving technologies. *Acad. Strateg. Manag. J.* **2019**, *18*, 1–6.
12. Environmental and Energy Study Institute. Energy Efficiency. Available online: https://www.eesi.org/topics/energy-efficiency/description (accessed on 7 March 2020).
13. Pérez-Lombard, L.; Ortiz, J.; Velázquez, D. Revisiting energy efficiency fundamentals. *Energy Effic.* **2013**, *6*, 239–254. [CrossRef]
14. Bhardwaj, K.; Gupta, E. Analyzing the "energy-efficiency gap". *Indian Growth Dev. Rev.* **2017**, *10*, 66–88. [CrossRef]
15. Chai, K.-H.; Baudelaire, C. Understanding the energy efficiency gap in Singapore: A motivation, opportunity, and ability perspective. *J. Clean. Prod.* **2015**, *100*, 224–234. [CrossRef]
16. Johansson, M.T.; Thollander, P. A review of barriers to and driving forces for improved energy efficiency in swedish industry—Recommendations for successful in-house energy management. *Renew. Sustain. Energy Rev.* **2018**, *82*, 618–628. [CrossRef]
17. Fresner, J.; Morea, F.; Krenn, C.; Uson, J.A.; Tomasi, F. Energy efficiency in small and medium enterprises: Lessons learned from 280 energy audits across Europe. *J. Clean. Prod.* **2017**, *142*, 1650–1660. [CrossRef]
18. Jaffe, A.B.; Stavins, R.N. The energy-efficiency gap. What does it mean? *Energy Policy* **1994**, *22*, 804–810. [CrossRef]
19. Solnørdal, M.T.; Foss, L. Closing the energy efficiency gap—A systematic review of empirical articles on drivers to energy efficiency in manufacturing firms. *Energies* **2018**, *11*, 518. [CrossRef]
20. Hasan, A.S.M.; Rokonuzzaman, M.; Tuhin, R.A.; Salimullah, S.M.; Ullah, M.; Sakib, T.H.; Thollander, P. Drivers and barriers to industrial energy efficiency in textile industries of Bangladesh. *Energies* **2019**, *12*, 1775. [CrossRef]
21. Zobel, T.; Malmgren, C. Evaluating the management system approach for industrial energy efficiency improvements. *Energies* **2016**, *9*, 774. [CrossRef]
22. Marchi, B.; Zanoni, S. Supply chain management for improved energy efficiency: Review and opportunities. *Energies* **2017**, *10*, 1618. [CrossRef]
23. Wang, C.N.; Ho, H.X.T.; Hsueh, M.H. An integrated approach for estimating the energy efficiency of seventeen countries. *Energies* **2017**, *10*, 1597. [CrossRef]
24. Lin, J.; Xu, C. The impact of environmental regulation on total factor energy efficiency: A cross-region analysis in China. *Energies* **2017**, *10*, 1578. [CrossRef]
25. Cagno, E.; Trianni, A.; Spallina, G.; Marchesani, F. Drivers for energy efficiency and their effect on barriers: Empirical Evidence from Italian manufacturing enterprises. *Energy Effic.* **2016**, *10*, 855–869. [CrossRef]
26. Worrell, E.; Bernstein, L.; Roy, J.; Price, L.; Harnisch, J. Industrial energy efficiency and climate change mitigation. *Energy Effic.* **2009**, *2*, 109. [CrossRef]
27. Thollander, P.; Backlund, S.; Trianni, A.; Cagno, E. Beyond barriers—A case study on driving forces for improved energy efficiency in the foundry industries in Finland, France, Germany, Italy, Poland, Spain, and Sweden. *Appl. Energy* **2013**, *111*, 636–643. [CrossRef]
28. Cai, W.; Lai, K.-H.; Liu, C.; Wei, F.; Ma, M.; Jia, S.; Jiang, Z.; Lv, L. Promoting sustainability of manufacturing industry through the lean energy-saving and emission-reduction strategy. *Sci. Total Environ.* **2019**, *665*, 23–32. [CrossRef]
29. Avgerinou, M.; Bertoldi, P.; Castellazzi, L. Trends in data centre energy consumption under the european code of conduct for data centre energy efficiency. *Energies* **2017**, *10*, 1470. [CrossRef]
30. Yuan, X.; Ma, R.; Zuo, J.; Mu, R. Towards a sustainable society: The status and future of energy performance contracting in China. *J. Clean. Prod.* **2016**, *112*, 1608–1618. [CrossRef]

31. Tang, T.; Hill, H. Implementation and impacts of intergovernmental grant programs on energy efficiency in the USA. *Curr. Sustain. Renew. Energy Rep.* **2018**, *5*, 59–66. [CrossRef]

32. Guerrini, A.; Romano, G.; Indipendenza, A. Energy efficiency drivers in wastewater treatment plants: A double bootstrap DEA analysis. *Sustainability* **2017**, *9*, 1126. [CrossRef]

33. Gawin, B.; Marcinkowski, B. Business intelligence in facility management: Determinants and benchmarking scenarios for improving energy efficiency. *Inf. Syst. Manag.* **2017**, *34*, 347–358. [CrossRef]

34. Backlund, S.; Thollander, P.; Palm, J.; Ottosson, M. Extending the energy efficiency gap. *Energy Policy* **2012**, *51*, 392–396. [CrossRef]

35. Schulze, M.; Nehler, H.; Ottosson, M.; Thollander, P. Energy management in industry—A systematic review of previous findings and an integrative conceptual framework. *J. Clean. Prod.* **2016**, *112*, 3692–3708. [CrossRef]

36. Ameli, N.; Brandt, N. Determinants of households' investment in energy efficiency and renewables: Evidence from the OECD survey on household environmental behaviour and attitudes. *Environ. Res. Lett.* **2015**, *10*, 044015. [CrossRef]

37. Harris, J.; Anderson, J.; Shafron, W. Investment in energy efficiency: A survey of Australian firms. *Energy Policy* **2000**, *28*, 867–876. [CrossRef]

38. Riehle, D.M.; Fleischer, S.; Becker, J. A web-based information system to evaluate different versions of IT artefacts in online experiments. *AIS Trans. Enterp. Syst.* **2019**, *4*, 10. [CrossRef]

39. Stal, J.; Paliwoda-Pekosz, G. Fostering development of soft skills in ICT curricula: A case of a transition economy. *Inf. Technol. Dev.* **2019**, *25*, 250–274. [CrossRef]

40. March, S.T.; Storey, V.C. Design science in the information systems discipline: An introduction to the special issue on design science research. *Mis. Q.* **2008**, *32*, 725–730. [CrossRef]

41. Mendling, J. From scientific process management to process science: Towards an empirical research agenda for business process management. In Proceedings of the 8th ZEUS Workshop, Vienna, Austria, 27–28 January 2016; pp. 1–4.

42. Offermann, P.; Blom, S.; Schönherr, M.; Bub, U. Artifact types in information systems design science—A literature review. *Lect. Notes Comput. Sci.* **2010**, *6105*, 77–92. [CrossRef]

43. Hevner, A.R.; March, S.T.; Park, J.; Ram, S. Design science in information systems research. *Mis. Q.* **2004**, *28*, 75–105. [CrossRef]

44. Murphy, A.; Tucker, H.; Coyne, M.; Touryalai, H. *GLOBAL 2000. The World's Largest Public Companies*; Forbes: Jersey City, NJ, USA, 2020.

45. Coulter, L.M. *Energy for a better future. Annual Report and Accounts for the Year Ended December 31, 2019*; Royal Dutch Shell PLC: Hague, The Netherlands, 2020.

46. Fowlie, M.; Greenstone, M.; Wolfram, C. Do energy efficiency investments deliver? Evidence from the weatherization assistance program. *Q. J. Econ.* **2018**, *133*, 1597–1644. [CrossRef]

47. Zivin, J.G.; Novan, K. Upgrading efficiency and behavior: Electricity savings from residential weatherization programs. *Energy J.* **2016**, *37*, 1–24. [CrossRef]

48. Themistocleous, M.; Soja, P.; Cunha, P.R. The same, but different: Enterprise systems adoption lifecycles in transition economies. *Inf. Syst. Manag.* **2011**, *28*, 223–239. [CrossRef]

49. Soja, P.; Cunha, P.R. ICT in transition economies: Narrowing the research gap to developed countries. *Inf. Technol. Dev.* **2015**, *21*, 323–329. [CrossRef]

Publisher's Note: MDPI stays neutral with regard to jurisdictional claims in published maps and institutional affiliations.

Article

How to Foster the Adoption of Electricity Smart Meters? A Longitudinal Field Study of Residential Consumers

Anna Kowalska-Pyzalska [1,*, Katarzyna Byrka [2] and Jakub Serek [2]**

[1] Department of Operations Research and Business Intelligence, Faculty of Computer Science and Management, Wroclaw University of Science and Technology, Wyb. Wyspianskiego 27, 50-370 Wroclaw, Poland

[2] Wrocław Faculty of Psychology, SWPS University of Social Science and Humanities, 50-505 Wrocław, Poland; kbyrka@swps.edu.pl (K.B.); jserek@st.swps.edu.pl (J.S.)

* Correspondence: anna.kowalska-pyzalska@pwr.edu.pl; Tel.: +48-713202524

Received: 20 August 2020; Accepted: 7 September 2020; Published: 11 September 2020

Abstract: The objective of this research was to explore correlates and predictors that play a role in the process of adopting and withdrawing from using a smart metering information platform (SMP). The SMP supports energy monitoring behaviors of the electricity consumers. The literature review shows, however, that not every customer is ready to the same extent to adopt novel solutions. Adoption requires going through stages of readiness to monitor energy consumption in a household. In a longitudinal field experiment on Polish residential consumers, we aimed to see whether messages congruent with the stage of readiness in which participants declared to be at a given moment will be more effective in prompting participants to progress to the next stage than a general message or a passive control condition. We also tested the effect of attitude and knowledge about energy monitoring on phase changes. Our study reveals that what affects the phase change is the participation in the study. The longer the participants were engaged in the usage of SMP, the more willing they were to monitor their energy consumption in the future. This result sheds light on the future educational and marketing efforts of the authorities and energy suppliers.

Keywords: energy monitoring; electricity smart meters; smart metering information platforms; knowledge; longitudinal study; consumers

---

1. Introduction

Recently, many countries, for environmental and political reasons, have been striving to increase the energy efficiency of production, distribution, and consumption of energy. The goal of increasing energy efficiency is closely correlated with the new approach to the power system, namely the concept of smart grids (SG). Intelligent networks use modern communication technologies to exchange information between market agents (producers, market operators, and end users) to improve production efficiency and energy consumption [1–4]. One of the milestone steps in the transformation of the traditional power system to SG is the extensive implementation of the smart meters (SM) among electricity end users [5–7]. A smart meter is an electronic device that measures energy use and sends this information automatically over wireless networks to the energy supplier. The consumer can benefit from SM in multiple ways—firstly, by receiving a much more accurate billing; secondly, by gaining an opportunity to control one's energy consumption in real time. The information collected by SM can provide consumers with a feedback on current energy consumption and energy efficiency via an SM information system (platform, SMP) that is a website or mobile application connected with SM [4,8–10].

Global roll-outs of SM are usually initiated by pilot programs and local deployment of SM in a given region or city [11–15]. A good example of such practices is Wroclaw—a capital city of Lower Silesia in Poland, with nearly 630,000 inhabitants. Since 2015 Tauron Dystrybucja S.A., the local electricity distribution system operator (DSO) has been running a project AMIPlus Wrocław, which aimed at installing a smart meter at each household and enabling access to the SM platform (both an Internet website and a mobile app called e-licznik).

As SM is still a novelty on the Polish energy consumer market, and most of the electricity consumers are not fully aware of the potential of the installed devices [3,4], we have taken this opportunity to better understand the process of adopting novel electricity solutions. Our longitudinal field study was performed to explore individual variables that foster or hinder progression in the stages of readiness to adopt using a smart meter platform: e-licznik. The originality of this contribution relies on using the stage model approach, so far not explored thoroughly in the energy related studies.

The remainder of the paper is as follows. In Section 2, we provide the literature review of variables having an impact on SM and SMP adoption and energy monitoring. We also discuss the theoretical background of the study. Next, in Section 3, we present the methodology of the survey and its design. In Section 4, the obtained results are presented and discussed. Finally, in Section 5, the outcomes of the survey are concluded and some practical recommendations are provided.

2. Literature Review

2.1. Barriers in Using Smart Meter Platforms

The worldwide roll-outs of SM and the access to the information about the real-time energy consumption create some new opportunities for consumers and suppliers [12,16,17]. The literature provides a number of findings from the recent studies in which: (i) willingness to monitor energy in general, and by means of SMP is investigated [4,8,10,18–20], (ii) factors influencing the acceptance of SM and SMP by end-users are studied [3,6,15,21–25].

There is a great number of barriers to SM acceptance that limit users' willingness to use the enabling technologies, such as smart metering information systems (platforms, SMP) [26]. The barriers include among others distrust in the industry, lack of familiarity, a sense of procedural fairness, and concerns related to privacy and cost [7,23,25]. To focus on benefits of using SM, customers must be willing to accept this technology. Various aspects of community and social SM acceptance have been already explored [3,6,7,23–25,27–29]. As in the case of any other energy technology, the lack of acceptance may lead to slowing or a halting of the development [7,30]. Evidence from SM roll-outs run in various countries all over the world have shown that the widespread implementation of SM is unlikely to be successful unless it adequately addresses the perspectives and needs of the consumers [5,7,11,16,17,31].

Table 1 summarizes the most common incentives and barriers to SM and SMP adoption.

2.2. Monitoring of Energy Consumption

Many studies emphasize that the introduction of smart grids and a broader implementation of SM may open new perspectives for consumers in terms of their awareness and control of energy consumption [5,31,32]. The question is, however, if they are interested and ready to control energy consumption and, if yes, what motivates them most: savings, environmental attitudes, social influence, or maybe something else.

The impact of information and feedback about energy consumption on consumers' habits and behaviors have been already studied [17,33,34]. Especially computerized feedback, by means of SM devices, mobile apps, and smart metering platforms have been widely explored [8,9,18,35]. These studies reveal that computerized feedback may lead to some reduction of energy consumption by leading to habitual changes and/or prompting investments in smart and energy efficient home appliances and smart devices (e.g., smart plugs). At the same time, the user-friendliness and ease of access to the information is emphasized [18,36]. These are critical conditions that the computerized feedback must fulfill to engage consumers, especially as the general level of consumers' interest and knowledge is low [3,4,15,36].

Table 1. Incentives and barriers of SM and SMP adoption.

Factor	Description	References
Privacy concerns	These concerns originate from consumers' beliefs that using SM may lead to a loss of privacy by providing detailed information about household behaviors. Data collected by SM may reveal the activities of people inside of their home (i.e., their habits, usage, and type of home appliances they possess, etc.) In case of improper cyber security, SM data can be misused by authorized and unauthorized parties.	[7,25,37,38]
Procedural fairness	It refers to access to and control in the decision-making process. It indicates whether one has control over a certain process or procedure—in this case, SM data transmission and usage.	[7,39]
Trust	Both previous factors connect with the issue of trust in energy suppliers (whether they will secure the personal information and will not share it with third parties). Trust is especially vital in situations where familiarity with a technology is low, as it influences perceptions of risks and benefits.	[7,25]
Financial aspects	Some consumers are afraid that, due to SM installation, their cost of energy will increase (more adequate readings). On the other hand, some of them may expect immediate savings from SM, which is rather unrealistic.	[7,21,40,41]
Familiarity & knowledge	Familiarity of SM technology is still low. Consumers mistake SM with some other smart home devices. To some extent, knowledge and exposure to SM may be associated with increased concerns about negative attributes of these technologies. However, at the same time, it may increase interest and willingness to monitor energy consumption.	[8,21,22,42]
Environmental concern	The impact of environmental beliefs and concerns on SM acceptance is ambiguous. Generally, people who are aware of climate change are supposed to be more willing to accept SM as a useful and energy efficient technology.	[7]
Acceptance & engagement	There is some empirical evidence indicating an impact of SM acceptance on SM related behaviors, i.e., energy saving and monitoring.	[8,10]

The first step in monitoring energy consumption is its measurement [8], based on the traditional electricity bills and/or SMP. The second step includes observations of the measurements and its comparative analysis [4,8]. Energy monitoring behavior may increase the general awareness of one's energy usage, or the energy consumption of certain home appliances [43]. However, the possibility of monitoring energy by means of SMP may still not be enough to create a habitual behavior. Consumers may need some additional incentives, such as customized feedback [43], or some combination with demand side management and demand response tools, such as dynamic electricity tariffs [26].

2.3. Phase Changes of Behaviors

The acceptance and use of SM platforms is a phase process, as in the case of other eco-innovations, e.g., the use of ecological forms of transportation [44] or green energy [45]. Our study has been motivated and inspired by the stage model of self-regulated behavioral change (SSCB), proposed by Bamberg [46].

This model draws from a classic action phases model proposed by Heckhausen and Gollwitzer [47,48]. Accordingly, behavioral change, such as adoption of novel solutions, is a goal-directed and deliberate process in which individuals take gradual steps to the goal. In the first stage (pre-decisional), an individual has to choose a given behavior from competing options. In the second stage (pre-actional), an individual forms an intention to perform a behavior. He or she weighs the pros and cons of engaging in a certain behavior and specifies how the behavior will be performed. In the third stage (actional), an individual implements an intention. The fourth stage (post-actional) focuses on the evaluation of an action.

The model of innovation diffusion (DOI) proposed by Rogers [49] is another example of a phase model. The SSCB model refers to the diffusion stages of DOI, but it focuses more on individual determinants such social norms, attitudes, and perceived behavioral control as determinants of people's engagement in the following phases.

To illustrate four phases of behavior change in the context of using SMP, the example would be as follows: (1) Predecisional phase—when consumers choose to use SMP or to engage in an energy monitoring behavior; (2) Preactional phase—when consumers specify their intention to use SMP or to perform an energy monitoring behavior, (3) Actional phase—when consumers regularly use SMP or monitor energy consumption, and (4) Postactional phase—when consumers evaluate the satisfaction of using SMP or monitoring energy consumption [44,46]. As different phases of behavior change involve different psychological processes, past research has shown that consumers at different stages of the process need different methods to encourage them to move on to the next phase [44,46].

Literature shows that consumers' ecological behavior is strongly associated not only with professed values and opinions, but also with norms, barriers, and difficulties with accepting new behaviors, social norms, and legal regulations [31,50–53]. The SSCB model has been successfully used thus far to explore behaviors related to green public transportation [46]. In the context of energy market, phase models have not been widely used. Recently, one study applied the SSCB model and analyzed whether German SM platforms are properly designed so that, through their use, energy consumers can move from one decision-making phase to another [35]. The conclusions of this work show that the SSCB model is suitable for assessing consumer behavior related to energy saving.

2.4. Specific Research Goals

Although the acceptance of the SM and SMP acceptance and diffusion have already been studied, we still see a need to explore which factors are responsible for the transition from one behavioral stage to another, in the process of creating awareness, acceptance, and regular usage of SMP or energy monitoring behavior.

Hence, we aimed to see whether messages congruent with stages in which participants declared to be at a given moment will be more effective in prompting participants to progress to the next stage than a general message or a passive control condition. We also tested the effect of attitude and knowledge about energy monitoring on phase changes.

Based on the current knowledge on factors enhancing SM and SMP adoption, within our survey, we wanted to check what may enhance consumers' willingness to regularly monitor energy consumption by means of SMP. Hence, we checked the impact of the following issues such as: knowledge about the energy market, participation and engagement in the longitudinal study, environmental attitudes and behaviors, positive attitudes towards energy monitoring, and, finally, computer skills. In particular, we tested: (i) an impact of messages (interventions), (ii) an effect of an attitude towards energy monitoring, and (iii) an effect of knowledge about energy market on phase change of regular energy monitoring by means of SMP.

3. Methodology of the Study

3.1. Study Design

To address our research questions, we conducted a longitudinal experiment with six points of measurement: pretest (T0), posttests after interventions on Monday (T1), Wednesday (T2), Friday (T3), and Sunday (T4), and the follow-up (T5) and two control groups: one active (C1), one passive (C2), and an experimental group (Ex).

3.2. Procedure

The study was established on an Internet platform designed for the purpose of the project. The data were collected between July 2018 and July 2019. Participants were recruited by research assistants from the general population as well as from the initial, preliminary study conducted in March 2018 on a sample of adult inhabitants of Wroclaw (see [4], for more details). The inclusion criteria were living in the agglomeration of Wrocław—a large city in the southwest of Poland, having smart meters installed in the household, being over 18 years old, and being responsible for paying electricity bills.

In the first stage of the study, participants registered on the platform and completed the base measurement (T0) that is a questionnaire containing socio-demographic variables, knowledge about the energy market, various items measuring attitudes, and behaviors related to energy monitoring and environmental issues, and the declaration in which phase stage towards a smart metering platform—e-licznik—participants were (see Appendix A for a detailed description). E-licznik is a free mobile application and Internet widget developed by the energy supplier Tauron Dystrybucja S.A and broadly available to customers. The application provides data based on consumption metering from a smart electricity meter.

At least seven days after completion of T0, participants took part in the main study (T1–T4). On Monday, they received a message (text message or email) with a request to log on the platform. Then, they were asked to get acquainted with the instruction regarding the e-licznik platform. Subsequently, participants were asked to report on the platform the readings from the application regarding their energy consumption from the previous day and to complete short questionnaires measuring the behavioral stage that the respondents were in, and attitudes towards monitoring, environmental issues, and behaviors. The same procedure was repeated on Wednesday (T2), Friday (T3) and Sunday (T4). On the last day, the participants also completed a post-test questionnaire, identical to the T0 one. The study framework with a timeline is presented in Figure 1.

At T1, participants were randomly assigned either to a passive control group (C1), to an active control group (C2), or to an experimental group (Ex). In the passive control group (C1), participants completed the questionnaires at T1, T2, T3, and T4 without any help or reminders from the research assistants. In the active control group (C2), they received instructions on how to log to an e-licznik platform and were asked to do it and report their energy consumption. In the experimental group, participants additionally received text messages adjusted to their behavioral stage (F1–F4) reported in the last questionnaire. In particular, the following messages were sent to participants at T1, T2, T3, and T4:

- Group C2: "Log into the https://inteligentnylicznik.pl and fill in information about your energy consumption."
- Group Ex, Stage F1: "Log into the https://inteligentnylicznik.pl. You probably think that monitoring energy consumption is time consuming, but it only takes 10 min."
- Group Ex, Stage F2: "Log into the https://inteligentnylicznik.pl. Load the attached instruction. It will help you start monitoring your energy consumption."
- Group Ex, Stage F3: "Log into the https://inteligentnylicznik.pl. Plan your day to find 10 min to monitor energy consumption. For example, after checking your email in the evening, log in to the e-licznik platform."

- Group Ex, Stage F4: "Log into the https://inteligentnylicznik.pl. You can organize your time so that you can continue to regularly monitor energy consumption for at least a month."

Pretest T1	Posttest T1-T4				Follow-up T5
	Monday T1	Wednesday T2	Friday T3	Sunday T4	
main questionnaire measurement of the behavioral stage (F1-F4)	measurement of the energy consumption of the previous day by means of e-licznik				main questionnaire measurement of the behavioral stage (F1-F4)
	short questionnaire with pro-environmental behavioral questions				
	measurement of the behavioral stage (F1-F4)				
				main questionnaire	
	Respondents are divided into 3 groups: C1: control, passive (no interventions) C2: control, active (basic interventions) Ex: experimental, active (interventions sent in text messages on each day of measurement; interventions are adjusted to the current behavioral stage (F1-F4) of the respondent				

| T0 | Phase T1 starts after T0 on the nearest Monday | T1-T4 | Phase T5 starts 4 weeks after the last measurement of T1 (Sunday) | T5 |

Figure 1. Survey framework with six measurement points: T0–T5.

Each behavioral stage (F1, F2, F3, F4) was measured with the following questions:

- pre-decisional stage F1: "I never use e-licznik web platform/application";
- pre-actional stage F2: "Currently, I sometimes use e-licznik web platform/application";
- actional stage F3: "My goal is to organize my week so that I can monitor my energy consumption regularly";
- post-actional stage F4: "I often monitor the energy consumption of my household using e-licznik platform/application".

Participants responded to these questions on a Likert scale from strongly disagree (1) to strongly agree (5).

In all groups apart from the control group C1, over the course of the study, participants were also receiving text messages and emails reminding them about the next measurement in the study. At least four weeks after the T4, in the last stage of the study (T5), we measured again the behavioral stage at which participants were at the moment. We also measured their satisfaction with using the e-licznik platform. Those participants who completed the whole survey were gifted with a smart plug or another small smart device worth ca. 50 PLN (c.a. 11 Euro).

4. Results of the Study

4.1. Statistical Analyses

The results section is organized as follows. First, we present descriptive statistics for the demographic and control variables. Second, we describe details on measures and materials used in a study, the measure of energy monitoring, the measures of an attitude towards environmental issues, and knowledge on the energy market.

Analyses were performed using statistical language R v. 3.4.3 (RCore Team, 2019) for logistic regression models and IBM SPSS Statistics v. 25 for the rest of the analyses performed. To construct measures of attitudes towards energy monitoring and pro-environmental issues, we reduced the number of the items into meaningful components by means of the Principle Components Analysis (PCA).

Then, we directly addressed stated hypotheses and explored whether the time of measurement (T0, T4, T5) and the experimental manipulation predicted phase changes (F1–F4). Next, we tested correlations between monitoring of energy consumption and attitude towards pro-environmental issues and a the level of education, and knowledge. All analyses were conducted in the frequentist approach with α-level set to 0.05.

4.2. Participants

In total, 289 respondents have been recruited to stage T0 and 142 (49%) completed all measurements (T0–T5). It is noteworthy that such an attrition rate is quite common in longitudinal studies, especially with strict inclusion criteria. The final sample's mean age was M = 35.5 years old (SD = 0.89). The sample was equally represented by men (50.7%) and women (49.3%) and over represented by participants with higher education (76.8%). Likely, the reason of such a distribution of age and education is that inhabitants, having access to e-licznik, need to have better computer skills and are more familiar with technology.

Participants were asked about their age, gender, income, the type of household, and the number of inhabitants in the household. Figure 2 presents demographics of the respondents who have completed the survey (all T0–T5 points of measurement). In terms of material situation, 10.5% of the respondents stated that it is lower than average, 57% that it is similar to average, and 26.8% that it is higher or much higher (1.4%) than average. Most of the respondents live either in blocks of flats or modern apartments, in families with 2 (35.9%), 3 (28.2%), or more (21.1%) members. Finally, the average monthly electricity bills did not exceed 50 PLN (11 Euro) in case of 5.6% of participants, were between 51–100 (12–22 Euro) PLN for 39.4%, between 101–200 PLN (23–45 Euro) for 41.5%, and are higher than 201 PLN (45 Euro) for 11.3%.

The majority of the respondents confirmed using a computer for at least an hour every day (95%, M = 4.68, SD = 0.59), using social media and applications for communication with friends and family (e.g., Facebook, Twitter, WhatsApp, Hangout, and others) (86%, M = 4.39, SD = 1.05), has at least one email address (97%, M = 4.67, SD = 0.62), can download a new application or program from the Internet to their computer or mobile phone (96%, M = 4.68, SD = 0.56).

Participants also indicated their attitudes towards SM and SMP. They expressed their willingness to receive information and reports on their current energy consumption in general and of individual electrical appliances in their household directly via the website or an application in their mobile phones. The lack of trust in the energy supplier appeared not to be an issue for participants. More than 75% of them believed that the energy consumption data collected by SM is safely stored by the energy supplier and will not be sold to third parties without consumer' permission. Only 11% stated that the energy supplier had an excessive knowledge of their habits thanks to SM, and 13% were afraid that the data provided by SM was not sufficiently secured and that unauthorized persons may have access to them. Thanks to the installation of SM and access to the data on current energy consumption via SMP more than 60% of participants expected to have more knowledge about the energy consumption of individual electrical devices in their households, and 25% believed to be able to change their habits and use more electricity when it is cheaper.

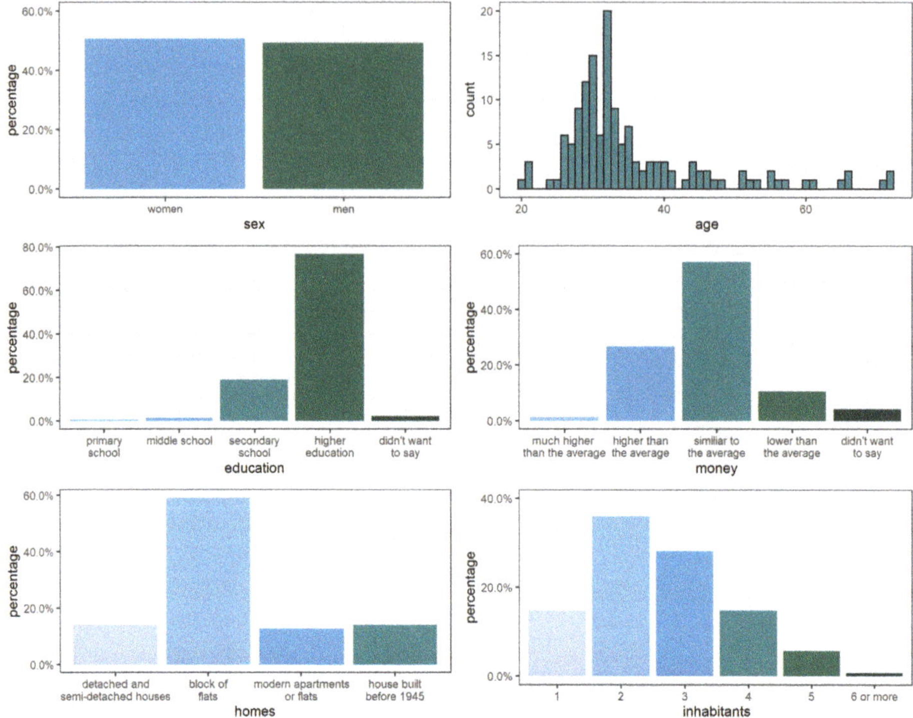

Figure 2. Frequencies of the demographics for participants who completed all measurement points of the study (n = 142).

Finally, we asked the participants what annual savings they expect thanks to the installation of SM in their household. Interestingly, 21% of them had no financial expectations. The rest of the participants expected a certain level of savings starting from 1–5% per year (21% of respondents), 6–10% (30%), 11–15% (12%), 16–20% (5%), 21–25% (5%), and more than 25% (6%).

4.3. Predicting the Phase Change

We applied multinomial logistic regression model from NNet package [54] to predict phase change (F1–F4) depending in which group a given person belonged to (C1, C2, or Ex), the time of measurement (T0, T4, T5), and the interaction of the group and the time of measurement. As a reference point, we took a null model with an intercept only and without any predictors entered. We performed null, group, time of measurement, and the interaction of group × time of measurement models, and we compared them against each other using AIC and ANOVA tests (Table 2).

Table 2. Model comparisons.

Model	AIC	Psuedo-R^2	df	LR	p-Value
Null	849.35	<0.01	-	-	-
Group	853.34	0.01	6	8.01	0.237
Time of measurement	838.16	0.03	6	23.20	<0.001
Group × Time of measurement	858.01	0.05	18	16.15	0.582

4.3.1. The Group Model

To test the effect of the manipulation (congruent vs. control messages) on phase changes, we included as predictors a passive control group (C1), an active control group (C2), and the experimental group (Ex) into the model. We performed the regression analysis, in which we compared these groups using the contrasts. Specifically, we compared C1 to Ex and C2 to Ex. Even though C1 and C2 had slightly different procedures, we also examined possible difference between joined control groups (C1 + C2) and the experimental group (Ex). The analysis yielded the following results. The omnibus group model was not significantly different from the null model, and the AIC value (853.54) was bigger than the value of null model (849.35). Based on these results, we inferred that experimental manipulation was not successful as assignments to the groups were not significant predictors of the phase change. Therefore, we do not report specific results for contrasts' analyses.

4.3.2. The Time of Measurement Model

To test the effect of the time of measurement, we entered T0, T4, and T5 measurements to the model (see 'time of measurement model' in Table 2). The reason for which we chose these three measurement points is that we were interested in possible long-term phase changes and not in day-to-day changes. Moreover, we wanted to keep the same number of points of measurement across most of the analyses performed. Once again, we set custom contrasts for the time of measurement model in which T0 was compared to T4, T0 to T5, and T4 to T5. The difference between the time of measurement model and the null model was statistically significant, and the AIC value for the time of measurement was lower (838.16) than that of the null model (849.35). It indicated that this model fits the collected data better than the null model.

The time of measurement model explained 3% of variance of the dependent variable (based on McFadden pseudo-R2). The outcomes of the Wald tests revealed that the difference between T0 and T4 and T0 and T5 for phase change from F1 to F2 were significant (see Table 3 for more details). More specifically, the change from T0 to T4 increased the odds of phase change from F1 to F2 by 1.37. In addition, the change from phase F1 to phase F2 was 1.48 odds higher on T5 when compared to T0. Similar results were obtained for change from phase F3 to phase F4. The significant predictors were contrasts between T0 and T4, and T0 and T5. Change from T0 to T4 increased the odds of phase change by 1.48, and change from T0 to T5 increased the odds by 1.86. The contrasts between the time of measurements did not predict the likelihood of changing from the phase F2 to the phase F3.

Table 3. Multinominal regression coefficients of the time of measurement model.

Odds	Effect	Estimate	SE	Wald	*p*-Value	Exp(β)
	Intercept	0.60	0.17	3.45	<0.001	1.82
$P(Y = F2)/P(Y = F1)$	T0–T4	0.31	0.13	2.37	0.018	1.37
	T0–T5	0.39	0.14	2.74	0.006	1.48
	T4–T5	0.08	0.15	0.54	0.588	1.08
	Intercept	0.42	0.18	2.34	0.019	1.52
$P(Y = F3)/P(Y = F2)$	T0–T4	0.04	0.13	0.32	0.750	1.04
	T0–T5	0.16	0.14	1.10	0.270	1.17
	T4–T5	0.12	0.16	.73	0.463	1.12
	Intercept	0.13	0.19	0.69	0.489	1.14
$P(Y = F4)/P(Y = F3)$	T0–T4	0.39	0.16	2.53	0.012	1.48
	T0–T5	0.62	0.16	3.86	<0.001	1.86
	T4–T5	0.23	0.16	1.46	0.145	1.25

4.3.3. The Interaction of the Group and the Time of Measurement

We compared the model with an interaction term (time of measurement × group) to the time of the measurement model. They were not significantly different. In conclusion, the best fitting model was the one with the time of measurement as the predictor of the phase change. It suggests that mere participation in the study independent of the group was the best predictor of changes from phase 1 (pre-decisional stage) to phase 2 (pre-actional stage) and from the phase 3 (actional stage) to phase 4 (post-actional stage), see Table 2.

4.4. The Effect of the Participation in the Study on Energy Monitoring and Attitude towards Environmental Issues

In the next step, we conducted three exploratory Principal Component Analyses (PCA), one for each point of measurement T0 (n = 274), T4 (n = 145), T5 (n = 142), for questions A1–A6 (pro-enviromental attitudes), B1–B5 (monitoring behaviors), and M1–M16 (attitudes towards monitoring) with the exclusion of items M1, M2, and M10 (see Table A1 in the Appendix A for a description of variables, their coding and scales used in the study). We excluded these items because they were referring to energy monitoring and environment protection at the same time, which caused ambiguity we wanted to avoid. Altogether, we included 24 items in conducted PCAs.

The results of Bartlett sphericity tests and KMO coefficients indicated that a reduction of dimensions may be useful with collected data (see Table 4 for details). We used eigenvalues above 1 as a criterion to select the number of components. In effect, for each measurement (T0, T4, and T5), the solution with two components best fitted the data. We based our selection of items for each component on item loading cut-off point, which was set to 0.3.

Table 4. Coefficients of Bartlett sphericity tests, KMO, eigenvalues, and percentage of explained variance for solutions with two components.

Measurement T	χ^2	df	p	KMO	Components	Eignevalue	%Variance
T0	2726.58	276	<0.001	0.86	1.EM	6.20	32.62
					2.EA	2.59	13.62
T4	1880.03	276	<0.001	0.86	1.EM	8.34	34.75
					2.EA	2.85	11.87
T5	1976.64	276	<0.001	0.88	1.EM	10.69	39.59
					2.EA	2.88	10.66

Note: df for Bartlett sphericity test are based upon the number of variables included in the analysis.

The first component was **energy monitoring (EM)**, which contained the following items: B1–B5, M4, M5, M6, M7, M8, M9, M11, M12, M13, M14, M15, and M16. This component explained respectively 32.62% (T0), 34.75% (T4), and 39.59% (T5) of variance. The exemplary items that best describe this component are: "I decided to use internet platforms/applications to monitor energy consumption in my household" (M11), "I check monthly energy consumption according to data from the electricity meter" (B2), "I believe that energy monitoring is good" (M9), "I feel bad when I don't control the energy consumption in my household" (M7).

In the second component, **attitude towards environmental issues (EA)**, we included items number A1, A2R, A3, A4, A5R, A6 (R—means negative loading). This component explained respectively 13.62% (T0), 11.87% (T4) and 10.66% (T5) of variance. The items that best describe this factor are: "In my opinion, reports about the ecological crisis are exaggerated" (A2R), "I am happy when the climate and environment protection plays an important role in politics" (A3), "In my opinion, every person has an impact on environmental protection through his own behavior" (A4), "Protecting the environment is particularly important to me (A1).

After exploring the results of PCA, we decided to remove item M3 from further analyses as it was causing problems with coherent components' interpretation. In T0, item M3 was loading the second component but did not suit it from the semantic point of view. In T4, component loading of M3 did not exceed 0.3, and, in T5, it was loading the first component. There was a small variance of item loadings between each point of measurement, hence we decided to apply the same two component solutions for each measurement.

Finally, we created two factors from 23 items and each factor was produced by calculating arithmetic mean scores, where high scores mean more favorable attitude towards environmental issues and more endorsement of energy monitoring. The reliability of these two components was examined using the Cronbach's alpha. Cronbach's alpha for each component at each time of measurement was at least on the level of 0.65. Internal reliability for monitoring of energy consumption was $\alpha = 0.90$ (T0), $\alpha = 0.92$ (T4), $\alpha = 0.93$ (T5) and for a pro-environmental attitude was $\alpha = 0.65$ (T0), $\alpha = 0.77$ (T4), $\alpha = 0.74$ (T5). These results indicate an acceptable consistency of the measurement items and construct reliability. Some more descriptive statistics and normality test for EM and EA at three points of measurement T0, T4, and T5 are presented in Appendix A in Table A2.

To explore the effect of the participation in the study on the energy monitoring and attitude towards environmental issues, we conducted a repeated measures ANOVAs with the group variable as a between group factor and time of measurement of energy monitoring as a dependent variable measured at T0, T4, and T5 ($n = 142$). The results of the analysis showed a statistically significant main effect of the time of measurement for energy monitoring, $F(1.72, 242.09) = 14.74$, $p < 0.001$, partial-$\eta^2 = 10\%$ see Table 5. The results of a post-hoc pairwise comparison with Sidak correction revealed that participants energy monitoring at T4 (M = 3.30, SD = 0.72) was significantly higher than in T0 (M = 3.16, SD = 0.71), $\triangle = 0.14$, $p = 0.009$, and the energy monitoring at T5 (M = 3.39; SD = 0.75) was significantly higher than at T0 and T4, respectively $\triangle = 0.23$, $p < 0.001$ and $\triangle = 0.09$, $p = 0.020$. This outcome means that participants' energy Monitoring (EM) was increasing with each point of measurement. We also performed the same analysis with the attitude towards environmental issues (EA). However, we found no significant effects of participation in the study on participants' attitudes towards environmental issues (see Table 5 and Figure 3).

Table 5. Results of the repeated measures ANOVA for energy monitoring (EM) and attitude towards environmental issues (EA).

Variables	Greenhouse-Geiser ϵ	p-Value	F	df	p-Value	Partial-η^2
EM	0.86	<0.001	14.74	1.72, 242.09	<0.001	0.10
EA	0.88	<0.001	2.65	1.76, 249.98	0.080	0.02

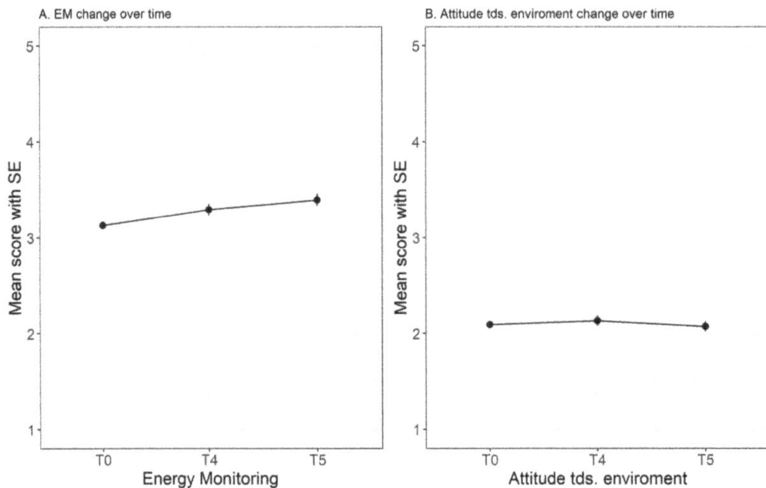

Figure 3. Mean scores with SE for repeated measurements of energy monitoring (EM) and attitude towards environmental issues (EA).

4.5. Knowledge and Education as Correlates of Energy Monitoring and Attitude towards Environmental Issues

In the last analysis, we explored relationships between energy monitoring (EM), attitude towards environmental issues (EA), and knowledge measured at T0, T4, and T5, as well as education level. To measure knowledge, we asked four questions (K1–K4) testing participant's familiarity with the following terms and issues: (K1) the concept of smart grid; (K2) the concept of smart metering; (K3) the opportunity to change the energy supplier; and (K4) the most energy-consuming home appliance. Each question had only one correct answer, so the sum of the collect answers might have ranged from 0 to 4. In the T5 point of measurement, the majority of respondents knew which of the home appliances is the most energy-intensive (91.5% correct answers). In addition, most of the respondents (83%) were aware that SM enables remote reading of energy consumption by the energy supplier. Less respondents were aware of who may change the electricity supplier or what smart grid means (62.7% and 30% of the correct answers, respectively).

We used the Spearman correlation coefficient as it is less susceptible to extreme cases, and allows for assessing the relationship for ordinal data (see Table 6). The results of the conducted analyses showed that energy monitoring (EM) at T0 was moderately negatively correlated with education level and positively correlated with knowledge at T0 and T4. Energy monitoring in T4 and T5 was positively correlated with knowledge in T0, T4, and T5. Surprisingly, we found no correlations with attitude towards environmental issues and knowledge or education.

Table 6. Correlation analysis coefficients for relationships between energy monitoring (EM), attitude towards environmental issues (EA), Education, and Knowledge in T0, T4, T5, and weekly attitude.

Variables	Coeff.	Education	Knowledge T0	Knowledge T4	Knowledge T5
EM in T0	rho Spearmana	−0.25	0.25	0.35	0.14
	p-value	<0.001	<0.001	<0.001	0.024
EM in T4	rho Spearmana	−0.16	0.34	0.31	0.26
	p-value	0.053	<0.001	<0.001	0.001
EM in T5	rho Spearmana	−0.15	0.31	0.35	0.28
	p-value	0.084	<0.001	<0.001	<0.001
EA in T0	rho Spearmana	−0.02	0.00	−0.07	−0.08
	p-value	0.716	0.940	0.411	0.212
EA in T4	rho Spearmana	0.03	−0.12	0.00	0.01
	p-value	0.743	0.156	0.959	0.925
EA in T5	rho Spearmana	0.06	−0.02	0.02	−0.05
	p-value	0.505	0.837	0.841	0.593

5. Discussion and Conclusions

Although the acceptance of smart meters has been studied in the literature, the consumers' readiness to use SM platform still warrants exploration.

Expecting that acceptance of SMP and involvement in energy monitoring is a phase process, we aimed to test whether messages congruent with behavioral stages in which participants declared to be are more effective in prompting participants to progress to the next stage than general messages or passive control conditions. Based on the current literature review, we have expected to observe that phase change as well as participants' attitudes to use SMP and monitor energy regularly will be affected by their environmental attitudes, energy monitoring behaviors, and knowledge on the energy market [18,36].

5.1. Summary of the Results

In summary, our results showed that the most important factor affecting phase change was the participation in the study. The longer the participants remained in the study, the higher was the chance that they progressed from the pre-decisional to pre-actional stage and from the actional to the post-actional stage. Moreover, the time of measurement affected energy monitoring.

We found no differences between the control groups and the experimental group. One explanation could be purely statistical, the power of the performed test was too low. That is, the effects we tested were too small to detect with the sample size we had. Another explanation, which seems more plausible, is that participation in such a demanding study even in the control group in which participants completed a number of questionnaires was an experience strong enough to affect changes. Numerous studies in psychology show that an investment of effort in some issues makes people value the given cause more [55]. In other words, effort invested could have given additional value to energy monitoring even in the control group. This interpretation could be additionally supported by the results showing that participants were more eager to engage in energy monitoring as the study progressed.

Participation in the study also affected attitudes towards environmental issues, but to a lesser extent. Thus, the participation in the study was more effective for a variable closer related to behaviors referring to the control of energy consumption.

Knowledge about energy market was correlated with participants' energy monitoring. This is quite an intuitive result as probably specific knowledge provided know-how for participants in the study. More surprising are the results that education was negatively related to energy monitoring. We may speculate that participants with higher education have more absorbing professional lives and spend more time in front of the computer. Therefore, they are less willing to control energy, using technology in their spare time.

For most of the participants, monitoring energy by means of SMP has similar pros and cons. The higher control over one's energy consumption and better energy management belonged to the biggest advantages of using SMP, whereas time consumption and low effectiveness in terms of financial savings were mentioned as the biggest disadvantages and barriers of regular SMP usage. For such consumers, the energy supplier should offer automatic transmission of e.g., daily reports on energy consumption or information on exceeding a given level of energy consumption (e.g., daily limit set by the energy consumer according to his own needs). Such services could increase the level of interest and engagement in SMP usage.

5.2. Limitations of the Study and Future Work

The main limitation of our study was a restricted sample size, relatively small, but also composed of volunteers. It is also possible that the study itself was overly time-consuming and difficult for our participants. This would explain why part of the participants resigned from the participation in the study.

Moreover, we focused on participants' declarations and not on real behaviors as indicators of energy consumption. We asked participants to report energy consumption, but we observed a large proportion of missing responses for this item.

Future work should focus on larger, more diverse samples and provide easier to use applications for participants. Ideally, some data could be collected directly from SMP and compared to survey data in collaboration with an energy provider.

Despite a few mentioned limitations, the originality of this contribution relies on using the stage model approach, so far not explored thoroughly in the energy related studies (see [35]). Moreover, we tested our hypotheses in a longitudinal field experiment, which allowed us to observe changes in the process.

5.3. Practical Recommendations

Based on our results, we may conclude that, while electricity smart meters are useful for the energy providers, they might not offer enough real benefits for the residential consumers. Even if SM are combined with smart metering information platforms, such as Internet widgets and mobile applications, their role in prompting energy monitoring is very limited. At the same time, we observed that mere participation in the study, independent on the group and getting acquainted with the e-licznik application, enhances the phase change and the readiness to monitor energy consumption. These findings suggest that using SMP without any prompts and instructions is unlikely to occur as there are no reasonable incentives that could convince respondents to monitor energy. Financial, social, or environmental benefits are probably too low and the effort too high to lead to a permanent behavioral change.

In conclusion, we believe that some good practices are needed. It is necessary to make monitoring of electricity consumption easy, intuitive and non time-consuming. Designers and suppliers of smart metering platforms should provide user friendly solutions. Smart meters should also be proposed with some additional enabling technologies, such as e.g., smart plugs or smart devices, as well some IT solutions enabling remote adjustment of energy consumption of home appliances or air conditioning to the current electricity prices. Moreover, to raise awareness, some educational campaigns would be helpful. Our results suggest, however, that the role of theoretical knowledge in the energy market should not be overestimated when it comes to energy monitoring and phase changes. Knowledge appears to affect attitudes on monitoring more than behaviors. Rather reasonable price polices, such as additional financial incentives for consumers to control energy consumption and to shift from pick to off-pick hours, would be more beneficial.

Author Contributions: Conceptualization of the study and Methodology: A.K.-P. and K.B., Introduction: A.K.-P.; Literature review: A.K.-P.; Theoretical background: K.B. and A.K.-P.; Methods and Survey design: K.B. and A.K.-P.; Data analysis: J.S. and K.B.; Results and discussion: K.B., J.S., and A.K.-P.; Conclusions: K.B. and A.K.-P.;

Writing and editing the paper: A.K.-P., K.B., and J.S. All authors have read and agreed to the published version of the manuscript.

Funding: Open access of this article was financed by the Ministry of Science and Higher Education in Poland under the 2019-2022 program "Regional Initiative of Excellence", project number 012/RID/2018/19.

Acknowledgments: This work was supported by the National Science Center (NCN, Poland) by Grant No. 2016/23/B/HS4/00650.

Conflicts of Interest: The authors declare no conflict of interest.

Abbreviations

The following abbreviations are used in this manuscript:

SG	Smart grids
SM	Electricity smart meters
SMP	Smart metering platform (SM information systems)
DOI	Diffusion of innovation model
SSCB	Stage model of self-regulated behavioral change

Appendix A

Table A1. Definitions of the variables, coding, and description.

Variable	Code	Description
Demographics	**D1–D7**	
Gender	D1	2 categories (nominal)
Age	D2	integer (ordinal)
Education	D3	5 categories (nominal)
Housing	D4	4 categories (nominal)
Material situation	D5	5 categories (ordinal)
Range of electricity bill (in PLN per month)	D6	4 categories (ordinal)
Inhabitants in the household	D7	6 categories (ordinal)
Pro-environmental attitudes	**A1–A6**	
Environmental protection is especially important to me	A1	
In my opinion, reports of the ecological crisis are exaggerated	A2	
I am glad that climate and environmental protection play an important role in politics	A3	
In my opinion, every person has an impact on environmental protection through their own behavior	A4	scale from 1 to 5
As an individual, I do not have much influence on environmental protection	A5	
I would be willing to pay higher taxes in order to protect the natural environment better and more effectively	A6	
Energy monitoring behaviors	**B1–B6**	
I check monthly energy consumption according to data from electricity bills	B1	
I check the monthly energy consumption according to the data from the electricity meter	B2	
I use a platform or web application to monitor energy consumption	B3	
I use an intelligent energy management system in my household (the so-called home area network)	B4	scale from 1 to 5
I have an electronic device installed in my household and can see my current electricity consumption	B5	
Do you use other methods of monitoring energy consumption? (open question)	B6	
Attitudes towards monitoring	**M1–M16**	
To care for the environment and increase energy efficiency, everyone should monitor the energy consumption of their household	M1	
Everyone can contribute to taking care of the environment by monitoring the energy consumed in the household using e.g., access to data from an energy meter	M2	scale from 1 to 5
To reduce energy consumption, I turn off the lights, avoid leaving appliances on stand-by, only turn on the washing machine and dishwasher when they are full	M3	

Table A1. *Cont.*

Variable	Code	Description
Attitudes towards monitoring	**M1–M16**	
Regardless of what others may think, my own rules oblige me to monitor household energy use	M4	
I know that some of my neighbors and friends reduce their energy consumption by regularly monitoring their energy consumption by accessing data from an energy meter. It motivates me to try to do the same	M5	
I feel good when I know I am in control of my energy consumption by regularly accessing consumption data from my energy meter, e.g., via a platform or web application	M6	
I feel bad not having control of the energy consumption in my household	M7	
I can see the possibility of regular energy monitoring, e.g., by accessing data from an intelligent energy meter via a platform/web application	M8	
I believe that monitoring energy consumption is good	M9	
I intend to contribute to the protection of the environment by regularly monitoring energy consumption, e.g., using a platform/web application	M10	scale from 1 to 5
I have decided to use a web platform/application to monitor my household energy consumption	M11	
I have decided to use a web platform/application to monitor my household energy consumption	M12	
I foresaw possible problems that may arise and prevent me from carrying out regular monitoring of energy consumption via the platform/web application	M13	
I have developed a way to avoid problems and obstacles in the implementation of regular monitoring of energy consumption and how to flexibly adapt the monitoring to a given situation	M14	
For the next 7 days, I am going to monitor energy consumption via the platform/web application	M15	
I intend to continue using the web platform/app to monitor my energy consumption even when it is inconvenient	M16	
Computer skills	**S1–S4**	
I use my computer for at least an hour every day	S1	
I use social media and applications to communicate with friends and family (e.g., Facebook, Twitter, Whatsapp, Hangout, and others)	S2	scale from 1 to 5
I have at least one email address	S3	
I can download a new application or program from the Internet to my computer or mobile phone	S4	
Knowledge about energy market	**K1–K4**	
How do we call an energy system that integrates the activities of all participants in the generation, transmission, distribution and use processes (1) smart metering; (2) smart grids; (3) advanced metering infrastructure; (4) I do not know	K1	
For energy consumers who have an intelligent energy meter installed, it is possible to: (1) Individual appointments of a collector to read energy consumption; (2) Remote reading of energy consumption by the seller and monitoring of energy consumption through the web portal; (3) Settlements based on forecasts of electricity consumption, made by the electricity supplier on the basis of (4) I do not know	K2	selection test (one answer is correct)
What is true: (1) In Poland, every energy consumer has the right to change the electricity supplier; (2) In Poland, only industrial and institutional customers have the right to change the electricity supplier; (3) In Poland, changing the electricity supplier requires the consent of the President of the Energy Regulatory Office; (4) I do not know	K3	
The most energy-intensive household electronics and household appliances include: (1) computer; (2) refrigerator; (3) home lighting; (4) I do not know	K4	

Table A1. *Cont.*

Variable	Code	Description
Preferences towards SM	**P1–P3**	
Access to information from e-licznik would be most useful to me for	P1	
My confidence in the energy supplier regarding data security is best described by the sentence	P2	selection test (option to choose one answer)
Thanks to the installation of an intelligent energy meter and access to data on my current energy consumption, I expect	P3	
Behavioral stages	**F1–F4**	
I never use e-licznik web platform /application	F1	
Currently, I sometimes use e-licznik web platform /application	F2	
My goal is to organize my week so that I can monitor my energy consumption regularly	F3	scale from 1 to 5
I often monitor the energy consumption of my household using e-licznik platform/application	F4	

Note: Likert scale from 1 (fully disagree) to 5 (fully agree).

Table A2. Descriptive Statistics with the Shapriro–Wilk normality test for EM and EA at T0, T4, and T5.

Variables	M	Me	SD	Sk.	Kurt.	Min	Max	W	p
EA T1	2.05	2.00	0.58	0.71	1.05	1.00	4.33	0.96	<0.001
EA T4	2.13	2.17	0.63	0.19	−0.41	1.00	3.83	0.98	0.016
EA T5	2.06	2.00	0.61	0.29	−0.56	1.00	3.67	0.97	0.006
EM T1	3.16	3.18	0.71	−0.06	−0.09	1.24	4.88	0.99	0.895
EM T4	3.30	3.41	0.72	−0.21	0.16	1.29	5.00	0.99	0.364
EM T5	3.39	3.41	0.75	−0.12	0.12	1.59	5.00	0.98	0.033

References

1. Rixen, M.; Weigand, J. Agent-based simulation of policy induced diffusion of smart meters. *Technol. Forecast. Soc. Chang.* **2014**, *85*, 153–167. [CrossRef]
2. Zhang, T.; Nuttall, J.W. Evaluating government's policies on promoting smart metering diffusion in retail electricity markets via agent-based simulation. *J. Prod. Innov. Manag.* **2011**, *28*, 169–186. [CrossRef]
3. Chawla, Y.; Kowalska-Pyzalska, A. Public awareness and consumer acceptance of smart meters among Polish social media users. *Energies* **2019**, *12*, 2759. [CrossRef]
4. Kowalska-Pyzalska, A.; Byrka, K. Determinants of the willingness to energy monitoring by residential consumers: A case study in the city of Wroclaw in Poland. *Energies* **2019**, *12*, 907. [CrossRef]
5. Ellabban, O.; Abu-Rub, H. Smart grid customers' acceptance and engagement: An overview. *Renew. Sustain. Energy Rev.* **2016**, *65*, 1285–1298. [CrossRef]
6. Verbong, G.P.; Beemsterboer, S.; Sengers, F. Smart grids or smart users? involving users in developing a low carbon electricity economy. *Energy Policy* **2013**, *52*, 1175–125. [CrossRef]
7. Bugden, D.; Stedman, R. A synthetic view of acceptance and engagement with smart meters in the United States. *Energy Res. Soc. Sci.* **2019**, *47*, 137–145. [CrossRef]
8. Foulds, C.; Robison, R.A.V.; Macrorie, R. Energy monitoring as a practice: Investigating use of the imeasure online energy feedback tool. *Energy Policy* **2017**, *104*, 194–202. [CrossRef]
9. Schleich, J.; Faure, C.; Klobasa, M. Persistence of the effects of providing feedback alongside smart metering devices on household electricity demand. *Energy Policy* **2017**, *107*, 225–233. [CrossRef]
10. Buchanan, K.; Russo, R.; Anderson, B. Feeding back about eco-feedback: How do consumers use and respond to energy monitors? *Energy Policy* **2014**, *73*, 138–146. [CrossRef]
11. Crispim, J.; Braz, J.; Castro, R.; Esteves, J. Smart grids in the EU with smart regulation: Experiences from the UK, Italy and Portugal. *Util. Policy* **2014**, *31*, 85–93. [CrossRef]
12. Zhou, S.; Brown, M.A. Smart meter deployment in Europe: A comparative case study on the impacts of national policy schemes. *J. Clean. Prod.* **2017**, *144*, 22–32. [CrossRef]

13. Chawla, Y.; Kowalska-Pyzalska, A.; Skowronska-Szmer, A. Perspectives of smart meters' roll-out in India: An empirical analysis of consumers' awareness and preferences. *Energy Policy* **2020**, *146*, 111798. [CrossRef]

14. Sovacool, B.K.; Kivimaa, P.; Hielscher, S.; Jenkins, K. Vulnerability and resistance in the United Kingdom's smart meter transition. *Energy Policy* **2017**, *109*, 767–781. [CrossRef]

15. Chawla, Y.; Kowalska-Pyzalska, A.; Widayat, W. Consumer Willingness and Acceptance of Smart Meters in Indonesia. *Resources* **2019**, *8*, 177. [CrossRef]

16. Biresselioglu, M.E.; Nilsen, M.; Demir, M.H.; Royrvik, J.; Koksvik, G. Examining the barriers and motivators affecting European decision makers in the development of smart and green energy technologies. *J. Clean. Prod.* **2018**, *198*, 417–429. [CrossRef]

17. Burchell, K.; Rettie, R.; Roberts, T.C. Householder engagement with energy consumption feedback: The role of community action and communications. *Energy Policy* **2016**, *88*, 178–186. [CrossRef]

18. Ma, G.; Lin, J.; Li, N. Longitudinal assessment of the behavior-changing effect of app-based eco-feedback in residential buildings. *Energy Build.* **2018**, *159*, 486–494. [CrossRef]

19. Hargreaves, T.; Nye, M.; Burgess, J. Making energy visible: A qualitative field study of how householders interact with feedback from smart energy monitors. *Energy Policy* **2010**, *38*, 6111–6119. [CrossRef]

20. Gangale, F.; Mengolini, A.; Onyeji, I. Consumer engagement: An insight from smart grid projects in Europe. *Energy Policy* **2013**, *60*, 621–628. [CrossRef]

21. Krishnamutri, T.; Schwartz, D.; Davis, A.; Fischoff, B.; de Bruin, W.B.; Lave, L.; Wang, J. Preparing for smart grid technologies: A behavioral decision research approach to understanding consumer expectations about smart meters. *Energy Policy* **2012**, *41*, 790–797. [CrossRef]

22. Kahma, N.; Matschoss, K. The rejection of innovations? rethinking technology diffusion and the non-use of smart energy services in Finland. *Energy Resour. Soc. Sci.* **2017**, *34*, 27–36. [CrossRef]

23. Hess, D.J. Smart meters and public acceptance: Comparative analysis and governance implications. *Health Risk Soc.* **2014**, *16*, 243–258. [CrossRef]

24. Chawla, Y.; Kowalska-Pyzalska, A.; Silveira, P.D. Marketing and communications channels for diffusion of electricity smart meters in Portugal. *Telemat. Inform.* **2020**, *50*, 101385. [CrossRef]

25. Chen, C.F.; Xu, X.; Arpan, L. Between the technology acceptance model and sustainable energy technology acceptance model: Investigating smart meter acceptance in the United States. *Energy Res. Soc. Sci.* **2017**, *25*, 93–104. [CrossRef]

26. Kowalska-Pyzalska, A. What makes consumers adopt to innovative energy sources in the energy market? A review of incentives and barriers. *Renew. Sustain. Energy Rev.* **2018**, *82*, 3570–3581. [CrossRef]

27. Wolsink, M. Distributed generation for sustainable energy as a common pool resource: Social acceptance in rural setting of smart (micro-) grid configurations. In *New Rural Spaces: Towards Renewable Energies, Multifunctional Farming, and Sustainable Tourism*; Frantal, B., Martiant, S., Eds.; Czech Academy of Sciences: Prague, Czech Republic, 2014.

28. Wuestenhagen, R.; Wolsink, M.; Buerer, M.J. Social acceptance of renewable energy innovation: An introduction to the concept. *Energy Policy* **2008**, *35*, 2683–2691. [CrossRef]

29. Chawla, Y.; Kowalska-Pyzalska, A.; Oralhan, B. Attitudes and opinions of social media users towards smart meters' rollout in Turkey. *Energies* **2019**, *3*, 732. [CrossRef]

30. Negro, S.O.; Alkemade, F.; Hekkert, M.P. Why does renewable energy diffuse so slowly? A review of innovation system problems. *Renew. Sustain. Energy Rev.* **2012**, *16*, 3836–3846. [CrossRef]

31. Avancini, D.B.; Rodriques, J.J.P.C.; Martins, S.G.B.; Rabelo, R.A.L.; Al-Mahtadi, J.; Solic, P. Energy meters evolution in smart grids: A review. *J. Clean. Prod.* **2019**, *217*, 702–715. [CrossRef]

32. Bellido, M.H.; Rosa, L.P.; Pereida, A.O.; Falcoa, D.M.; Ribeiro, S.K. Barriers, challenges and opportunities for microgrid implementation: The case of Federal University of Rio de Janeiro. *J. Clean. Prod.* **2018**, *180*, 203–216. [CrossRef]

33. Darby, S.; McKenna, E. Social implications of residential demand response in cool temperature climates. *Energy Policy* **2012**, *49*, 759–769. [CrossRef]

34. Bonino, D.; Corno, F.; De Russis, L. Home energy consumption feedback: A user survey. *Energy Build.* **2012**, *47*, 383–393. [CrossRef]

35. Nachreiner, M.; Mack, B.; Matthies, E.; Tampe-Mai, K. An analysis of smart metering information systems: A psychological model of self-regulated behavioral change. *Energy Res. Soc. Sci.* **2015**, *9*, 85–97. [CrossRef]

36. Buchanan, K.; Banks, N.; Preston, I.; Russo, R. The British public's perception of the UK smart metering initiative: Threats and opportunities. *Energy Policy* **2016**, *91*, 87–97. [CrossRef]

37. Hmielowski, J.D.; Boyd, A.D.; Harvey, G.; Joo, J. The social dimensions of smart meters in the United States: Demographics, privacy, and technology readiness. *Energy Res. Soc. Sci.* **2019**, *55*, 189–197. [CrossRef]

38. Razavi, R.; Gharipour, A. Rethinking the privacy of the smart grid: What your smart meter data can reveal about your household in Ireland. *Energy Res. Soc. Sci.* **2018**, *44*, 312–323. [CrossRef]

39. Good, N.; Ellis, K.A.; Mancarella, P. Review and classification of barriers and enablers of demand response in the smart grid. *Renew. Sustain. Energy Rev.* **2017**, *16*, 57–72. [CrossRef]

40. Lineweber, D.C. Understanding residential customer support for and opposition to smart grid investments. *Electr. J.* **2011**, *24*, 92–100. [CrossRef]

41. Mack, B.; Tampe-Mai, K.; Kouros, J.; Roth, F.; Diesch, E. Bridging the electricity saving intention-behavior gap: A German field experiment with a smart meter website. *Energy Res. Soc. Sci.* **2019**, *53*, 34–46. [CrossRef]

42. Raimi, K.T.; Carrico, A.R. Understanding and beliefs about smart energy technology. *Energy Res. Soc. Sci.* **2016**, *12*, 68–74. [CrossRef]

43. Podgornik, A.; Sucic, B.; Blazic, B. Effects of customized consumption feedback on energy efficient behavior in low-income households. *J. Clean. Prod.* **2016**, *130*, 25–34. [CrossRef]

44. Bamberg, S. Changing environmentally harmful behaviors: A stage model of self-regulated behavioral change. *J. Environ. Psychol.* **2013**, *34*, 151–159. [CrossRef]

45. Ozaki, R. Adopting sustainable innovation: What makes consumers sign up to green electricity? *Bus. Strategy Environ.* **2011**, *20*, 1–17. [CrossRef]

46. Bamberg, S. Applying the stage model of self-regulated behavioral change in a car use reduction intervention. *J. Environ. Psychol.* **2013**, *33*, 68–75. [CrossRef]

47. Gollwitzer, P.M.; Heckhausen, H.; Steller, B. Deliberative and implemental mind-sets: Cognitive tuning toward congruous thoughts and information. *J. Personal. Soc. Psychol.* **1990**, *59*, 1119–1127. [CrossRef]

48. Heckhausen, H.; Gollwitzer, P.M. Thought contents and cognitive functioning in motivational versus volitional states of mind. *Motiv. Emot.* **1987**, *11*, 101–120. [CrossRef]

49. Rogers, E.M. *Diffusion of Innovations*, 5th ed.; Free Press: New York, NY, USA, 2003.

50. Gadenne, D.; Sharma, B.; Kerr, D.; Smith, T. The influence of consumers' environmental beliefs and attitudes on energy saving behaviors. *Energy Policy* **2011**, *39*, 7684–7694. [CrossRef]

51. Nolan, J.M.; Schultz, P.W.; Cialdini, R.B.; Goldstein, N.J.; Griskevicius, V. Normative social influence is underdetected. *Personal. Soc. Psychol. Bull.* **2008**, *34*, 913–923. [CrossRef]

52. Allcott, H. Social norms and energy conservation. *J. Public Econ.* **2011**, *95*, 1082–1095. [CrossRef]

53. Byrka, K.; Jędrzejewski, A.; Sznajd-Weron, K.; Weron, R. Difficulty is critical: The importance of social factors in modeling diffusion of green products and practices. *Renew. Sustain. Energy Rev.* **2016**, *62*, 723–735. [CrossRef]

54. Venables, W.N.; Ripley, B.D. *Modern Applied Statistics with S*, 4th ed.; Springer: New York, NY, USA, 2002.

55. Inzlicht, M.; Shenhav, A.; Olivola, C.Y. The effort paradox: Effort is both costly and valued. *Trends Cogn. Sci.* **2018**, *22*, 337–349. [CrossRef] [PubMed]

Article

An Assessment of Corporate Average Fuel Economy Standards for Passenger Cars in South Korea

Seungho Jeon [1], Minyoung Roh [1], Almas Heshmati [2] and Suduk Kim [1,*]

[1] Department of Energy Systems Research, Ajou University, Suwon 16449, Korea; sngho.jeon@gmail.com (S.J.); rohmin9122@gmail.com (M.R.)

[2] Jönköping International Business School (JIBS), Jönköping University, SE-551 11 Jönköping, Sweden; almas.heshmati@ju.se

* Correspondence: suduk@ajou.ac.kr; Tel.: +82-31-219-2689

Received: 30 June 2020; Accepted: 25 August 2020; Published: 1 September 2020

Abstract: The shift in consumer preferences for large-sized cars has increased the energy intensity (EI) of passenger cars, while growth in battery electric vehicle (BEV) sales has decreased EI in recent years in South Korea. In order to lower passenger cars' EI, the South Korean government has implemented the Corporate Average Fuel Economy (CAFE) standards with a credit system, in which the sale of one energy-efficient car (for example, a BEV) can get multiple credits. This study analyzes CAFE standards in terms of both the EI improvement sensitivity scenarios and the degree of credits for BEVs and fuel cell electric vehicles (FCEVs) by using the Global Change Assessment Model (GCAM). In this study, passenger cars include small, medium, and large sedans, sport utility vehicles (SUVs) of internal combustion engine vehicles (ICEVs), BEVs, and FCEVs. The findings of this study are as follows: First, from the policy design perspective, a proper setting of the credit system for BEVs and FCEVs is a very important variable for automakers to achieve CAFE standards. Second, from the technology promotion perspective, active promotion of fuel efficiency improvements through CAFE standards are important since Better-EI and Best-EI scenarios are found to achieve CAFE standards even when a BEV or a FCEV receives a credit of one car sale in 2030.

Keywords: Corporate Average Fuel Economy standards; Global Change Assessment Model

1. Introduction

In 2017, the transportation sector accounted for 28.9% of the total energy consumption in South Korea [1]. Due to a heavy reliance on petroleum products, the transportation sector was the most CO_2-emitting sector among all end-use sectors in the country [2]. The South Korean government has implemented various policies for reducing energy consumption and greenhouse gas emissions in the transportation sector. Since the road sector accounted for 79.7% of the total transportation energy consumption in 2017, excluding that of bunkering [3], many policies focus on the road sector in South Korea. As many countries have adopted the Corporate Average Fuel Economy (CAFE) standards, South Korea also implemented CAFE standards in 2008. However, CAFE standards only deal with a tank-to-wheels analysis which is a part of a comprehensive analysis of vehicle energy use and emissions [4], thus restricting the annual average fuel economy (km/L) or greenhouse gas (GHG) emissions (g/km) of automobiles for automakers. To facilitate automakers in selling fuel-efficient cars and satisfying CAFE standards, there is a credit system in the CAFE standards in South Korea. According to this credit system, a sale of one fuel-efficient car can earn multiple credits. For example, the sale of one battery electric vehicle (BEV) is counted as three car sales by calculating the annual average fuel efficiency performance and annual average GHG emission performance. Even the sale of one gasoline vehicle which has a fuel efficiency of more than 23.4 km/L is counted as two car sales. It is worth introducing the U.S. CAFE standards here in which automakers are able to trade credits [5].

For instance, an automaker with a CAFE performance lower than what is required can opt to buy some credits in the credit market (that is, from other automakers). This flexibility in the U.S. CAFE standards allows automakers to lower costs for achieving CAFE standards. U.S. CAFE standards are calculated based on the wheelbase (length) and footprint (area) [6], causing larger cars to be less affected than smaller cars, while the South Korea CAFE standards consider a vehicle's curb weight.

Fuels emit GHG emissions through their life cycle—well-to-wheels process (WtW)—which can be disaggregated into well-to-tank (WtT) (extraction, refining, and transportation) and tank-to-wheels (TtW) (combustion). CAFE standards regulate only TtW emissions. The results of Khan et al.'s [7] WtW study in Pakistan show that TtW emissions accounted for 73–86% of the life cycle of GHG emissions for internal combustion vehicles (ICEVs). Song et al.'s [8] WtW study in Macau showed that for a gasoline vehicle, TtW emissions accounted for 87% of its life-cycle GHG emissions. Jang and Song's [9] WtW study in South Korea showed that TtW emissions accounted for 82.8% and 83.4% of the life cycle GHG emissions for gasoline and diesel vehicles, respectively. Previous studies have found that TtW GHG emissions are a major contributor to life-cycle GHG emissions. Hence, this study focuses on an analysis of TtW GHG emissions and assesses CAFE standards in South Korea using the Global Change Assessment Model (GCAM) with a sensitivity analysis.

2. Current Status of Passenger Cars in South Korea

As shown in Figure 1, the total number of cars in South Korea has increased rapidly. This increase has primarily been led by sales of passenger cars. Over the last ten years, the number of passenger cars and trucks has increased by 47.2% and 13.5%, with a current total of 19.17 million passenger cars and 3.59 million trucks, while vans sales decreased by 24.9% with a total of 0.81 million vans in 2019. That is, passenger cars will be a crucial target for reducing transportation energy consumption and GHG emissions in the road sector.

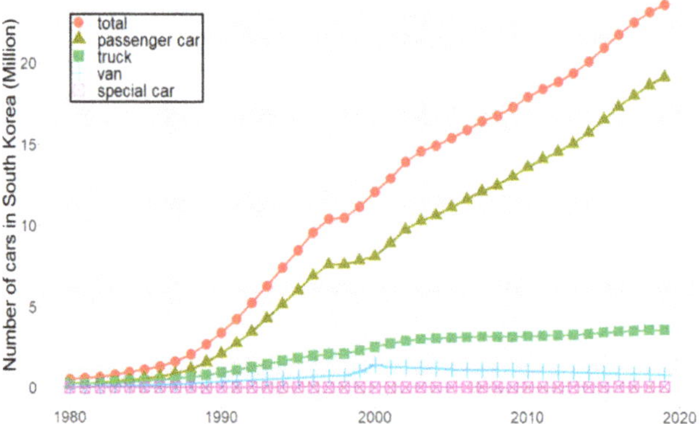

Figure 1. Number of cars in South Korea.

Figure 2 shows historical trends of the share of passenger cars by engine size and average energy intensity (EI). During 2000–2019, the share of large-sized passenger cars (more than 2000 cc) noticeably increased from 8.6% to 28.5%, while the percentage of small-sized passenger cars (1000 cc~1600 cc) sharply decreased from 49.3% to 21.2%. Considering that the energy intensity of large-sized cars is usually higher than that of small-sized cars [10], this shift in consumer preferences for large-sized cars must have had a negative impact on the overall energy intensity of passenger cars. The increase in average energy intensity from 2.71 MJ/km to 3.37 MJ/km during 2001–2016 (Figure 2) provides empirical evidence of such an impact.

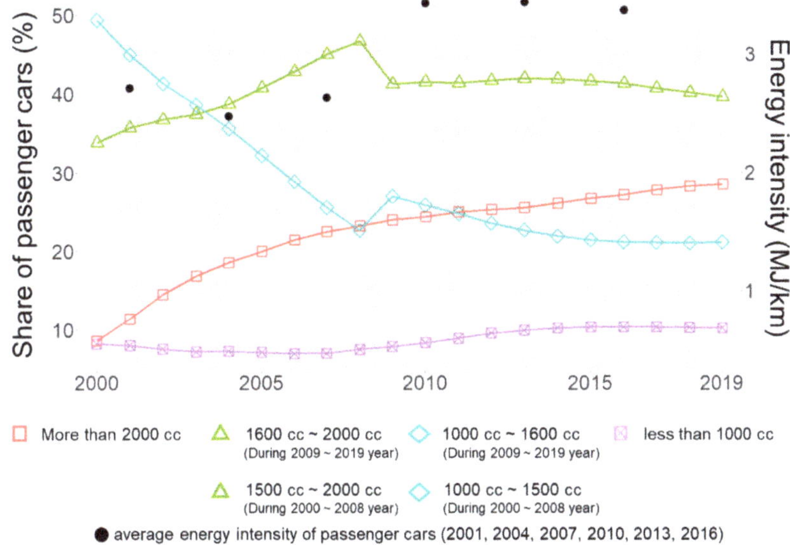

Figure 2. Share of passenger cars by engine size and weighted average energy intensity (EI) (source: Ministry of Land, Infrastructure and Transport [11], Korea Transport Institute [12], Korea Energy Economics Institute [13]). Note: MOLIT [11] changed the statistical classification of passenger cars in 2009.

While the shift in consumer preferences for large-sized passenger cars tends to increase energy intensity, the promotion of energy-efficient cars such as BEVs and fuel cell electric vehicles (FCEVs) could lower passenger cars' energy intensity. South Korea has implemented various policies for promoting BEVs, for example, providing a subsidy for buying a BEV. As a result of this policy, for the transportation sector, BEVs' market share in South Korea increased from 0.05% to 1.95% during 2013–2018 (Figure 3). Even though the number of BEVs in 2018 was small, it is expected that BEVs can be a primary technology for improving energy intensity in the transportation sector in the near future.

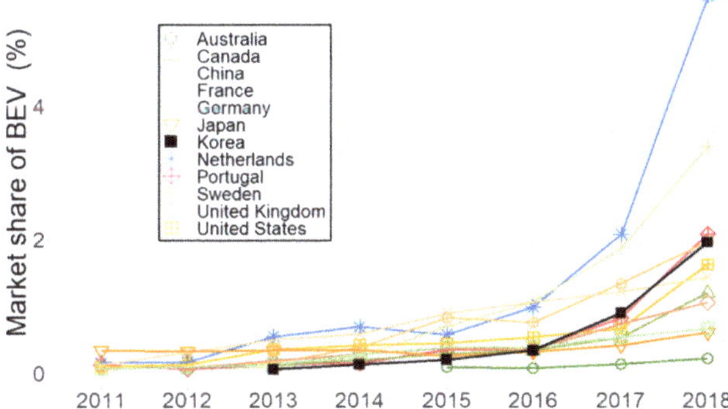

Figure 3. Market share of battery electric vehicles (BEVs) (source: International Energy Agency [14]). Note: Market share means share of new BEV registrations as a percentage of total new passenger car registrations.

3. CAFE Standards in South Korea and the Objective of this Study

In South Korea, the implementation of CAFE standards is based on the Energy Use Rationalization Act [15]. During the initial period of the implementation of the standards (2008–2011), the target was to achieve 12.4 km/L for mini and small-sized cars and 9.6 km/L for medium and large cars by 2011. During the second period (2012–2015), no specific targets for the types of cars were given and the overall target was to achieve 17 km/L or 140 g/km by 2015. During the current period (2016–2020), the target is to achieve 24.3 km/L or 97 g/km by 2020. A new target of achieving 28.1 km/L or 84 g/km by 2030 [16] has also been announced. Under the current CAFE standards, the performance and standards of annual average fuel efficiency are calculated as [17] (see Supplementary Materials).

$$f = \frac{N}{\sum\limits_{i=1}^{N} \frac{q_i}{f_i}} \tag{1}$$

$$f^s = \frac{N}{\sum\limits_{i=1}^{N} \frac{q_i}{f_i^s}} \tag{2}$$

where

f: average fuel efficiency performance,
f^s: average fuel efficiency standard,
N: total sales of cars,
i: car model i,
q_i: sales of car model i,
f_i: fuel efficiency performance of car model i,
f_i^s: fuel efficiency standard of car model i.

$$f_i^s = \begin{cases} \alpha + \beta m_i & if\ m_i > 1070\ \text{kg} \\ \delta & if\ m_i \leq 1070\ \text{kg} \end{cases} \tag{3}$$

where

$\alpha, \beta,$ and δ: given parameters,
m_i: the curb weight of car model i.

Fuel efficiency standard of car model i is directly given as δ regardless of its curb weight if the curb weight of model i is less than 1070 kg and the performance and standard of annual average GHG emissions is calculated as

$$e = \frac{\sum\limits_{i=1}^{N} e_i q_i}{N} \tag{4}$$

$$e^s = \frac{\sum\limits_{i=1}^{N} e_i^s q_i}{N} \tag{5}$$

where e is average GHG emission performance, e^s is the average GHG emission standard, e_i denotes the GHG emission performance of car model i, and e_i^s denotes the GHG emission standard of car model i, which is calculated as

$$e_i^s = \begin{cases} \alpha' + \beta'(m - w) & if\ m > 1070\ \text{kg} \\ \delta' & if\ m \leq 1070\ \text{kg} \end{cases} \tag{6}$$

where w is an additionally given parameter compared to the calculation of fuel efficiency standards. Until 2011, fuel efficiency was regulated by Federal Test Procedure-75 (FTP-75), a driving test known as the city driving test. Since 2012, fuel efficiency is regulated in a combined mode considering both city driving and highway driving tests. Under current CAFE standards, the sale of one BEV or one FCEV is counted as three car sales by calculating the annual average fuel efficiency performance and annual average GHG emission performance. For calculating the average GHG emission performance of BEVs or FCEVs, e_i is counted as zero.

Figure 4 shows historical fuel efficiency and carbon emissions by passenger cars with their CAFE standards. In Figure 4, neither fuel efficiency nor carbon emissions show much improvement during 2013–2018. Considering these unfavorable historical trends in passenger cars' fuel efficiency and carbon emissions, it would be meaningful to assess CAFE standards.

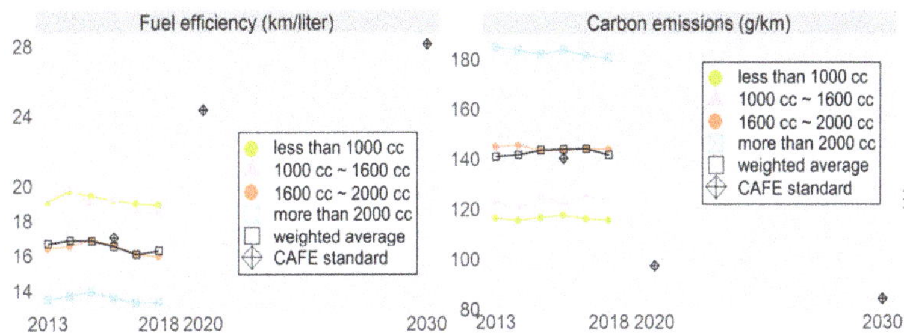

Figure 4. Observed versus Corporate Average Fuel Economy (CAFE) standards in terms of fuel efficiency and carbon emissions in the passenger car sector (source: Korea Energy Agency [17]).

4. Methodology and Data

4.1. Global Change Assessment Model

GCAM was chosen as an integrated assessment model for creating representative concentration pathways for the Intergovernmental Panel on Climate Change (IPCC)'s Fifth Assessment Report (AR5) [18]. GCAM represents various sectors including energy systems, agriculture, land use, land use change and forestry (LULUCF), economy, water, and climate for an analysis of their interactions. GCAM runs in five-year time steps, solving for market equilibrium. At the equilibrium, supply equals demand in all markets. The transportation sector is one of the end-use sectors in GCAM's energy system. One of the advantages of using GCAM is a well-represented hierarchical structure of the sector (for example, passenger road sector), mode (for example, small, medium, large car, and sport utility vehicle (SUV)), and technology (for example, ICEV, BEV, and FCEV) [19]. Mishra et al. [20] explain the methodological details of the GCAM transportation module. Kyle and Kim [21] and Yin et al. [22] can also be used as reference studies for an analysis of the transportation sector using GCAM.

The passenger transportation service demand at time t is given as

$$D_t = \sigma(Y_t)^{\alpha}(P_t)^{\beta}(N_t) \tag{7}$$

where

D: Passenger transportation demand (passenger kilometers travelled or PKT),

Y: Per capita income ($\$$),

P: Price of transportation service ($\$$/PKT),

N: Population,

α: Income elasticity,

β: Price elasticity,

t: Year in five-year time steps (for example, 2010 for calibration, 2015, 2020).

The price of transportation services (P) is calculated from the weighted average cost of sector, mode, and technology as

$$P_t = \sum_i S_{i,t} P_{i,t} \tag{8}$$

$$P_{i,t} = \sum_s S_{s,i,t} P_{s,i,t} \tag{9}$$

$$P_{s,i,t} = \sum_j S_{j,s,i,t} P_{j,s,i,t} + \frac{W}{SP_{s,i,t}} \delta_i \tag{10}$$

$$P_{j,s,i,t} = \frac{FP_{j,s,i,t} EI_{j,s,i,t} + NFP_{j,s,i,t}}{L_{j,s,i,t}} \tag{11}$$

where

i: Sector (for example, passenger road sector, passenger rail sector),
s: Mode (for example, small car, medium car),
j: Technology (for example, ICEV, BEV),
W: Hourly wage ($/h),
SP: Speed of mode (km/h),
δ: A parameter for the calculation of value of time,
FP: Fuel price ($/joule),
EI: Energy intensity (joule/VKT),
NFP: Non-fuel price ($/VKT),
L: Load factor (PKT/VKT),
S: Market share.

For example, the share of technology j in mode s is determined as

$$S_{j,s,i,t} = \frac{(SW_{j,s,i,t} P_{j,s,i,t})^{\lambda_i}}{\sum\limits_j (SW_{j,s,i,t} P_{j,s,i,t})^{\lambda_i}} \tag{12}$$

where SW means share-weight as a parameter for calibration and λ denotes the logit exponent.

Figure 5 shows the structure of the transportation sector used in this study. The passenger car sector includes four different modes—small sedan, medium sedan, large sedan, and SUV. Each mode has three technology options—ICEV, BEV and FCEV. The input data for modeling the transportation sector are based on Jeon and Kim [23], Jeon et al. [24], Korea Energy Economics Institute [3], Korea Transport Institute [12], Korea Energy Agency [17], and Korea Transportation Safety Authority [25].

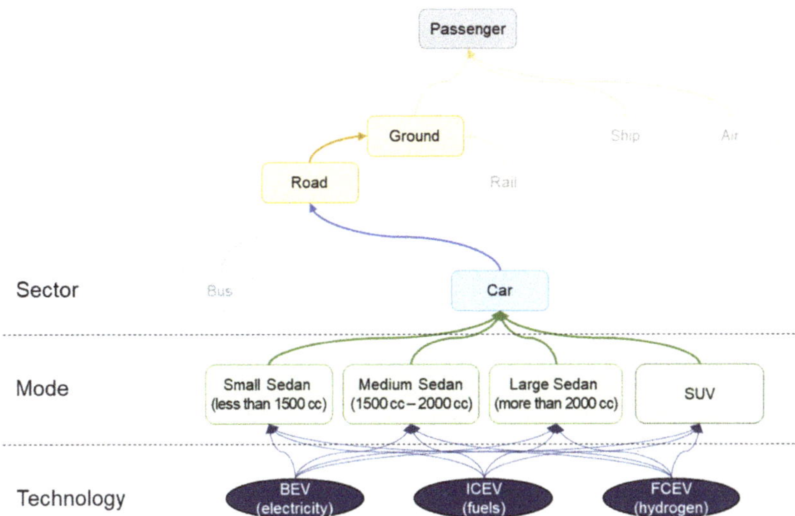

Figure 5. Representation of the transportation sector's structure used in this study.

4.2. Vehicle Cost Assumptions

This study uses the default data in GCAM [26] only for the composition ratio of technology costs. For the assumption of the year of cost parity (ICEV versus BEV), previous studies are referred to. Lutsey and Nicholas [27] expect that 2025 is the earliest when BEVs will reach cost parity with ICEVs. This study points out that mass production of BEVs could lower their costs, especially because of their lower battery costs. Likewise, Soulopoulos et al. [28] expect substantial cost reductions in BEVs because of improvements in battery technology and economies of scale. In their study, the cost parity of ICEVs and BEVs will be realized around 2022–2026, depending on vehicle size and a BEV's range (for example, smaller cars will reach cost parity earlier). As shown in Figure 6, this study assumes that ICEV costs are constant over all periods, while BEV costs are assumed to reach cost parity with ICEV costs in 2025, as referred to in references [27–29]. Then, BEV costs will be 85% of ICEV costs from 2030.

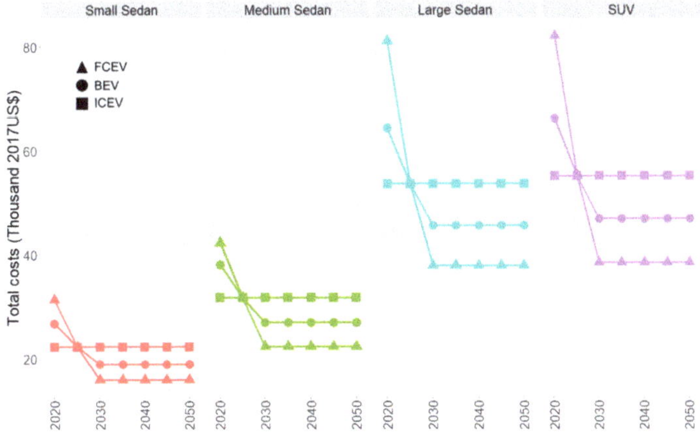

Figure 6. Assumption of total costs (see Supplementary Materials).

We refer to Morrison et al.'s [30] research on cost competitiveness between BEVs and FCEVs for the assumption of the year of cost parity (BEV versus FCEV). In particular, it is found that a BEV with a 150-mile range will reach cost parity with a FCEV around 2025, and after 2025 a FCEV's costs will be lower than a BEV's costs (for more details and numerical values, please refer to the tables in the Supplementary Materials).

4.3. Scenario Description

First, this study adopts Shared Socioeconomic Pathway 2 (SSP2) for socioeconomic assumptions (see Supplementary Materials). Table 1 gives the scenario settings in terms of the relative energy intensity of cars based on the reference case. Three additional scenarios are analyzed in this study, depending on the degree of improvements in energy intensity in all passenger cars (Figure 5). The reference assumes no improvement in energy intensity in the future. The 'Mod-EI' scenario is based on the energy intensity improvement rate applied by Ruffini and Wei [29]. The 'Better-EI' and the 'Best-EI' scenarios have an additional 10 and 20 percentage points of energy intensity improvement compared to the 'Mod-EI' scenario in 2050, respectively.

Table 1. Energy intensity improvement scenarios (unit: normalized).

Scenario	Description	Tech	Relative Energy Intensity of Passenger Cars			
			2020	2030	2040	2050
Reference	No improvement	ICEV	1	1	1	1
		BEV	1	1	1	1
		FCEV	1	1	1	1
Mod-EI	Moderate improvement in energy intensity	ICEV	1	0.889	0.800	0.727
		BEV	1	0.881	0.838	0.821
		FCEV	1	0.842	0.825	0.816
Better-EI	High improvement in energy intensity	ICEV	1	0.849	0.720	0.627
		BEV	1	0.841	0.758	0.721
		FCEV	1	0.802	0.745	0.716
Best-EI	Very high improvement in energy intensity	ICEV	1	0.809	0.640	0.527
		BEV	1	0.801	0.678	0.621
		FCEV	1	0.762	0.665	0.616

Since GCAM does not account for the number of vehicles explicitly, a conversion of transportation service demand into the number of cars is required. As shown in Equation (13), the number of vehicles (*Veh*) can be calculated from transportation demand (*D*) multiplied by the inverse of the load factor (L^{-1}) and the inverse of the VKT per vehicle (V^{-1}). The load factor (*L*) and VKT per vehicle (*V*) are assumed to refer to references [12,25], respectively, for all technologies in this study.

$$Veh \equiv PKT \times \frac{VKT}{PKT} \times \frac{Veh}{VKT}$$
$$= D \times L^{-1} \times V^{-1} \tag{13}$$

5. Results

Figure 7 shows the simulation results of service demand by passenger cars along with historical trends in demand in selected OECD countries in conjunction with GDP per capita. In Figure 7, the simulation results are from the reference case. It is found that the simulation results of service demand by passenger cars are consistent across all scenarios. At maximum across all scenarios and periods, the Best-EI scenario increases service demand for passenger cars 0.16% more than the reference in 2050. That is, all the scenarios do not have a tangible impact on total service demand by passenger cars. It should be noted that all the scenarios assume income elasticity of passenger transportation demand (α in Equation (7)) as 1.1 for all periods. For example, Dunkerley et al. show that the income elasticity of passenger transportation demand is in the range 0.5 to 1.4 [31].

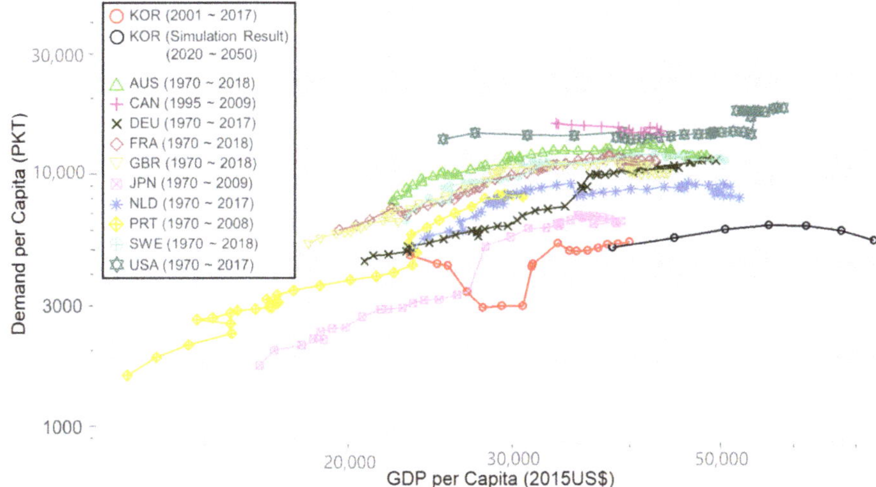

Figure 7. GDP per capita versus demand per capita for passenger cars (source: OECD [32,33]).

Figure 8, which is used for calculating CAFE's performance, shows the market share of ICEVs, BEVs, and FCEVs in the reference case. Since all scenarios do not have a big impact on the overall trend in market share, only the reference results are given in Figure 8. At maximum across all scenarios and periods, market share in 2050 is as follows: Reference (ICEV 9.6%; BEV 26.4%; FCEV 64.0%), Best-EI (ICEV 11.1%; BEV 22.5%; FCEV 66.4%). After the cost parity point (2025), ICEVs' market share is expected to decline rapidly from 70.2% to 21.1% during 2025–2030. Over the same period, FCEVs' market share will grow sharply from 9.8% to 52.1%. A slight increase of BEVs' market share, from 20.0% to 26.8%, will also contribute to shrinking ICEVs' market share.

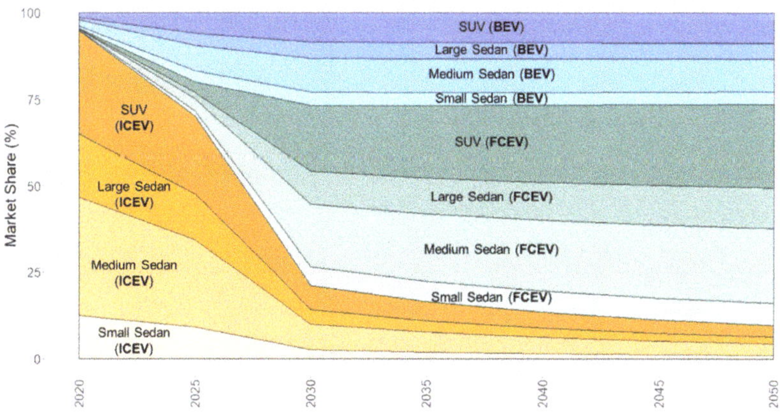

Figure 8. Market share of ICEVs, BEVs and FCEVs; new car penetrations as a percentage of total new passenger car penetrations.

Using their mix scenario, Krause et al. [34] expect that the market share of conventional vehicles will be below 20% in 2050 in the European Union road transportation sector. Bloomberg New Energy

Finance [35] foresees South Korea achieving a high level of electric vehicle adoption, representing around 60% of the market share in 2040.

The credit system in South Korea's CAFE standards provides incentives for selling BEVs and FCEVs. However, the credit system can change depending on market circumstances. For example, a sale of one FCEV was counted as five car sales by calculating CAFE's performance until 2017. Now, the credit for one FCEV sale has decreased to three car sales. This means the credit system could be an important variable in assessing CAFE standards. Figure 9 assesses CAFE standards in terms of the energy intensity sensitivity scenarios and also the degree of BEV and FCEV credits. In 2025, if the credit for a BEV or a FCEV sale is one car sale, only the Best-EI scenario will achieve CAFE standards. If the credit for a BEV or an FCEV sale is two car sales, only the reference will not achieve CAFE standards. If the credit for a BEV or a FCEV sale is three car sales, all scenarios will achieve CAFE standards. In 2030, if the credit for one BEV or FCEV sale is one car sale, Better-EI and Best-EI scenarios will achieve CAFE standards. However, if the credit for a BEV or a FCEV is more than two car sales, all scenarios will achieve CAFE standards.

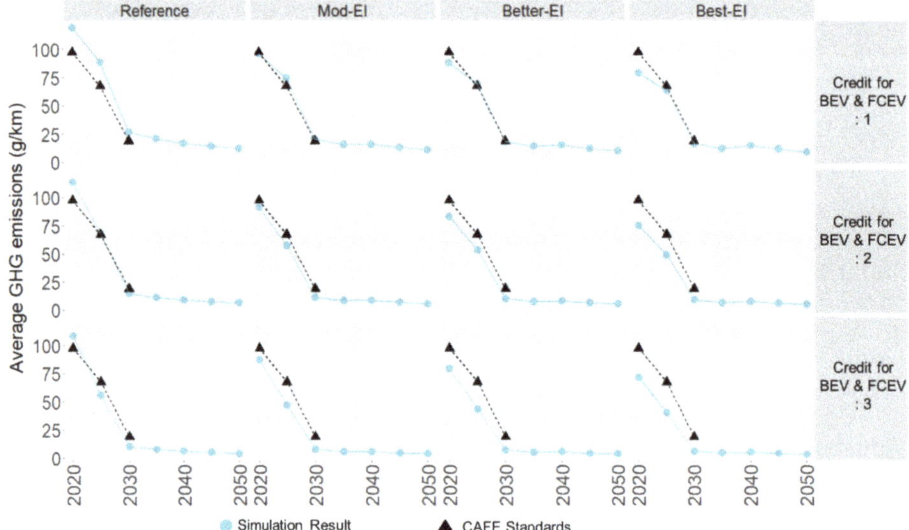

Figure 9. Average greenhouse gas (GHG) emissions by scenarios and the degree of credit for BEVs and FCEVs.

6. Conclusions

Energy intensity is one of the important factors that influence CO_2 emissions [36,37]. In this respect, this study introduced the current status of South Korea's passenger cars from two perspectives. First, the shift in consumer preferences towards large-sized cars is deteriorating the overall energy intensity of passenger cars. Second, the increasing promotion of BEVs could improve the energy intensity of passenger cars. CAFE standards were implemented for improving the energy intensity of passenger cars in South Korea. This study assessed CAFE standards by doing a sensitivity analysis of energy intensity improvement scenarios using GCAM. In addition to an analysis of the scenarios the study also assessed the credit system of CAFE standards.

The results are summarized as follows. First, all scenarios have a negligible impact on total service demand by passenger cars and on the overall trends of market share among ICEVs, BEVs, and FCEVs. According to the CAFE performance analysis, all scenarios will achieve CAFE standards in 2030 if the current credit system of three credits for one BEV or FCEV sale or at least two credits for one BEV or

FCEV sale is kept. However, in the case of no additional credits for one BEV or FCEV sale (that is, one credit for a BEV or an FCEV sale), the reference and the Mod-EI scenario will not achieve CAFE standards in 2030.

Some other findings of this study are as follows: First, from the policy design perspective, a proper setting of the credit system for BEVs and FCEVs will be a very important variable for automakers to achieve CAFE standards. Second, from the technology promotion perspective, active promotion of fuel efficiency improvements through CAFE standards is important since the Mod-EI scenario does not achieve CAFE standards when the credit for a BEV is one car sale in 2030.

The current study can be extended to include the following issues: First, various types of hybrid electric vehicles can be explicitly considered by extending this study. Second, various types of costs not included in vehicle costs, such as user costs, inconvenience costs, and time costs of refueling, can be modeled by applying this study if additional information becomes available.

Supplementary Materials: The following are available online at http://www.mdpi.com/1996-1073/13/17/4533/s1.

Author Contributions: Conceptualization, S.J. and M.R.; methodology, S.J. and M.R.; software, S.J. and M.R.; data curation, S.J.; visualization, S.J. and M.R.; writing—original draft preparation, S.J.; writing—review and editing, S.K.; and A.H.; supervision, S.K.; funding acquisition, S.K. All authors have read and agreed to the published version of the manuscript.

Funding: This work was supported by the Technology Development Program to Solve Climate Changes of the National Research Foundation (NRF) funded by the Ministry of Science, ICT and Future Planning (NRF-2017M1A2A2081253), and the Korea Ministry of Environment (MOE) as Graduate School specialized in Climate Change.

Acknowledgments: The authors are indebted to Jaeick Oh for his valuable comments and advice. Two anonymous referees' comments and suggestions on an earlier version of this paper are gratefully acknowledged.

Conflicts of Interest: The authors declare no conflict of interest.

References

1. International Energy Agency (IEA). World Energy Balances. 2019. Available online: https://www.iea.org/subscribe-to-data-services/world-energy-balances-and-statistics (accessed on 23 June 2020).
2. International Energy Agency (IEA). CO_2 Emissions from Fuel Combustion. Available online: https://www.iea.org/subscribe-to-data-services/co2-emissions-statistics (accessed on 23 June 2020).
3. Korea Energy Economics Institute (KEEI). *2018 Yearbook of Energy Statistics*; Korea Energy Economics Institute: Ulsan, Korea, 2018; Available online: http://www.keei.re.kr/keei/download/YES2018.pdf (accessed on 18 February 2020).
4. Curran, S.J.; Wagner, R.M.; Graves, R.L.; Keller, M.; Green, J.B., Jr. Well-to-wheel analysis of direct and indirect use of natural gas in passenger vehicles. *Energy* **2014**, *75*, 194–203. [CrossRef]
5. Durrmeyer, I.; Samano, M. To rebate or not to rebate: Fuel economy standards versus feebates. *Econ. J.* **2018**, *128*, 3076–3116. [CrossRef]
6. National Highway Traffic Safety Administration (NHTSA). Corporate Average Fuel Economy. Available online: https://www.nhtsa.gov/laws-regulations/corporate-average-fuel-economy (accessed on 19 July 2019).
7. Khan, M.I.; Shahrestani, M.; Hayat, T.; Shakoor, A.; Vahdati, M. Life cycle (well-to-wheel) energy and environmental assessment of natural gas as transportation fuel in Pakistan. *Appl. Energy* **2019**, *242*, 1738–1752. [CrossRef]
8. Song, Q.; Wu, Y.; Li, J.; Wang, Z.; Yu, D.; Duan, H. Well-to-wheel GHG emissions and mitigation potential from light-duty vehicles in Macau. *Int. J. Life Cycle Assess.* **2018**, *23*, 1916–1927. [CrossRef]
9. Jang, J.J.; Song, H.H. Well-to-wheel analysis on greenhouse gas emission and energy use with petroleum-based fuels in Korea: Gasoline and diesel. *Int. J. Life Cycle Assess.* **2015**, *20*, 1102–1116. [CrossRef]
10. Yanni, T.; Paul, J. Impact and sensitivity of vehicle design parameters on fuel economy estimates. No. 2010-01-0734. *SAE Tech. Pap.* **2010**. [CrossRef]
11. Ministry of Land, Infrastructure and Transport (MOLIT). Total Registered Motor Vehicles. Available online: https://stat.molit.go.kr/portal/main/portalMain.do (accessed on 24 June 2020).

12. Korea Transport Institute (KOTI). Korea Transportation Statistics (Various Years), Korea Transport Institute. Available online: https://www.ktdb.go.kr/www/index.do (accessed on 23 June 2020).
13. Korea Energy Economics Institute (KEEI). Energy Consumption Survey (Various Years), Korea Energy Economics Institute. Available online: http://www.kesis.net/sub/sub_0002.jsp?M_MENU_ID=M_M_002&S_MENU_ID=S_M_010 (accessed on 24 June 2020).
14. International Energy Agency (IEA). *Global EV Outlook 2019*; International Energy Agency: Paris, France, 2019; Available online: https://webstore.iea.org/global-ev-outlook-2019 (accessed on 24 June 2020).
15. International Energy Agency (IEA). Energy Use Rationalization Act. Available online: https://www.iea.org/policies/392-energy-use-rationalization-act (accessed on 28 August 2020).
16. Ministry of Trade, Industry and Energy. Energy Efficiency Innovation Strategy, South Korea. 2019. Available online: http://www.motie.go.kr/motie/ne/presse/press2/bbs/bbsView.do?bbs_cd_n=81&bbs_seq_n=161993 (accessed on 28 June 2020).
17. Korea Energy Agency (KEA). Vehicle Fuel Economy and CO_2 Emissions: Data and Analyses (Various Years), Korea Energy Agency. Available online: http://bpms.kemco.or.kr/transport_2012/pds/month_pds.aspx (accessed on 24 June 2020).
18. Van Vuuren, D.P.; Edmonds, J.; Kainuma, M.; Riahi, K.; Thomson, A.; Hibbard, K.; Hurtt, G.; Kram, T.; Krey, V.; Lamarque, J.-F.; et al. The representative concentration pathways: An overview. *Clim. Chang.* **2011**, *109*, 5. [CrossRef]
19. Kim, S.H.; Edmonds, J.; Lurz, J.; Smith, S.J.; Wise, M. The ObjECTS framework for integrated assessment: Hybrid modeling of transportation. *Energy J.* **2006**. [CrossRef]
20. Mishra, G.S.; Kyle, P.; Teter, J.; Morrison, G.M.; Kim, S.; Yeh, S. *Transportation module of Global Change Assessment Model (GCAM): Model Documentation*; Institute of Transportation Studies, University of California: Davis, CA, USA, 2013; Available online: https://trid.trb.org/view.aspx?id=1262949 (accessed on 28 August 2020).
21. Kyle, P.; Kim, S.H. Long-term implications of alternative light-duty vehicle technologies for global greenhouse gas emissions and primary energy demands. *Energy Policy* **2011**, *39*, 3012–3024. [CrossRef]
22. Yin, X.; Chen, W.; Eom, J.; Clarke, L.E.; Kim, S.H.; Patel, P.L.; Yu, S.; Kyle, G.P. China's transportation energy consumption and CO_2 emissions from a global perspective. *Energy Policy* **2015**, *82*, 233–248. [CrossRef]
23. Jeon, S.; Kim, S. Modeling Domestic Transportation Sector Using Global Change Assessment Model. *J. Korean Soc. Transp.* **2017**, *35*, 91–104. [CrossRef]
24. Jeon, S.; Roh, M.; Oh, J.; Kim, S. Development of an Integrated Assessment Model at Provincial Level: GCAM-Korea. *Energies* **2020**, *13*, 2565. [CrossRef]
25. Korea Transportation Safety Authority (TS). Automobile Mileage Analysis (Various Years), Korea Transportation Safety Authority. Available online: https://www.kotems.or.kr/app/kotems/forward?pageUrl=/kotems/ptl/bbs/KotemsPtlBbsStatsLs&topmenu1=06&topmenu2=03&topmenu3=03 (accessed on 25 June 2020).
26. Joint Global Change Research Institute (JGCRI). GCAM v5.1 Documentation: Global Change Assessment Model (GCAM). Available online: http://jgcri.github.io/gcam-doc (accessed on 14 April 2020).
27. Lutsey, N.; Nicholas, M. Update on Electric Vehicle Costs in the United States through 2030. *Int. Counc. Clean Transp.* **2019**, 1–12. Available online: https://theicct.org/sites/default/files/publications/EV_cost_2020_2030_20190401.pdf (accessed on 19 July 2020).
28. Soulopoulos, N. When will electric vehicles be cheaper than conventional vehicles. *Bloom. New Energy Financ.* **2017**, *12*. Available online: https://www.blogmotori.com/wp-content/uploads/2017/07/EV-Price-Parity-Report_BlogMotori_COM_MobilitaSostenibile_IT.pdf (accessed on 19 July 2020).
29. Ruffini, E.; Wei, M. Future costs of fuel cell electric vehicles in California using a learning rate approach. *Energy* **2018**, *150*, 329–341. [CrossRef]
30. Morrison, G.; Stevens, J.; Joseck, F. Relative economic competitiveness of light-duty battery electric and fuel cell electric vehicles. *Transp. Res. Part C Emerg. Technol.* **2018**, *87*, 183–196. [CrossRef]
31. Dunkerley, F.; Rohr, C.; Daly, A. Road Traffic Demand Elasticities: A Rapid Evidence Assessment. 2014. Available online: https://www.rand.org/pubs/research_reports/RR888.html (accessed on 28 June 2020).
32. Organization for Economic Co-operation and Development (OECD). Road Passenger Transport by Passenger Cars. Available online: https://data.oecd.org/transport/passenger-transport.htm (accessed on 27 June 2020).
33. Organization for Economic Co-Operation and Development (OECD). Level of GDP per Capita and Productivity. Available online: https://stats.oecd.org/Index.aspx?DataSetCode=PDB_LV (accessed on 27 June 2020).

34. Krause, J.; Thiel, C.; Tsokolis, D.; Samaras, Z.; Rota, C.; Ward, A.; Prenninger, P.; Coosemans, T.; Neugebauer, S.; Verhoeve, W. EU road vehicle energy consumption and CO_2 emissions by 2050–Expert-based scenarios. *Energy Policy* **2020**, *138*, 111224. [CrossRef]
35. Bloomberg New Energy Finance (BNEF). Electric Vehicle Outlook 2020. *Bloomberg Finance Limited Partnership*. 2020. Available online: https://about.bnef.com/electric-vehicle-outlook/ (accessed on 20 July 2020).
36. Parker, S.; Bhatti, M.I. Dynamics and drivers of per capita CO_2 emissions in Asia. *Energy Econ.* **2020**, 104798. [CrossRef]
37. World Bank. CO^2 Emissions (Metric Tons per Capita). Available online: https://data.worldbank.org/indicator/EN.ATM.CO2E.PC (accessed on 25 August 2020).

Article

Modeling Air Pollutant Emissions in the Provincial Level Road Transportation Sector in Korea: A Case Study of the Zero-Emission Vehicle Subsidy

Minyoung Roh [1], Seungho Jeon [1], Soontae Kim [2], Sha Yu [3], Almas Heshmati [4] and Suduk Kim [1,*

[1]   Department of Energy Systems Research, Ajou University, Suwon 16449, Korea; rohmin9122@gmail.com (M.R.); sngho.jeon@gmail.com (S.J.)
[2]   Department of Environmental Engineering, Ajou University, Suwon 16449, Korea; soontaekim@ajou.ac.kr
[3]   Joint Global Change Research Institute, Pacific Northwest National Laboratory, College Park, MD 20740, USA; sha.yu@pnnl.gov
[4]   Jönköping International Business School (JIBS), Jönköping University, P.O. Box 1026, SE-551 11 Jönköping, Sweden; almas.heshmati@ju.se
*   Correspondence: suduk@ajou.ac.kr; Tel.: +82-31-219-2689

Received: 30 June 2020; Accepted: 2 August 2020; Published: 3 August 2020

Abstract: South Korea has been suffering from high $PM_{2.5}$ pollution. Previous studies have contributed to establishing $PM_{2.5}$ mitigation policies but have not considered provincial features and sector-interactions. In that sense, the integrated assessment model (IAM) could complement the shortcomings of previous studies. IAM, capable of analyzing $PM_{2.5}$ pollution levels at the provincial level in Korea, however, has not been developed yet. Hence, this study (i) expands on IAM which can represent provincial-level spatial resolution in Korea (GCAM-Korea) with air pollutant emissions modeling which focuses on the road transportation sector and (ii) examines the zero-emission vehicles (ZEVs) subsidy policy's effects on $PM_{2.5}$ mitigation using the expanded GCAM-Korea. Simulation results show that $PM_{2.5}$ emissions decrease by 0.6–4.1% compared to the baseline, and the Seoul metropolitan area contributes 38–44% to the overall $PM_{2.5}$ emission reductions. As the ZEVs subsidy is weighted towards the light-duty vehicle 4-wheels (LDV4W) sector, various spillover effects are found: ZEVs' share rises intensively in the LDV4W sector leading to an increase in its service costs, and at the same time, driving bus service costs to become relatively cheaper. This, in turn, drives an increase in bus service demand and emissions discharge. Furthermore, this type of impact of the ZEVs subsidy policy does not reduce internal combustion engine vehicles (ICEVs) in freight trucks, although diesel freight trucks are a major contributor to $PM_{2.5}$ emissions and also to NO_x.

Keywords: integrated assessment model; subsidy policy; air quality improvement; zero-emission vehicles; fine particulate matter

---

1. Introduction

1.1. Background

In recent years, South Korea has been suffering from deteriorating air quality because of high particulate matter (PM) levels. In capital Seoul, $PM_{2.5}$ (PM of 2.5 μm or less in diameter) concentration is nearly two times higher than what is prescribed by the World Health Organization (WHO) guidelines [1]. According to WHO, $PM_{2.5}$ exposure leads to an increase in mortality because of respiratory and cardiovascular diseases [2]. In Korea, a total of 11,924 deaths were attributable to $PM_{2.5}$ in 2015 [3].

$PM_{2.5}$ can be directly emitted from human activities such as power plants, business facilities, and internal combustion engine vehicles (ICEVs). $PM_{2.5}$ can also be produced secondarily by photochemical reaction with $PM_{2.5}$ precursor species of nitrogen oxides (NO_x), sulfur oxides (SO_x), ammonia (NH_3), and volatile organic carbons (VOC) in the atmosphere [4]. Regarding sources of domestic $PM_{2.5}$ emissions, of all $PM_{2.5}$ in the atmosphere in Korea, half of it comes from secondary formation. Business facilities are the largest $PM_{2.5}$ emitters nationwide while diesel vehicles are the largest emitters in the Seoul metropolitan area (Seoul, Incheon, and Gyeonggi), where almost half the Korean population resides [5,6].

The Korean government has set up a series of countermeasures to control $PM_{2.5}$ emissions including a Comprehensive Plan on Fine Dust Management (CPFDM). CPFDM is a comprehensive plan for cutting $PM_{2.5}$ between 2020 and 2024 which aims to decrease the annual average $PM_{2.5}$ concentration by 35% below the 2016 levels (26 $\mu g/m^3$) by 2024. Also, domestic emission reduction target per year with reduction rate for $PM_{2.5}$ (noted in the parenthesis) was set at 3300 tonnes (8%) for the industry sector; 2000 tonnes (63%) for the power generation sector; 8600 tonnes (35%) for the transportation sector; and 5200 tonnes (17%) for everyday surroundings such as road-cleaning, illegal incineration, and introduction of domestic low-NO_x boilers (the term of "everyday surroundgins" is used in the official document) [7]. That is, the transportation sector, especially the road transportation sector, has the largest reduction target of $PM_{2.5}$ emissions, although the power sector is facing a stricter emission reduction target in terms of the reduction rate. The road transportation sector accounted for around 70% of the total $PM_{2.5}$ emissions from both road and non-road transportation sectors including fugitive road dust (FRD) in 2016. FRD is generated by tire wear, brake wear, and road wear. It is one of the major emitters accounting for 7% of the overall local emissions of $PM_{2.5}$ in 2016. At that time, emissions from the road transportation sector were 11% [8].

Zero-emission vehicles (ZEVs) such as electric battery vehicles (BEVs) and fuel cell electric vehicles (FCEVs), are globally promoted for improving air quality and reducing oil consumption [9]. In Korea, ZEVs have been strongly promoted as one of $PM_{2.5}$ mitigation measures for the transportation sector. In 2018, the government spent $757 million to carry forward $PM_{2.5}$ mitigation measures for the transportation sector, which accounted for 56% of the total budget for domestic $PM_{2.5}$ mitigation measures. In particular, budget spending on the subsidizing ZEVs' purchase accounted for 71% of the $PM_{2.5}$ mitigation budget for the transportation sector [10].

In addition to the ZEV purchase subsidy, the government also offers tax incentives (for example, tax breaks for special consumption tax, educational tax, acquisition tax, and automobile tax) for ZEVs' buyers [7], and mandates automakers to supply a certain percentage of ZEVs including low-emission vehicles (hybrid electric vehicles (HEVs) and plug in HEVs) without any incentive. Instead, the amount of mandatory supply can be deducted if automakers invest in charging station installations as a contribution to infrastructure construction [11]. Unlike automakers, owners of apartment houses, business facilities, and large car parks get a subsidy for charging station installations [12].

1.2. Main Objectives of This Study

Studies have widely used an integrated assessment model (IAM) for analyzing environmental policy within inter-related systems such as the economy, energy, land-use, agriculture, and climate [13]. IAM has also been used for emission projections, mortality costs, and air quality management for $PM_{2.5}$ (Table 1).

CPFDM was established based on the following studies but the studies have some shortcomings. Kim et al. [14] prioritized PM reduction policies using the Analytic Hierarchy Process (AHP) and suggested 'Mandatory reduction of air pollution in the manufacturing industry and the suspension of such factories operation' as the top priority. Since they did not consider provincial emissions patterns, their suggestion may not be applicable to some provinces. For example, policies associated with diesel vehicle reduction might have been given a higher priority than the suggested policy in the Seoul metropolitan area if Kim et al. [14] had taken into account provincial emission patterns. In this sense, our study can make up the gap in Kim et al.'s study.

For the computation of $PM_{2.5}$, NO_x, and SO_x concentrations at monthly and grid levels, the Community Multiscale Air Quality (CMAQ) model was used with the national emissions inventory [15–17]. Anthropogenic emissions control is constrained in socioeconomics assumptions such as population and economic growth, as well as technology development assumptions [18]. However, since some studies are based on the point of view of atmospheric chemical reactivity, they do not consider socioeconomics assumptions. Besides, there is also a study which estimates social costs of $PM_{2.5}$ [19].

Table 1. Previous studies on anthropogenic PM2.5 emissions using IAM.

Research Topic	Pollutant							Region	Spatial Scope	IAM	Reference
	$PM_{2.5}$	PM_{10}	NO_x	SO_x	BC	OC	CO_2				
Emissions projections	✓		✓	✓				USA	US States	GCAM-USA [1]	[18]
Mortality costs	✓		✓	✓			✓	USA	US States	GCAM-USA	[20]
Emissions projections	✓	✓						Europe	National	RAINS [2]	[21]
Emissions projections and policy impact analysis			✓	✓			✓	USA	US Census Division	EPA-MARKAL [3]	[22]
Global emissions aspect	✓	✓			✓	✓		Global	25 Global Regions	GAINS [4]	[23]
Energy efficiency measures' impacts on emissions in the cement industry	✓	✓	✓	✓			✓	China	China Provinces	GAINS	[24]

[1] Global Change Assessment Model-USA developed by PNNL/JGCRI; [2] Regional Air Pollution INformation and Simulation developed by IIASA; [3] Environmental Protection Agency-MARKet Allocation developed by EPA; [4] and Greenhouse gas Air pollution Interactions and Synergies developed by IIASA.

An analysis of cross-sectoral dynamics is a pre-requisite for preventing unexpected harm of the spillover effects in multiple sectors, but there are few studies on how the policy impact of $PM_{2.5}$ emissions changes in multiple sectors. Hence, using IAM can remedy the shortcomings of previous studies. However, to the best of our knowledge IAM has not been applied for tackling $PM_{2.5}$ issues in Korea. Moreover, IAM is also capable of analyzing $PM_{2.5}$ at the provincial level and this too has not been developed yet by researchers.

Hence, the first goal of this study is modeling air pollutant emissions using IAM that represents Korean province partial resolution (GCAM-Korea). This study focuses on the road transportation sector in GCAM-Korea as the first step. Pollutant coverage is primary $PM_{2.5}$ as well as the precursors NO_x, SO_x, VOC, and NH_3. The second goal is assessing the ZEV subsidy policy's impact on air pollutant emissions across the road transportation sector and provinces.

2. Methodology and Data

2.1. Global Change Assessment Model and GCAM-Korea

GCAM is a community model which has been managed by the Joint Global Change Research Institute (JGCRI) for over 30 years. As a community model, GCAM is a fully open source code and model data on Github [25]. GCAM can investigate human-earth system dynamics alongside detailed representation of technology. The system consists of the economy, energy systems, agriculture and land-use, water, and the physical Earth system. As a partial equilibrium model based on a given socioeconomic pathway, GCAM finds equilibrium in the supply and demand of goods and services in each market and then determines market-clearing quantity and price [26,27]. GCAM models technology competition using the logit type of share equation based on the relative costs developed by McFadden [28]. The share of technology in each sector and period is changed smoothly by costs or policy changes [29]. That is, the logit share equation can prevent the winner-takes-all phenomenon which can be caused by an abrupt and slight price change in linear programming optimization [30,31].

Population and GDP (Gross Domestic Product) are exogenous inputs and driving forces for determining final energy service demand in conjunction with the cost of energy services and sector-specific energy services' price elasticity. The model is calibrated for energy consumption and pollutant emissions at the base year. In GCAM, GDP can affect future emissions of air pollutants. Smith et al. [32] examined the relationship between sulfur dioxide emission reduction and GDP per capita in Purchasing Power Parity (PPP) in 17 world regions from 1850 to 2000. Their study developed an income-based parameterization for an IAM to control sulfur dioxide emissions. Based on their study, GCAM adopted the income-based emission control function for NO_x and SO_x. As a result, fast economic growth tends to implement emission reduction rapidly. In GCAM, anthropogenic air pollutant emissions are driven not only by fuel consumption but also GDP per capita.

While GCAM's energy-economy system presents 32 regions globally including South Korea as a separate region, the recent GCAM represents various spatial resolutions for capturing the heterogeneity of certain regions which have not been modeled separately. As an example of a country-specific GCAM, which was not modeled as a separate region before, GCAM-Ethiopia was developed by separating Ethiopia from Eastern Africa that is one of the 32 global regions to go over biomass policy effects on Ethiopian energy consumption [33]. GCAM-Gujarat is a bit more detailed country GCAM. GCAM-Gujarat is an extended version of GCAM-India and was used for assessing building energy policies in Gujarat state in India [34]. GCAM-China has a higher resolution, which represents 31 provinces in China with other global regions. GCAM-China was used for examining the role of technologies such as carbon dioxide capture, utilization, and storage (CCUS) [35] and nuclear power plants [36] in China at the provincial level. Another example of higher spatial GCAM is GCAM-USA, which subdivided the USA region into 50 US states and D.C. and was also used as a $PM_{2.5}$ analysis tool for US states and D.C. Shi et al. [17] projected NO_x, SO_2, and $PM_{2.5}$ emissions, and Ou et al. [20] estimated $PM_{2.5}$ mortality costs.

GCAM-Korea is developed based on GCAM-USA ver. 5.1.3 for investigating the South Korean energy system at the provincial level. GCAM-Korea subdivides South Korea into 16 provinces except for Sejong (Figure 1). As Sejong is a relatively new city established in 2012 and it has only 0.5% of South Korea's residents not enough information is available on the region as yet. In GCAM-Korea, 31 global regions outside South Korea interact with 16 provinces in South Korea. Socioeconomics and energy systems are represented at the provincial level, while land-use and water systems adopt the default GCAM system. Although GCAM-Korea operates in 5-year periods from 2010 to 2040, the operation period can be extended through further modeling work. The base year is 2010 for calibration of energy and emissions. Input data for GCAM-Korea is available at GitHub (https://github.com/rohmin9122/gcam-korea-release) [37].

GCAM-Korea exhibits the provincial features of the energy sector. Electricity from coal power plants is mostly generated in Chungnam and Gyeongnam. Electricity is mainly consumed by the building and industrial sectors which are mostly located in the Seoul metropolitan area, Gyeonggi, Chungnam, and Jeonbuk; 77% of the national industrial energy is consumed in four provinces: Jeonam, Chungnam, Ulsan, and Gyeongbuk. Energy consumption in the building and transportation sectors is intensive in the Seoul metropolitan area which accounted for 52% and 44% of the total energy consumption in the building and transportation sectors, respectively, in 2015.

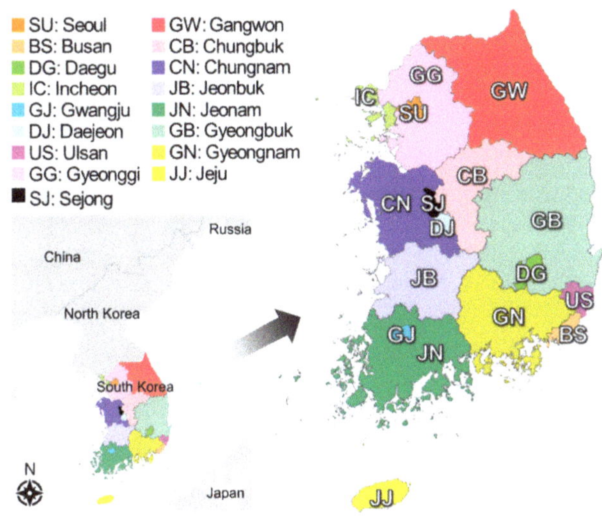

Figure 1. Administration divisions in Korea [37].

2.2. Modeling Air Pollutant Emissions in GCAM-Korea

As the current GCAM-Korea is modeled only for socioeconomics and energy systems, air pollutant emissions modeling is a new feature which requires to be augmented. Hence, this study further develops GCAM-Korea by using air pollutant emissions data from the national air pollutant emissions inventory.

2.2.1. National Air Pollutant Emissions Inventory

The National Institute of Environmental Research (NIER) provides estimated annual emissions through the Clean Air Policy Support System (CAPSS). The source classification code (SCC) is based on the classification of CORe INventory AIR (CORINAIR) published by the European Environment Agency (EEA), and it is adjusted by NIER to fit Korean activity classifications. The sources of fugitive dust and biomass burning were not included in the annual information till 2015. SCC comprises of 13 large categories—energy production combustion, non-industrial combustion, manufacturing industry, industrial processes, energy transport and storage, solvent use, road transportation, non-road transportation, waste, agriculture, other sources and sinks, fugitive dust, and biomass burning; 56 medium categories; and more than 200 small categories (activity sources) at the district level since 2016. In the national emissions inventory, air pollutants consist of CO, NO_x, SO_x, TSP, PM_{10}, $PM_{2.5}$, VOC, NH_3, and BC [38,39].

Road transportation is composed of eight vehicles by fuel type (gasoline, LPG, diesel and compressed natural gas (CNG)). The vehicle types are: passenger cars, taxis, vans, buses, freight cars, recreational vehicles, two-wheeled vehicles, and special vehicles. In the emissions inventory, emissions from road transportation are estimated using total vehicle kilometers traveled (VKT), the statistics of total registered motor vehicles, and emission factors by each type of vehicle, fuel, and species [38]. The total VKT is a sum of measured VKT and unmeasured VKT. Measured VKT is calculated using traffic volume and road length by road sections. Unmeasured VKT is estimated using vehicle type, vehicle age, and average driving speed on a district basis. All provinces are assumed to have the same emission factors for each vehicle type and species. Emission factors, however, are known to deteriorate with high driving speed and a vehicle's age [40].

Fugitive dust in the emissions inventory is composed of eight sub-categories including paved road dust, unpaved road dust, and construction. However, this study only focuses on paved and unpaved road fugitive dust (from now on referred to as FRD). FRD's estimation is based on total

VKT and emission factor for wear (tire wear, brake wear, and road wear). The emission factor for tire and brake wear is calculated using data measured by the mandatory vehicle inspection. The road wear emission factor is calculated using vehicle weight and measured silt loading. Silt loading means resuspended road dust per road surface [41,42].

2.2.2. Applying Air Pollutant Emissions Data in GCAM-Korea

Air pollutant emissions data obtained from the National Air Pollutant Emissions Service [43] was reclassified to match NIER's activity sources to road transportation modes in GCAM-Korea (see Appendix A), fuels, and provinces in GCAM-Korea; 276 districts excluding Sejong are merged into 16 provinces. Gasoline, LPG, and diesel are aggregated into refined liquids, and CNG is mapped to gas in GCAM-Korea. FRD sources are sub-divided based on their share of energy use that is calculated using the energy consumption survey [44], VKT [45], and fuel efficiency [46] because the sub-classification of FRD sources is aggregated across vehicle type, vehicle size, and fuel type in the emissions inventory. As FRD emissions data for BEVs and FCEVs is currently not available in the emissions inventory, these emissions models are ignored in this study.

The calibration year for GCAM-Korea is 2010. However, emissions data for various years is used for the model's calibration (see Table 2) on account of missing data or data which contradicts energy use as illustrated in Figure 2.

Table 2. Year of air pollutant emissions data used for the calibration.

Year	NH3	NOx	PM2.5		SOx	VOC
			Primary Emissions	FRD		
2010	✓	✓				
2013					✓	✓
2016			✓	✓		

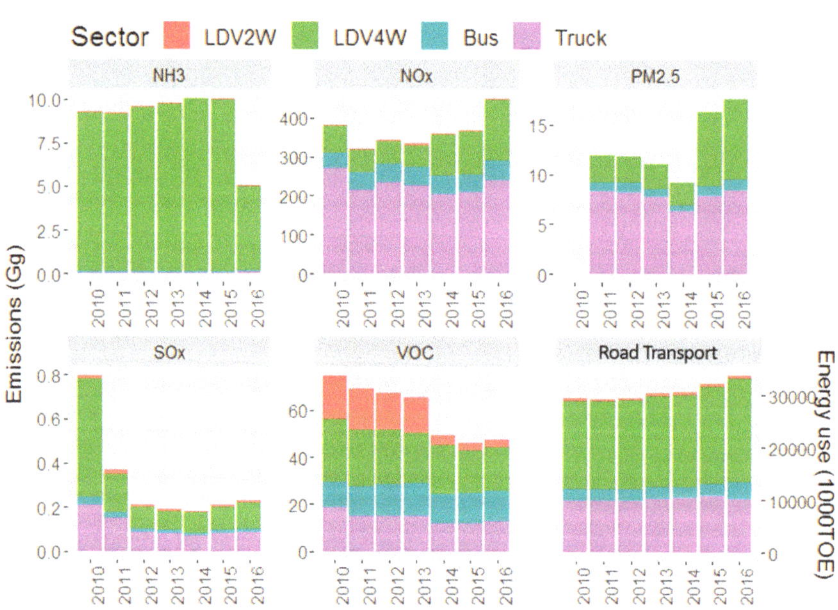

Figure 2. Reclassified emissions and energy use in road transportation. Note: Total energy use of road transportation is from the energy balance table in the Yearbook of Energy Statistics [44], and the sectoral share [47–49] is applied to total energy use.

For example, SO_x emissions from light-duty 4-wheel vehicles (LDV4W) were notably high only in 2010, while its energy use increased steadily during the study period. As VKT is closely related to energy use, comparing emissions to energy use instead of VKT is suitable under the assumption that there is no big change of technology development or regulations on SO_x emissions. In actual fact, there is no big change. The main reason that the calibration year's data is not used is for avoiding overestimation or underestimation of future emissions. If SO_x emissions in 2010 are used for calibration, future emissions will be overestimated. Another reason for using different years' emissions data is the absence of data in the calibration year. FRD and $PM_{2.5}$ emissions from light-duty vehicle 2-wheels (LDV2W) were newly released in 2015 and 2016, respectively.

3. Scenario Design

The ZEVs purchase subsidy is provided for cars (LDV4W) and buses for both BEVs and FCEVs. Subsidy for motorcycles (LDV2W) and freight trucks (less than 1 tonne) is available only for BEVs. Subsidy is provided not only by the national government but also by the local government. Subsidy from the national government is the same everywhere, but subsidy from the local government is different. For example, local government subsidy for LDV4W's BEVs range from $4100 in Seoul to $10,000 in Ulleung-gun, Gyeongbuk. The national government subsidy for LDV4W is between $5600 and $7500, depending on the vehicle model. On the other hand, local government subsidies for LDV4W's FCEVs are available only in eight provinces, ranging from $9100 in Incheon to $18,000 in Goseong-gun, Gangwon. The government subsidy for one of the FCEVs, NEXO, manufactured by Hyundai, is $20,500. Note that subsidy for LDV2W, buses, and trucks is equally supported by all local governments. Although the national government offers tax incentives for ZEV buyers, this study considers only the ZEV purchase subsidy.

To apply subsidy to GCAM-Korea, vehicle models are first classified into vehicle types. Then, the average subsidy of each vehicle type is calculated for each province. The calculated subsidy for BEVs and FCEVs is given in Appendices B and C respectively.

Second, future subsidy scenarios are developed (Table 3). According to CPFDM, subsidy for passenger cars will gradually be phased out, although the exact information on expiration has not been announced. A 'Sunset' scenario, therefore, is assumed in which subsidy for only LDV4W's BEVs will be phased out by 2040. In this scenario, the subsidy declines linearly to zero by 2040. A 'NoSunset' scenario is assumed for comparison. In both the scenarios, ZEVs subsidy is available from 2020. For the baseline analysis without any subsidy, a 'REF' or a reference case for the projected emissions of the baseline is prepared.

Table 3. Description of scenarios.

Scenario	Assumption
REF	Baseline without any subsidies
Sunset	Phaseout on subsidy for electric passenger cars only by 2040
NoSunset	Maintaining current subsidies till 2040

In GCAM-Korea, new technologies such as hydrogen buses, electric buses, and electric freight trucks (less than 1 tonne), have not been modeled yet. Hence, these new technologies are added to the nesting structure of the transportation sector in GCAM-Korea for an analysis (Figure 3). As future technology cost estimations largely depend on the scope of research, a relative cost approach is adopted. Purchase costs are obtained from various sources, and maintenance costs are calculated by applying the ratio of maintenance costs to the present value of purchase costs from previous studies (see Appendix D). Infrastructure costs such as charging stations and hydrogen production facilities are not considered. Future cost trends of electric buses and trucks are assumed based on the decreasing rate of cost of electric passenger cars in GCAM version 5.1.3. Likewise, the trend of hydrogen buses is assumed based on the trend of hydrogen passenger cars in GCAM.

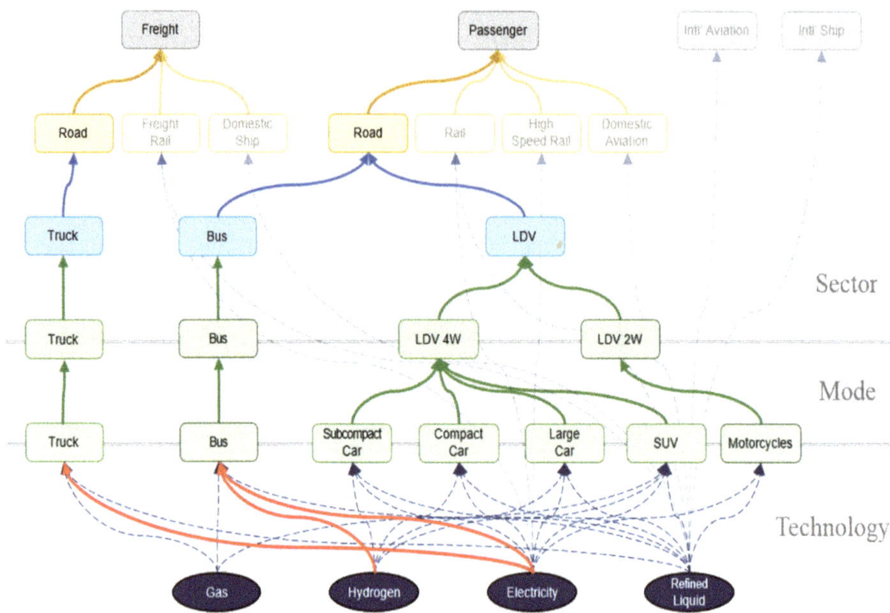

Figure 3. New nesting structure for the transportation sector in GCAM-Korea. Note: Red line indicates a newly added technology and blurry figures denote non-road transportation sectors.

Finally, the subsidy is subtracted from the total cost of each vehicle technology. According to our calculations, the total cost of an electric bus and a hydrogen bus is about 2.1 times and 4.0 times that of a diesel bus respectively. The total cost of an electric freight truck is about 1.8 times that of a one-tonne diesel truck.

4. Results

4.1. Projected Emissions at the Baseline

Table 4 summarizes projected emissions at the baseline (REF). It compares emissions from GCAM-Korea and those from the national emissions inventory. The projected emissions are captured fairly well in terms of sectors and provinces. The NH_3, NO_x, $PM_{2.5}$, SO_x, and VOC emissions in 2015 are projected as 80%, 94%, 97%, 81%, and 129% respectively, compared to emissions in the emissions inventory. In REF, the LDV4W and truck sectors are the main contributors to $PM_{2.5}$ emissions. The truck sector in particular accounted for 71% of NO_x emissions while the LDV4W sector accounted for 98% of the total NH_3 emissions.

Projected emissions by year and province are given in Figure 4. Sectoral emissions are projected between 72% and 119% compared to emissions in the emissions inventory except for VOC for LDV2W (456%) and NO_x for LDV4W (53%). VOC emissions for LDV2W are overestimated because of the abrupt decrease in emissions reported in the national emissions inventory. Its emissions in 2015 (2.96 Gg) fell by 80% as compared to emissions in 2013 (15.25 Gg), whereas energy use for LDV2W increased slightly from 484 KTOE in 2013 to 514 KTOE in 2015. For a similar reason, NO_x emissions for LDV4W cannot be captured well. Its emissions in the emissions inventory have dramatically increased since 2014, when it was more than two times the emissions in 2010.

Table 4. Comparison of emissions from the national inventory and those from GCAM-Korea (2015).

(Unit: Gg)		LDV2W	LDV4W	Bus	Truck	Total
	REF	0.04	7.84	0.02	0.09	7.99
NH_3	Inventory	0.05	9.88	0.02	0.09	10.04
	REF/Inventory	0.84	0.79	0.80	1.00	0.80
	REF	2.82	58.01	38.45	247.91	347.19
NO_x	Inventory	2.9	109.6	47.06	208.36	367.92
	REF/Inventory	0.97	0.53	0.82	1.19	0.94
	REF	0.06	6.89	1.11	7.78	15.84
$PM_{2.5}$	Inventory	0.07 [1]	7.46	0.99	7.87	16.39
	REF/Inventory	0.89	0.92	1.12	0.99	0.97
	REF	0.01	0.07	0.01	0.08	0.17
SO_x	Inventory	0.01	0.1	0.02	0.08	0.21
	REF/Inventory	0.82	0.72	0.85	0.92	0.81
	REF	13.49	17.9	14.04	13.75	59.18
VOC	Inventory	2.96	18.45	12.89	11.69	45.99
	REF/Inventory	4.56	0.97	1.09	1.18	1.29

[1] Note: $PM_{2.5}$ of LDV2W is indicated in 2016 because of no data for 2015.

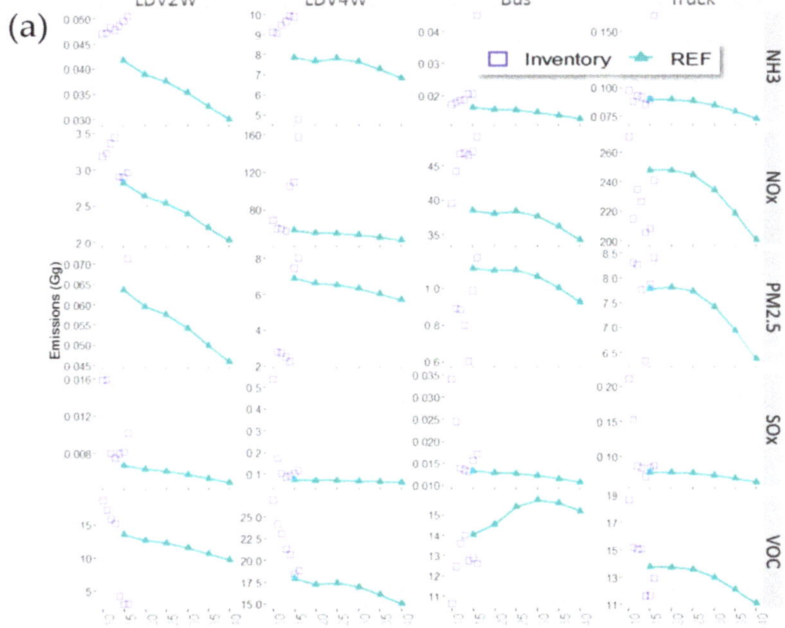

Figure 4. *Cont.*

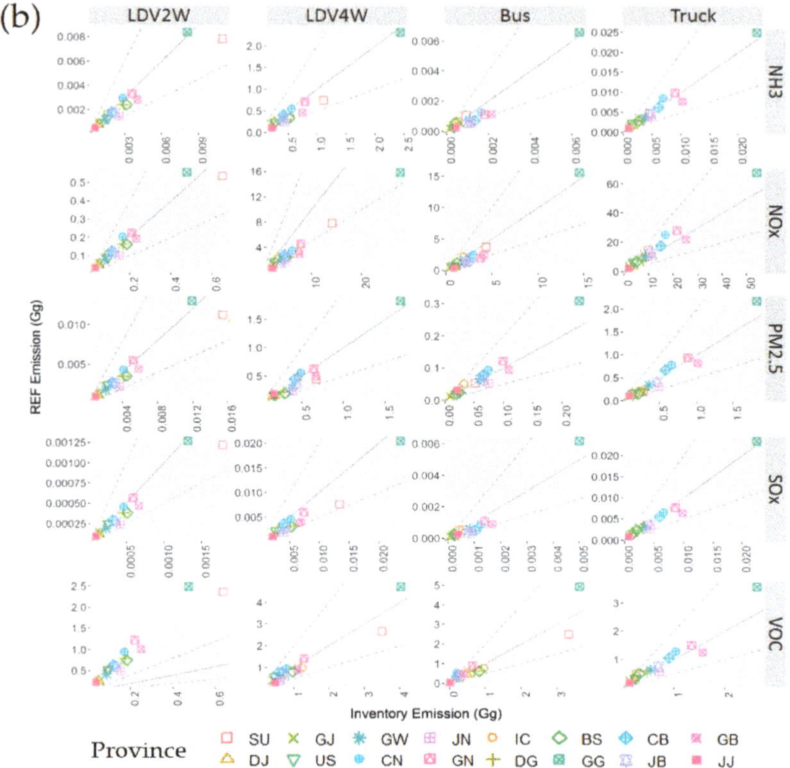

Figure 4. Projected emissions from the baseline (REF) compared to the national emissions inventory. Note: (**a**) Trend of the projected emissions and (**b**) projected emissions by provinces in 2015.

The REF scenario shows that projected emissions in the future are on a downward trajectory because of fuel switching from refined liquids to natural gas (NG), electricity, and hydrogen (see Figure 5b,c). The trends in NO_x and $PM_{2.5}$ emissions for the truck sector show a steeper decline than that for other sectors. The difference in emissions between 2020 and 2040 is 1446 tonnes of $PM_{2.5}$ and 47,050 tonnes of NO_x. The LDV4W sector, a major contributor to NH_3 emissions, is expected to reduce 860 tonnes of NH_3 emissions from 2020 till 2040.

As mentioned earlier, Seoul and Gyeonggi, a populous urban area with the highest number of vehicles [50], are expected to have most of the air pollutant emissions from all road transportation sectors. The truck sector in particular produces large emissions in Gyeonggi. In this province, annual VKT of trucks is the highest among all provinces because of the massive road freight volume due to the manufacture of plastics and synthetic rubber [51].

The second most polluted area is Gyeongsang province (Gyeongbuk and Gyeongnam), since this province is the second most populous province next to the Seoul metropolitan area which accounted for 12% of the whole population of South Korea in 2015. In this province, energy consumption by trucks accounted for 16% of the total truck energy consumption, serving a huge industrial complex in this region.

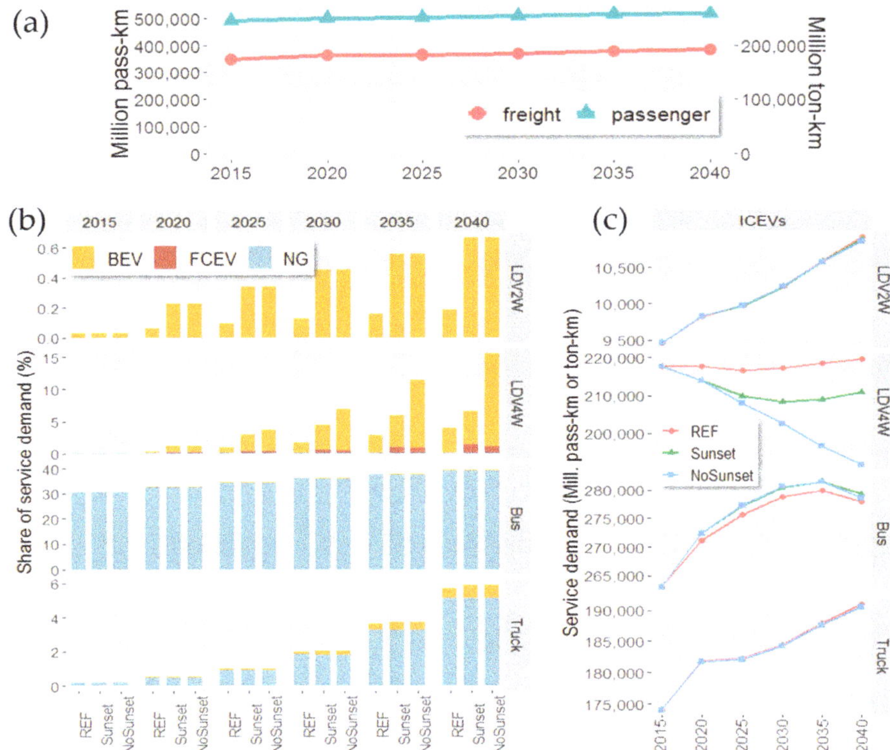

Figure 5. Service demand in the road transportation sector. Note: Total service demand: (**a**), Technology share (**b**), Service demand of ICEVs (**c**). The rest of the percentage of bars in (**b**) is the share of service demand for refined liquid vehicles.

4.2. ZEVs Promotion Using the Subsidy Policy

The ZEV subsidy increases ZEVs' service demand in all the sectors (Figure 5b) while total transportation service demand is kept almost the same (Figure 5a), showing only around 0.1% difference depending on the scenarios. BEVs' service demand noticeably increases in the LDV4W sector, since LDV4W is the main target of subsidy support. In the Sunset scenario, the share of service demand for BEVs and FCEVs is expected to be 2.6% and 0.2% respectively in 2025. In 2040, the share of BEVs and FCEVs increases to 5.3% and 1.2% respectively. REF's share is 0.8% for BEVs and 0.03% for FCEVs in 2040. The share for BEVs rises further to 14.4% in 2040 if the current subsidy is maintained till 2040 (the NoSunset case), while the share of FCEVs starts decreasing to around 1% despite the same amount of subsidy. Even if FCEVs receive the same subsidy, their market entry is disturbed by the introduction of BEVs considering the total service demand, which does not change significantly.

On the other hand, other vehicles excluding LDV4W, show minor effects on service demand change. As NG vehicles dominate service demand in the truck and bus sectors, ZEVs' share is less than 1% even in 2040. Besides, ICEVs' service demand increases in the bus sector with the ZEV subsidy, that is, there are intensive share increases in ZEVs' share in the LDV4W sector leading to an increase in its service demand and average service costs at the same time, while bus service costs become relatively cheaper. The reason for the increase in LDV4W sector's service costs is high-cost technologies (BEVs and FCEVs) being introduced in this sector. In 2040, the relative service cost of the bus sector is 0.80 in the Sunset case and 0.81 in the NoSunset case based on the LDV service cost of 1. By the price response, bus service demand increases by 0.3–0.7% compared to REF and increases further in the Sunset case (Figure 5c).

Demand for electricity and hydrogen increases with the growth of BEVs and FCEVs' service demands. In the Sunset case, electricity demand increases from 9.7PJ (REF case) to 12.2PJ in 2025, and from 16.1PJ (REF case) to 18.8PJ in 2040. In the NoSunset case, it further increases to 13.2PJ in 2025 and 31.3PJ in 2040. In the case of hydrogen demand, while the REF case shows the demand at 0.05PJ even in 2040, demands increases to 0.4PJ in 2025 and 2.3PJ in 2040 in the Sunset case. The NoSunset case shows the demand decreasing rather than increasing as compared to the Sunset case, which is 0.4PJ in 2025 and 1.9PJ in 2040, because of a decrease in FCEVs' service demand. Changes in the prices of electricity and hydrogen are negligible ranging between 0.0% and 0.3% during the period.

Figure 6 shows the estimates of a cumulative number of ZEVs and a comparison with the government's target for ZEV promotion. According to CPFDM, the goal is to have 850,000 BEVs and 150,000 FCEVs by 2024. In the Sunset case, the total number of vehicles is estimated to be approximately 313,000 BEVs and 22,000 FCEVs in 2025. BEVs and FCEVs are expected to be 3 times and 44 times more than the REF case respectively. In the NoSunset case, it is estimated at 399,000 BEVs and 21,000 FCEVs, which is a 22% increase and 3% decrease respectively compared to the Sunset case. But both scenarios fail to achieve the government's target of ZEV promotion.

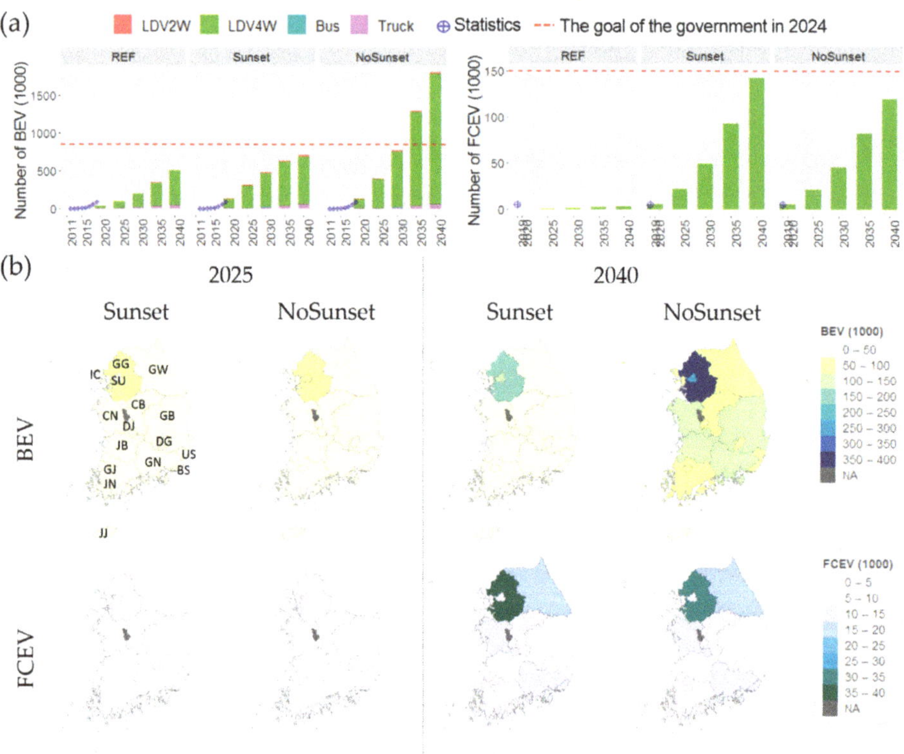

Figure 6. Estimates of the cumulative number of ZEVs by sector (**a**) and by province (**b**).

In the Seoul metropolitan area, the cumulative number of BEVs is estimated at 128,000 in 2025, which accounts for 41% of the total BEVs. In 2040, BEVs in this area are estimated at 307,000 in the Sunset case and 738,000 in the NoSunset case. FCEVs are mostly promoted in provinces where local subsidies are provided and not in the Seoul metropolitan area. Accordingly, Gangwon, where the largest subsidy for FCEVs is provided, is the second most diffused province. In Gangwon, the estimated number of FCEVs is 2900 in 2025 in the Sunset scenario. Chungnam and Gyeongnam follow with 1900 FCEVs each.

Table 5 summarizes the required subsidy for meeting the ZEVs scenarios from 2020 to 2040. It is estimated that total subsidy required during the period will be approximately $9.6 billion in the Sunset case and $24.7 billion in the NoSunset case. Above all, around 90% of the total subsidy spending is concentrated in the LDV4W sector.

Table 5. Required subsidy by scenarios (unit: Million $).

Year	Sunset					NoSunset				
	LDV2W	LDV4W	Bus	Truck	Total	LDV2W	LDV4W	Bus	Truck	Total
2020	10	1604	21	46	1682	10	1604	21	46	1682
2025	6	1991	27	59	2083	6	3484	27	59	3576
2030	6	1678	33	177	1894	6	4854	33	177	5071
2035	7	1662	39	309	2016	7	6894	39	309	7248
2040	7	1472	45	387	1911	7	6715	44	387	7153
Total	37	8406	165	978	9586	37	23,551	164	978	24,730

Note: Required subsidy is calculated as the average purchase subsidy multiplied by the increase in the number of vehicles during the 5-year period.

4.3. Effects of ZEV Promotion on Air Pollution

As shown in Figure 7, most emission reductions are expected from the LDV4W sector because ZEVs' dissemination is mostly expected in this sector. In general, emissions slightly reduce for all pollutants. In the Sunset case, emission reduction rates of NH_3, NO_x, $PM_{2.5}$, SO_x, and VOC are expected to be 3.7% (254 tonnes), 0.5% (1488 tonnes), 1.2% (155 tonnes), 1.5% (2 tonnes), and 0.9% (462 tonnes) respectively in 2040. In the NoSunset case, the NH_3 emission reduction rate is expected to be relatively higher due to the increase in ZEVs' share in the LDV4W sector—the LDV4W sector has high NH_3 emissions. On the other hand, emissions from the bus sector rise for all pollutants compared to REF with an increase in its service demand (Figure 5c). Estimates of $PM_{2.5}$ emission reduction are smaller than autonomously reduced emissions over time without any policy (the REF case).

According to CPFDM, the government has set a target of reducing NO_x, $PM_{2.5}$, SO_x, and VOC emissions in the transportation sector by 65%, 36%, 71%, and 44% of the emissions in 2024 respectively below those in 2016. In case of NH_3, there is no reduction target for the transportation sector. To compare the simulation results, the emission reduction target in the transportation sector is divided into emissions targets for the road transportation sector and the non-road transportation sector according to their proportion in base year 2016. As the simulation results are represented in a 5-year step, projected emissions are linearly interpolated.

Table 6 gives a comparison of the simulation results for emission reduction targets for the road transportation sector. The SO_x emission reduction target can be seen to be intended for the non-road transportation sectors considering the SO_x emissions portion in road transportation (0.6%). The simulation results show that NO_x, $PM_{2.5}$, and VOC emission reduction targets can be achieved as much as 4.0%, 11.5%, and 4.8% respectively in the Sunset case. According to a report released by the National Assembly Budget Office [10], the ZEV subsidy policy does not have a significant impact on reducing $PM_{2.5}$. $PM_{2.5}$ emission reductions by the ZEV subsidy policy accounted for only 3% of the overall emission reductions by $PM_{2.5}$ mitigation measures for the road transportation sector, whereas 76% of the overall budget for them was spent on the ZEVs subsidy in 2018 according to the report.

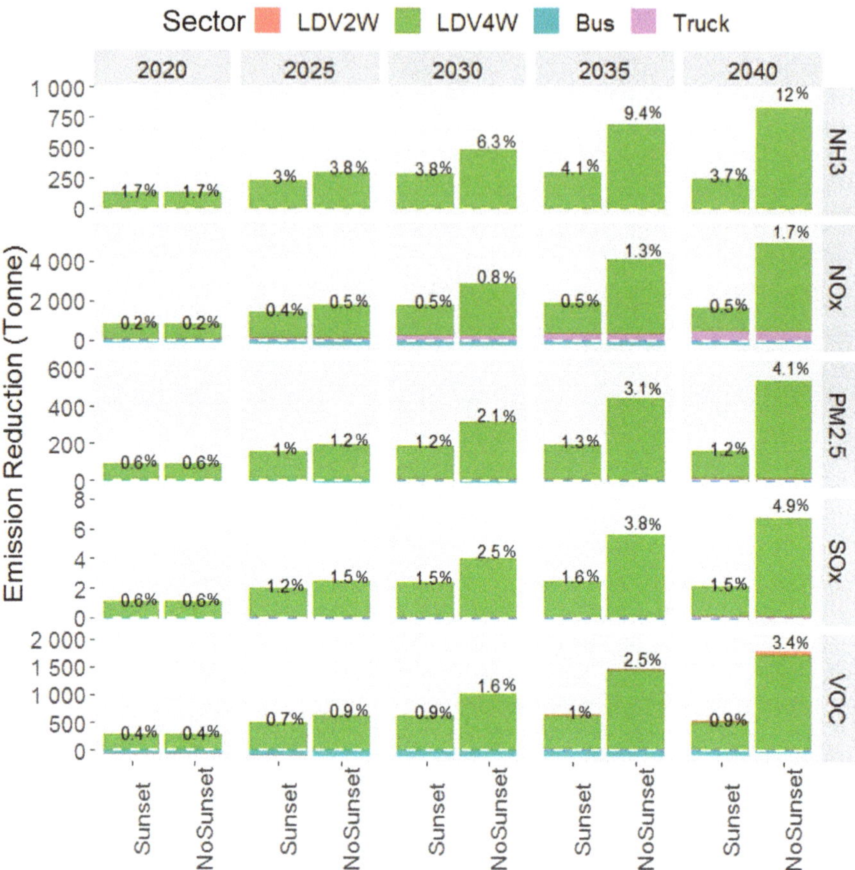

Figure 7. Emission reductions compared to emissions in REF (each scenario minus REF). Note: The values above the bars represent the percentage of emission reductions compared to REF.

Table 6. Expected emission reductions in 2024 compared to 2016.

	Emission Projection in 2016 (Tonne)	Emission Reduction Target (%)	Proportion of Road Transportation's Emissions [1] (%)	Road Transportation's Emission Reduction Target (Tonne)	2024 Expected Emission Reduction (Tonne)(Achievement Rate, %)	
	(A)	(B)	(C)	(A × B × C)	Sunset	NoSunset
NH_3 [2]	7910	-	99.5	-	236 (-)	288 (-)
NO_x	341,056	65	71.9	159,437	6380 (4.0)	6652 (4.2)
$PM_{2.5}$	15,459	36	70.9	3949	456 (11.5)	488 (12.4)
SO_x	164	71	0.6	1	6 (862)	6 (925)
VOC	58,489	44	66.5	17,126	818 (4.8)	916 (5.3)

[1] The base year is 2016.; [2] In CPFDM, the NH_3 emission reduction target was not set for the transportation sector.

Figure 8 illustrates expected emission reductions by province. The pattern of emission reductions is similar to expected ZEVs' dissemination (Figure 6b). For example, the Seoul metropolitan area has the highest transportation activities, showing the biggest emission reductions in all the scenarios. In the Sunset scenario, the emission reductions expected in 2040 are 130 tonnes of NH_3 (51% of national emission reductions); 711 tonnes of NO_x (48%); 68 tonnes of $PM_{2.5}$ (44%); 1 tonne of SO_x (50%);

and 244 tonnes of VOC (53%). Chungnam and Chungbuk, which provide the highest subsidy for BEVs in LDV4W, show the second and third-largest emission reductions, following the Seoul metropolitan area. The expected emission reductions in these two provinces are 40 tonnes of NH_3 (16%), 238 tonnes of NO_x (16%), 32 tonnes of $PM_{2.5}$ (21%), 0.3 tonnes of SO_x (15%), and 57 tonnes of VOC (12%) in the Sunset case.

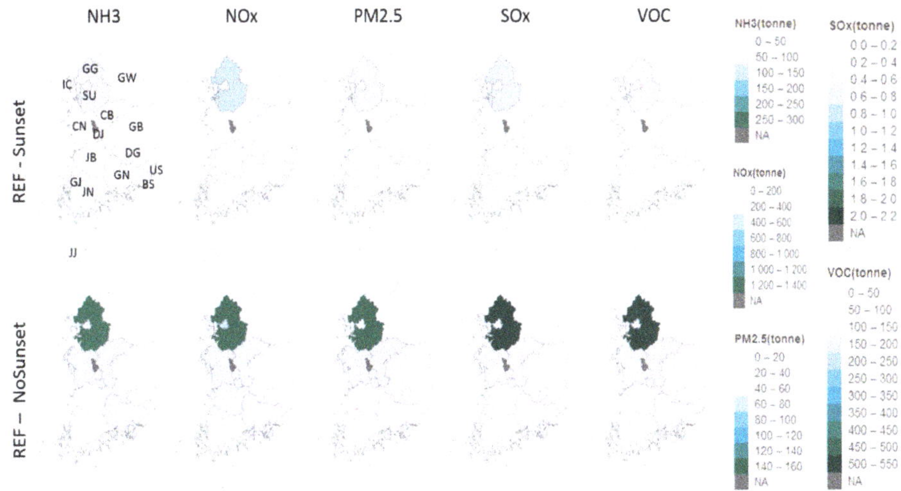

Figure 8. Emission reductions compared to REF in 2040.

5. Conclusions and Policy Implications

This study modeled air pollutant emissions using GCAM-Korea focusing on the road transportation sector. The projected emissions compared to the national emissions inventory using GCAM-Korea works fairly well with empirical data across sectors and provinces except for VOC from LDV2W in which the reported emissions in the emissions inventory contradict energy use.

The study applied the extended GCAM-Korea with air pollutant emissions modeling for examining the ZEV subsidy's effects on emission reductions for $PM_{2.5}$ as well as its precursors. Subsidy scenarios based on the current policy are found to have a major impact on the LDV4W sector in terms of change in service demand and emission reduction, whereas it is expected to have a minor impact on the other sectors. In all the scenarios, the government's target of ZEVs' dissemination is expected to be not attainable. The resulting expected emission reductions of $PM_{2.5}$ are 0.6–1.2% in the Sunset case and 0.6–4.1% in the NoSunset case compared to the baseline. The Seoul metropolitan area contributes 38–44% of the total emission reductions. Chungcheong province is the second most mitigated province next to the Seoul metropolitan area because of the second and third largest subsidy for BEVs in the LDV4W sector, even though this province has relatively low traffic and a small population compared to metropolitan areas. Its emission reduction accounts for 17–21% and 17–20% of the overall emission reductions in the Sunset and the NoSunset cases respectively. NH_3 is the most mitigated pollutant, for which the emission reduction rate is 1.7–3.7% in the Sunset case and 1.7–12% in the NoSunset case. On the other hand, NO_x emissions are expected to reduce very slightly with an emission reduction rate of 0.2–0.5% and 0.2–1.7% in the Sunset and NoSunset cases respectively.

As the ZEVs subsidy is weighted towards the LDV4W sector, as is shown in Table 5, various spillover effects are found: ZEVs' share rises intensively in the LDV4W sector, which leads to an increase in its service costs, while this drives the bus service costs to become relatively cheaper. This whole process, in turn, drives an increase in bus service demand and emissions. In other words, an imbalanced ZEVs subsidy distribution may dampen the subsidy's effect on air pollution improvements. Furthermore, the ZEVs subsidy is not expected to reduce ICEVs in the truck sector, although diesel freight trucks are a major contributor to $PM_{2.5}$ emissions as also NO_x. This means targeting emission reduction by promoting ZEVs might be misleading without explicit consideration of ICEVs in the truck sector. Another finding is that the decline in emissions over time without any policy is more than the ZEV subsidy's effects.

As this analysis does not cover uncertainty in the total costs of ZEVs, this should be considered in a future study. While infrastructure costs increase ZEVs' total costs, incentives for charging station installations and tax incentives for buyers decrease costs. Moreover, total costs can vary under future trends of efficiency and costs. The amount of ZEV purchase subsidy for the future is also uncertain because the government has not decided on this as yet. The uncertainty around cost eventually influences ZEVs' service demand, which changes the effects of the ZEV subsidy policy on air quality mitigation. In addition, emissions caused by increasing electricity and hydrogen consumption for ZEVs should also be considered from the perspective of the entire energy system. Emissions modeling for other sectors such as power generation and industry sectors will be conducted which is expected to provide more meaningful implications for cross-sector and cross-province aspects in the future.

Author Contributions: Conceptualization, M.R., S.J., S.K. (Suduk Kim), S.K. (Soontae Kim), and S.Y.; Methodology, M.R. and S.J.; Software, M.R. and S.J.; Data curation, M.R. and S.J.; Visualization, M.R.; Writing—Original Draft Preparation, M.R.; Writing—Review & Editing, S.K. (Suduk Kim), S.Y., S.K. (Soontae Kim) and A.H.; Supervision, S.K. (Suduk Kim); Funding acquisition, S.K. (Suduk Kim). All authors have read and agreed to the published version of the manuscript.

Funding: This work was supported by the Technology Development Program to Solve Climate Changes of the National Research Foundation (NRF) funded by the Ministry of Science, ICT & Future Planning (NRF-2017M1A2A2081253), and the Korea Ministry of Environment (MOE) as Graduate School specialized in Climate Change.

Acknowledgments: We would like to thank Jaeick Oh, Jaesung Jung, and three anonumous reviewers for their valuable comments.

Appendix A

Table A1. The Mapping of Vehicle Type in the National Inventory to Vehicle Mode in GCAM-Korea.

Classification of National Emissions Inventory		GCAM-Korea
Medium Category	**Small Category**	**Mode**
Passenger car	Compact	Subcompact Car
Passenger car	Small	Subcompact Car
Passenger car	Medium	Compact Car
Passenger car	Large	Large Car
Taxi	Medium	Compact Car
Taxi	Large	Large Car
Van	Compact	Bus
Van	Small	Bus
Van	Medium	Bus
Van	Large	Bus
Van	Special purpose	Bus

Table A1. *Cont.*

Classification of National Emissions Inventory		GCAM-Korea
Medium Category	**Small Category**	**Mode**
Bus	Chartered bus	Bus
Bus	City bus	Bus
Bus	Intercity bus	Bus
Bus	Express bus	Bus
Freight car	Compact	Truck
Freight car	Small	Truck
Freight car	Medium	Truck
Freight car	Large	Truck
Freight car	Special purpose	Truck
Freight car	Dump truck	Truck
Special vehicle (SV)	Recovery vehicle	Truck
Special vehicle (SV)	Wrecker car	Truck
Special vehicle (SV)	Others	Truck
Recreational vehicle (RV)	Small	Light Truck and SUV
Recreational vehicle (RV)	Medium	Light Truck and SUV
Two-wheeled vehicle	Less than 50 cc	Motorcycle
Two-wheeled vehicle	50 cc~99 cc	Motorcycle
Two-wheeled vehicle	100 cc~259 cc	Motorcycle
Two-wheeled vehicle	More than 260 cc	Motorcycle

Appendix B

Table A2. Average Subsidy for Battery Electric Vehicles by Province in 2020 (Unit: Thous.$).

Province	LDV2W	LDV4W				Bus	Truck
	Motor-Cycle	Subcompact	Compact	Large	SUV		
SU	2.1	6.2	10.8	10.7	11.5	74.9	15.5
IC	2.1	6.1	11.9	11.8	12.7	74.9	14.6
DJ	2.1	6.4	13.0	12.8	13.8	74.9	15.2
DG	2.1	5.5	11.3	11.1	12.0	74.9	13.9
GJ	2.1	5.9	11.9	11.8	12.7	74.9	13.7
BS	2.1	6.4	11.3	11.1	12.0	74.9	14.1
US	2.1	6.4	12.1	12.0	12.9	74.9	21.8
GG	2.1	5.9	11.3	11.2	12.0	74.9	15.3
GW	2.1	6.4	12.8	12.7	13.6	74.9	18.7
CB	2.1	8.1	13.8	13.7	14.7	74.9	23.2
CN	2.1	7.0	13.7	13.5	14.6	74.9	20.0
JB	2.1	5.9	14.7	14.5	15.6	74.9	18.4
JN	2.1	6.0	13.6	13.4	14.5	74.9	23.4
GB	2.1	6.4	12.3	12.1	13.1	74.9	19.2
GN	2.1	5.6	12.5	12.3	13.3	74.9	18.7
JJ	2.1	7.3	11.3	11.1	12.0	74.9	15.2

Note: The subsidy is calculated based on information obtained from [52].

Appendix C

Table A3. Average Subsidy for Fuel Cell Electric Vehicles by Province in 2020 (Unit: Thous.$).

| Province | LDV2W | LDV4W | | | | Bus | Truck |
	Motor-Cycle	Subcompact	Compact	Large	SUV		
SU	-	20.5	20.5	20.5	-	136.4	-
IC	-	15.8	29.5	29.5	-	136.4	-
DJ	-	20.5	20.5	20.5	-	136.4	-
DG	-	20.5	20.5	20.5	-	136.4	-
GJ	-	20.5	20.5	20.5	-	136.4	-
BS	-	16.8	31.4	31.4	-	136.4	-
US	-	20.5	20.5	20.5	-	136.4	-
GG	-	15.8	29.5	29.5	-	136.4	-
GW	-	20.6	38.6	38.6	-	136.4	-
CB	-	15.8	29.5	29.5	-	136.4	-
CN	-	16.6	31.1	31.1	-	136.4	-
JB	-	20.5	20.5	20.5	-	136.4	-
JN	-	17.0	31.8	31.8	-	136.4	-
GB	-	20.5	20.5	20.5	-	136.4	-
GN	-	16.1	30.1	30.1	-	136.4	-
JJ	-	20.5	20.5	20.5	-	136.4	-

Note: The subsidy is calculated based on information obtained from [52].

Appendix D

Table A4. Assumptions for an Electric Truck, Electric Bus, and Hydrogen Bus.

| Sector | Bus | | | Truck | |
Fuel Type	Electricity	Hydrogen	CNG	Electricity	Diesel
Fuel intensity (MJ/VKT)	5.3 [1]	12.9 [1]	5.8 [1]	1.2 [2]	1.5 [2]
Purchase cost ($/vehicle)	408,500 [3]	83,000 [4]	168,290 [5]	50,000 [2]	20,000 [6]
VKT [2] (miles/vehicle-year)	34,053	34,053	34,053	13,116	13,116
Lifetime [2] (year)	8	8	8	8	8
Non-energy cost ($/VKT)	0.23 [1]	0.22 [1]	0.26 [1]	0.1 [2]	0.17 [2]
Total cost ($/VKT-year)	2.48	4.78	1.18	0.815	0.456
Relative price	2.1	4.0	1	1.8	1

Note: A 10% the discount rate is applied for calculating the present value of future vehicle purchase costs. [1] Eudy et al. [53]; [2] Dana incorporated [54]; [3] Edison motors [55]; [4] Ministry of Environment [56]; [5] Daewoo bus [57]; and [6] Kia motors [58].

References

1. Trnka, D. *Policies, Regulatory Framework and Enforcement for Air Quality Management: The Case of Korea*; OECD: Paris, France, 2020; pp. 12–22. [CrossRef]

2. World Health Organization (WHO). *Health Effects of Particulate Matter. Policy Implications for Countries in Eastern Europe, Caucasus and Central Asia*; WHO: Copenhagen, Denmark, 2013; pp. 6–7. Available online: https://www.euro.who.int/__data/assets/pdf_file/0006/189051/Health-effects-of-particulate-matter-final-Eng.pdf?ua=1 (accessed on 20 March 2020).

3. Han, C.; Kim, S.; Lim, Y.; Bae, H.; Hong, Y. Spatial and temporal trends of number of deaths attributable to ambient PM2.5 in the Korea. *J. Korean Med. Sci.* **2018**, *33*, e193. [CrossRef]

4. National Institute of Environmental Research (NIER). *Sourcebook of PM2.5 Emission Factor*; NIER: Incheon, Korea, 2014; pp. 1–2. Available online: http://webbook.me.go.kr/DLi-File/NIER/09/019/5568844.pdf (accessed on 15 October 2019).

5. Ministry of Land, Infrastructure and Transport (MOLIT). Status of the Metropolitan Area and Provincial Area. Available online: http://www.index.go.kr/potal/main/EachDtlPageDetail.do?idx_cd=2729 (accessed on 19 March 2020).

6. Ministry of Environment (MOE). *2017 White Paper of Environment*; MOE: Sejong, Korea, 2017; pp. 126–128. Available online: http://webbook.me.go.kr/DLi-File/091/025/006/5636222.pdf (accessed on 18 March 2020).

7. Pan-government. *Comprehensive Plan on Fine Dust Management (2020–2024)*; Pan-government: Korea, 2019. Available online: https://www.me.go.kr/home/file/readDownloadFile.do?fileId=168738&fileSeq=5 (accessed on 21 April 2020).

8. NIER. Air Pollutants Emission Statistics in 2016. Available online: https://airemiss.nier.go.kr/common/downLoad.do?siteId=airemiss&fileSeq=512 (accessed on 17 February 2020).

9. Lutsey, Nic. *Transition to a Global Zero-Emission Vehicle Fleet: A Collaborative Agenda for Governments*; International Council on Clean Transportation: Washington DC, USA, 2015; pp. 1–2. Available online: http://theicct.org/sites/default/files/publications/ICCT_GlobalZEVAlliance_201509.pdf (accessed on 24 May 2020).

10. National Assembly Budget Office (NABO). *Analysis of Projects about Particulate Matter Measure*; NABO: Seoul, Korea, 2019; pp. 17–20. Available online: https://www.nabo.go.kr/system/common/JSPservlet/download.jsp?fCode=33315609&fSHC=&fName=%EB%AF%B8%EC%84%B8%EB%A8%BC%EC%A7%80+%EB%8C%80%EC%9D%91+%EC%82%AC%EC%97%85+%EB%B6%84%EC%84%9D.pdf&fMime=application/pdf&fBid=19&flag=bluene (accessed on 1 April 2020).

11. Ministry of Trand, Industry and Energy (MOTIE). Road Map for Regulatory Preemption of Eco-Friendly Vehicles. Available online: http://www.motie.go.kr/common/download.do?fid=bbs&bbs_cd_n=81&bbs_seq_n=162871&file_seq_n=1 (accessed on 22 July 2020).

12. EV. Support for Charging Station Installation. Available online: https://www.ev.or.kr/portal/buyersGuide/slowInstallSupport?pMENUMST_ID=21628 (accessed on 22 July 2020).

13. Kriegler, E.; Petermann, N.; Krey, V.; Schwanitz, V.J.; Luderer, G.; Ashina, S.; Bosetti, V.; Eom, J.; Kitous, A.; Mejean, A.; et al. Diagnostic indicators for integrated assessment models of climate policy. *Technol. Forecast. Soc. Chang.* **2015**, *90*, 45–61. [CrossRef]

14. Kim, B.; Lee, W.; Jo, H.; Lee, B. Making primary policies for reducing particulate matter. *J. Soc. E-Bus. Stud.* **2020**, *25*, 109–121. [CrossRef]

15. Kim, E.; Kim, H.; Kim, B.; Kim, S. PM2.5 Simulations for the Seoul Metropolitan Area: (VI) Estimating Influence of Sectoral Emissions from Chumngcheongnam-do. *J. Korean Soc. Atmos. Environ.* **2019**, *35*, 226–248. [CrossRef]

16. Kim, H.; Kim, S.; Lee, S.; Kim, B.; Lee, P. Fine-Scale Columnar and Surface NOx Concentrations over South Korea: Comparison of Surface Monitors, TROPOMI, CMAQ and CAPSS Inventory. *Atmosphere* **2020**, *11*, 101. [CrossRef]

17. Kim, O.; Bae, M.; Kim, S. Evaluation on Provincial NO_x and SO_2 Emission in CAPSS 2016 Based on Photochemical Model Simulation. *J. Korean Soc. Atmos. Environ.* **2020**, *36*, 64–82. [CrossRef]

18. Shi, W.; Ou, Y.; Smith, S.J.; Ledna, C.M.; Nolte, C.G.; Loughlin, D.H. Projecting state-level air pollutant emissions using an integrated assessment model: GCAM-USA. *Appl. Energy* **2017**, *208*, 511–521. [CrossRef] [PubMed]

19. Jeon, H.; Lee, H. *Valuing Air Pollution Using the Life Satisfaction Approach*; Korea's Allied Economic Associations Annual Meeting: Sejong, Korea, 2019; pp. 139–154. Available online: http://210.101.116.28/W_files/kiss6/33001315_pv.pdf (accessed on 23 March 2020).

20. Ou, Y.; West, J.J.; Smith, S.J.; Nolte, C.G.; Loughlin, D.H. Air pollution control strategies directly limiting national health damages in the US. *Nat. Commun.* **2020**, *11*, 1–11. [CrossRef] [PubMed]

21. Amann, M.; Johansson, M.; Lükewille, A.; Schöpp, W.; Apsimon, H.; Warren, R.; Warren, R.; Gonzales, T.; Tsyro, S. An integrated assessment model for fine particulate matter in Europe. *Water Air Soil Pollut.* **2001**, *130*, 223–228. [CrossRef]

22. Loughlin, D.H.; Benjey, W.G.; Nolte, C.G. ESP v1. 0: Methodology for exploring emission impacts of future scenarios in the United States. *Geosci. Model Dev.* **2011**, *4*, 287–297. [CrossRef]

23. Klimont, Z.; Kupiainen, K.; Heyes, C.; Purohit, P.; Cofala, J.; Rafaj, P.; Borken-Kleefeld, J.; Schöpp, W. Global anthropogenic emissions of particulate matter including black carbon. *Atmos. Chem. Phys. Discuss.* **2017**, *17*, 8681–8723. [CrossRef]

24. Zhang, S.; Worrell, E.; Crijns-Graus, W. Mapping and modeling multiple benefits of energy efficiency and emission mitigation in China's cement industry at the provincial level. *Appl. Energy* **2015**, *155*, 35–58. [CrossRef]

25. Joint Global Change Research Institute (JGCRI). GCAM v5.2 Documentation: Global Change Assessment Model (GCAM). Available online: http://jgcri.github.io/gcam-doc/index.html (accessed on 25 March 2020).

26. Edmonds, J.; Wise, M.; Pitcher, H.; Richels, R.; Wigley, T.; MacCracken, C. An integrated assessment of climate change and the accelerated introduction of advanced energy technologies-an application of MiniCAM 1.0. *Mitig. Adapt. Strateg. Glob. Chang.* **1997**, *1*, 311–339. [CrossRef]

27. Kim, S.; Edmonds, J.; Lurz, J.; Smith, S.J.; Wise, M. The ObjECTS framework for integrated assessment: Hybrid modeling of transportation. *Energy J.* **2006**, *27*, 63–91. [CrossRef]

28. McFadden, D. Conditional logit analysis of qualitative choice behavior. In *Frontiers in Econometrics*; Zarembka, P., Ed.; Academic Press: New York, NY, USA, 1973; pp. 105–142. Available online: https://eml.berkeley.edu/reprints/mcfadden/zarembka.pdf (accessed on 16 May 2020).

29. Brenkert, A.L.; Smith, A.J.; Kim, S.H.; Pitcher, H.M. *Model. Documentation for the MiniCAM.*; Pacific Northwest National Laboratory: Richland, WA, USA, 2003.

30. Keepin, B.; Wynne, B. Technical analysis of IIASA energy scenarios. *Nature* **1984**, *312*, 691–695. [CrossRef]

31. Loulou, R.; Goldstein, G.; Noble, K. Documentation for the MARKAL Family of Models. October 2004, pp. 65–73. Available online: http://iea-etsap.org/MrklDoc-I_StdMARKAL.pdf (accessed on 16 May 2020).

32. Smith, S.J.; Pitcher, H.; Wigley, T.M.L. Future sulfur dioxide emissions. *Clim. Chang.* **2005**, *73*, 267–318. [CrossRef]

33. Yurnaidi, Z.; Kim, S. Reducing Biomass Utilization in the Ethiopia Energy System: A National Modeling Analysis. *Energies* **2018**, *11*, 1745. [CrossRef]

34. Yu, S.; Tan, Q.; Evans, M.; Kyle, P.; Vu, L.; Patel, P.L. Improving building energy efficiency in India: State-level analysis of building energy efficiency policies. *Energy Policy* **2017**, *110*, 331–341. [CrossRef]

35. Yu, S.; Horing, J.; Liu, Q.; Dahowski, R.; Davidson, C.; Edmonds, J.; Liu, B.; Mcjeon, H.; McLeod, J.; Patel, P.; et al. CCUS in China's mitigation strategy: Insights from integrated assessment modeling. *Int. J. Greenh. Gas Control* **2019**, *84*, 204–218. [CrossRef]

36. Yu, S.; Yarlagadda, B.; Siegel, J.E.; Zhou, S.; Kim, S. The role of nuclear in China's energy future: Insights from integrated assessment. *Energy Policy* **2020**, *139*, 111344. [CrossRef]

37. Jeon, S.; Roh, M.; Oh, J.; Kim, S. Development of an Integrated Assessment Model at Provincial Level: GCAM-Korea. *Energies* **2020**, *13*, 2565. [CrossRef]

38. NIER. *National Air Pollutants Emission in 2016*; NIER: Incheon, Korea, 2018. Available online: https://airemiss.nier.go.kr/user/boardList.do?command=view&page=1&boardId=74&boardSeq=526&id=airemiss_040100000000 (accessed on 17 February 2020).

39. Yeo, S.; Lee, H.; Choi, S.; Seol, S.; Jin, H.; Yoo, C.; Lim, J.; Kim, J. Analysis of the National Air Pollutant Emission Inventory (CAPSS 2015) and the Major Cause of Change in Republic of Korea. *Asian J. Atmos. Environ.* **2019**, *13*, 212–231. [CrossRef]

40. Ahn, S.J.; Kim, L.; Kwon, O. Korea's social dynamics towards power supply and air pollution caused by electric vehicle diffusion. *J. Clean. Prod.* **2018**, *205*, 1042–1068. [CrossRef]

41. NIER. *Improvement of Methodology of Fugitive Dust Estimation and Development of Real-Time Road Fugitive Dust Measurement Method*; Inha University: Incheon, Korea, 2008. Available online: http://www.prism.go.kr/homepage/entire/retrieveEntireDetail.do;jsessionid=FD36972EFA936BA01CFCBEF3889EFBFF.node02?cond_research_name=&cond_research_start_date=&cond_research_end_date=&research_id=1480000-200800517&pageIndex=2833&leftMenuLevel=160 (accessed on 30 March 2020).

42. Han, S.; Youn, J.; Jung, Y. Characterization of PM10 and PM2.5 source profiles for resuspended road dust collected using mobile sampling methodology. *Atmos. Environ.* **2011**, *45*, 3343–3351. [CrossRef]

43. NIER. National Air Pollutants Emission Service. Available online: https://airemiss.nier.go.kr/user/boardList.do?handle=160&siteId=airemiss&id=airemiss_030500000000 (accessed on 17 February 2020).

44. Korea Energy Economics Institute (KEEI). *2017 Energy Consumption Survey*; KEEI: Ulsan, Korea, 2017. Available online: http://www.keei.re.kr/keei/download/YES2017.pdf (accessed on 18 February 2020).

45. Korea Transportation Safety Authority (TS). *2016 Average Mileage of Automobile*; TS: Gyeongsangbuk-do, Korea, 2017; pp. 32–46. Available online: http://stat.molit.go.kr/portal/common/downLoadFile.do (accessed on 17 February 2020).

46. Korea Energy Management Corporation (KEMCO). *Development of Pool of Greenhouse Gas Reduction Means in Building and Transport Sector*; Korea Institute of Civil Engineering and Building Technology (KICT): Ulsan, Korea, 2013; pp. 32–46. Available online: http://www.kemco.or.kr/web/kem_home_new/_common/ac_downFile.asp?f_name=%5B%uCD5C%uC885%uBCF4%uACE0%uC11C%5D%uAC74%uBB3C%uC218%uC1A1%uBD80%uBB38%20%uC628%uC2E4%uAC00%uC2A4%20%uAC10%uCD95%uC218%uB2E8%20Pool%uAD6C%uCD95.pdf&cc=1466 (accessed on 17 February 2020).

47. KEEI. *The 11th Edition (2011) Energy Consumption Survey*; KEEI: Ulsan, Korea, 2012. Available online: http://www.keei.re.kr/keei/download/ECS2011_Revised.pdf (accessed on 17 February 2020).

48. KEEI. *The 12th Edition (2014) Energy Consumption Survey*; KEEI: Ulsan, Korea, 2015. Available online: http://www.keei.re.kr/keei/download/ECS2014.pdf (accessed on 17 February 2020).

49. KEEI. *2017 Energy Consumption Survey*; KEEI: Ulsan, Korea, 2018. Available online: http://www.keei.re.kr/keei/download/ECS2017.pdf (accessed on 17 February 2020).

50. MOLIT. Total Registered Motor Vehicles. Available online: http://stat.molit.go.kr/portal/cate/statFileView.do?hRsId=58&hFormId=1242&hSelectId=1242&hPoint=001&hAppr=1&hDivEng=&oFileName=&rFileName=&midpath=&month_yn=N&sFormId=1242&sStart=202001&sEnd=202001&sStyleNum=809 (accessed on 21 February 2020).

51. Korea Transport Database (KTDB). *Freight in Korea*; KTDB: Gyeonggi-do, Korea, 2012. Available online: https://www.ktdb.go.kr/www/selectPblcteWebView.do?key=40&pblcteNo=27&searchLclasCode=PBL04 (accessed on 6 April 2020).

52. EV. Purchase Subsidies. Available online: https://www.ev.or.kr/portal/localInfo?pMENUMST_ID=21637 (accessed on 10 March 2020).

53. Eudy, L.; Post, M. *Fuel Cell Buses in U.S. Transit Fleets: Current Status 2018*; National Renewable Energy Laboratory (NREL): Golden, CO, USA, 2018. Available online: https://www.nrel.gov/docs/fy19osti/72208.pdf (accessed on 29 February 2020).

54. Dana Incorporated. TCO Calculator. Available online: https://apps.dana.com/commercial-vehicles/tco/ (accessed on 12 March 2020).

55. Edison Motors. eFIBIRD Purchase Cost. Available online: http://www.edisonmotorsev.com/service/order1 (accessed on 12 March 2020).

56. Ministry of Environment (MOE). Press Release; The Pilot Project of a Hydrogen City Bus Will be Undertaken at Six Cities. Available online: http://www.me.go.kr/home/web/board/read.do?boardMasterId=1&boardId=920200&menuId=286 (accessed on 12 March 2020).

57. Daewoo Bus. Bus Price Catalog. Available online: http://www.daewoobus.co.kr/newsite/KR/showroom/BusPrice_20170308.pdf (accessed on 12 March 2020).

58. Kia Motors. Fuel Efficiency and Purchase Cost of Bongo III EV. Available online: https://www.kia.com/kr/shopping-tools/catalog-price.html (accessed on 12 March 2020).

 *energies*

Article

Energy Intensity and Long- and Short-Term Efficiency in US Manufacturing Industry

Oleg Badunenko [1,†] and Subal C. Kumbhakar [2,3,*,†]

1   Department of Economics and Finance, College of Business, Arts and Social Sciences, Brunel University London, Uxbridge UB8 3PH, UK; oleg.badunenko@brunel.ac.uk
2   Department of Economics, State University of New York at Binghamton, Binghamton, NY 13902-6000, USA
3   Department of Innovation, Management and Marketing, University of Stavanger Business School, University of Stavanger, N-4036 Stavanger, Norway
*   Correspondence: kkar@binghamton.edu; Tel.: +1-607-777-4762
†   These authors contributed equally to this work.

Received: 22 June 2020; Accepted: 27 July 2020; Published: 1 August 2020

Abstract: We analyze energy use efficiency of manufacturing industries in US manufacturing over five decades from 1960 to 2011. We apply a 4-component stochastic frontier model, which allows disentangling efficiency into a short- and long-term efficiency as well as accounting for industry heterogeneity. The data come from NBER-CES Manufacturing Industry Database. We find that relative to decade-specific frontiers, the overall efficiency of manufacturing industries, which is a product of transient and persistent efficiencies has deteriorated greatly in the 1970s and rebounded only in the 2000s. The industries are very efficient in the short-term and this has not changed over five decades. The high level of overall inefficiency is almost completely due to the structural inefficiency which can be explained by what is referred to as the "energy paradox". Finally, higher energy-intensive industries perform worse in terms of energy use efficiency than their low energy-intensity counterparts.

Keywords: energy efficiency; energy intensity; stochastic frontier; persistent efficiency; transient efficiency; US manufacturing; energy paradox

---

1. Introduction

According to the U.S. Energy Information Administration, manufacturing industries in the US consume about a third of total energy consumed (see also [1]). (https://www.eia.gov/consumption/) The analysis of energy demand for manufacturing has therefore important implications for energy policy, where energy efficiency and savings is an important agenda (see, e.g., [2]). Additionally, improving energy use efficiency seems to be a natural way to mitigate climate change (see, e.g., [3]). If manufacturing industries are not efficient, it puts strains on the whole economy in general and energy producers and distributors in particular. This is especially true for the energy-intensive industries where energy consumption relative to its output is large.

For a long time, a lot of effort has been made to develop energy-efficient technologies, not only to lessen environmental damage but also to bring down the monetary cost of production. However, Refernece [4] identify and discuss the wide-spread "energy-paradox", whereby energy-efficient technologies, that would have paid-off, are in reality not adopted. The authors of [5] find that the adoption of energy-efficient technologies may be boosted by involving managers, who are in a position close to operations. The existence of the "energy-paradox" may indicate that the industry remains inefficient. Indeed, [6] find that the mean plant-level efficiency in the United States over the time-period 1987–2012 ranges from 33% to 86% for plants in various manufacturing industries.

Such huge inefficiencies are a matter of concern. Enormous financial savings could have been achieved if manufacturing firms were more efficient. The performance of manufacturing in terms of energy use is heterogeneous both in time and cross-sectional dimensions. One common factor influencing energy consumption is the price of energy. The authors of [7] confirm that the biggest determinant of energy intensity is the price of electricity. The cross-sectional variation is further determined by technology, i.e., some industries require more energy than others. The time variation has many determinants. Most important is probably the change in macroeconomic conditions. The beginning of the 1970s was marked by the oil prices, which had a detrimental effect on costs related to energy in manufacturing. It was an expectation that there should have been a surge in adopting the new energy-saving technologies which would eventually improve energy efficiencies. However, [8] finds that the energy consumption did not rebound quickly implying that the response to a decline in real energy prices was slow.

In this paper, we investigate the energy use efficiency of US manufacturing from 1958 to 2011. More specifically, we conduct an analysis at an aggregated level, where the unit of observation is defined as NBER 6-digit NAICS (see [9]). We split the whole time period into five decades and assume a decade-specific technology. We define energy intensity as energy demand per measure of economic activity (see, e.g., [10]). In each decade, we consider 10 percent the most and least energy-intensive industries. Further, following [11], we decompose overall inefficiency into persistent or structural and transient inefficiencies. This has an advantage over for example [3] or [6] since we can identify if efficiency can be improved with relatively small effort, or structural approach is required. We find a significant drop in energy use efficiency in the 1970s, which has probably been caused by the oil crisis. The return to the pre-1970s levels was reached only in the 2000s, which is in agreement with slow rebound estimates of energy consumption (see, e.g., [8]). Remarkably, such low levels of overall energy use efficiency owing to very low levels of structural inefficiency that cannot be managed with ease. This finding goes in unison with the "energy-paradox" (see [4]). Finally, higher energy-intensive industries are characterized by lower levels of energy use efficiency than low-intensive counterparts.

The paper is organized as follows. Section 3 introduces models that are used to measure energy use efficiency and 4-component stochastic frontier model that accounts for heterogeneity and splits overall inefficiency into persistent and transient components. Section 4 describes data and variable construction. Empirical results are presented and discussed in Section 5. Section 6 concludes.

2. Literature Review

Analysis of the energy use efficiency is interesting from both academic and business perspective. More efficient use, especially by energy-intensive industries, would result in lower demand for energy as well as output (see, e.g., [12]). Examining efficiency estimates could also complement accounting for rebound effects ([13]) when making energy consumption forecasts. Energy subsidies could also be inappropriately targeted to support highly inefficient producers if inefficiency measurement is improper (see, e.g., [14]). Here we provide a brief review of methods used in measuring energy and technical or cost efficiency.

Depending on the available data, measurement of technical efficiency can be done by using either stochastic frontier (SF) methods (see [15]) or data envelopment analysis (DEA) approach (see [16]). For a cross-sectional data with fewer observations, one can opt for DEA to estimate the benchmark and then measure inefficiency as a deviation from the benchmark. SF in contrast defines the benchmark accommodating stochastic noise and decomposes the composed error (sum of noise and inefficiency) into inefficiency and statistical noise. The noise can be both positive and negative and can be seen as positive and negative shocks to the production process. In the panel-data context, there are different possibilities to decompose the composed error term. One way is to allow inefficiency to be persistent and hence time-invariant. This approach is referred to as the first-generation panel-data SF modeling. The second-generation SF models assume that the inefficiency is time-varying. The first and second-generation models assume an error term (the deviation from the frontier) that has two components. Applying DEA to a panel data would be comparable to a second-generation SF model, which would produce time-varying efficiency estimates without accounting for possible noise. The third-generation SF model considers an error term with three components. The two components are time and firm-specific, i.e., statistical noise and time-varying inefficiency. The third component is time constant. The authors of [17–20] propose to treat it as time-invariant inefficiency. The authors of [21] assume it is an individual effect or firm heterogeneity. Thus, Kumbhakar and co-authors model two types of inefficiency (persistent and transient) ignoring heterogeneity, while Greene models transient inefficiency and heterogeneity ignoring persistent inefficiency. The fourth-generation class of SF models is originally introduced by [22] and accounts for both types of inefficiency as well as heterogeneity. Incidentally, the fourth-generation SF models are also known as the 4-component SF models.

Traditionally, energy efficiency measurement is contemplated in terms of energy intensity. However, it is argued that other measures should also be considered, for example DEA ([23]). This was one of the first studies to consider the production theory framework as a base for energy efficiency measurement. The authors of [23] employ DEA for the manufacturing sector constructed by the U.S. Bureau of Labor Statistics (BLS). She finds quite high efficiency scores for aggregate manufacturing for the 1970–2001 time period. Recall, however, that DEA does not account for heterogeneity or persistent efficiency akin to the second generation SF models, which can be seen as a disadvantage of using DEA. Furthermore, she finds higher efficiency scores towards the end of the sample. But because she used an intertemporal frontier approach, she could not distinguish whether this is attributed to technical progress or not. This can be viewed as the second disadvantage of using DEA when panel data are available. Many other studies have used DEA to analyze energy efficiency. The authors of [24], for example, investigate the energy efficiency of the Indian manufacturing sector for the 1998–2004 time period. The authors of [25] apply DEA to measure economy-wide energy efficiency using aggregated data on the OECD countries. The authors of [26] investigate energy use efficiency of canola production in Iran. See the review of [27] for other studies that employ DEA.

SFA has also been used to measure energy efficiency and efficiency in the energy sector. The authors of [28,29] were the first to advocate using SFA to estimate efficiency in manufacturing sectors. However, he did not go beyond a cross-sectional analysis. The authors of [30] use the second-generation SF model to measure energy efficiency of different states in the US residential sector. The authors of [31] investigate energy efficiency in the automotive manufacturing sector using plant-level data. The authors again use the second generation model. The authors of [32] are the first to use the third generation SF model to analyze the efficiency of the Swiss electricity distribution sector. The authors of [33] used the fourth-generation model to aggregate frontier energy demand model and estimate economy-wide persistent and transient energy efficiency in the US. The authors confirm the findings and arguments of [23] that energy intensity is not a good indicator of energy efficiency. The authors of [33] as well as [34] emphasize the importance of accounting for heterogeneity as well as estimating two types of inefficiency. This is the approach, which we apply for the first time to this type of data using three different models. Our models are described in the next section.

3. Methodology

3.1. Models

In this paper, we apply three different models to investigate energy use efficiency. In all models, we assume that the production technology consists of one output Y and a vector of four inputs $X = (L, K, NEM, E)$, where L is the labor, K is the capital stock, NEM is the non-energy materials, and E is the energy. The production technology using multiple outputs (transformation function), can be written, in implicit form as,

$$\mathcal{A}\mathcal{F}(Y, X) = 1. \tag{1}$$

If the manufacturing process does not experience production shocks, $\mathcal{A} = 1$, and $\mathcal{F}(Y, X) = 1$. However, since both positive and negative shows hit the production, the transformation function is made stochastic by setting $\mathcal{A} = \exp(v)$, v can be both positive and negative. Besides, if inputs are not used with 100% efficiency, the transformation function in (1) can be expressed as

$$\mathcal{A}\mathcal{F}(Y, \theta X; \beta) = 1, \tag{2}$$

where $\theta < 1$ is the input technical efficiency (defined as the ratio of minimum of each input required and actual amount used) and β is the set of the technology parameters of the function $\mathcal{F}$. Since the transformation function is homogeneous of degree 1 in inputs (see [35]), so we can rewrite (2) as

$$\mathcal{A}\mathcal{F}(Y, \lambda\theta X; \beta) = \lambda, \quad \lambda > 0. \tag{3}$$

Further, we can set $\lambda = (E\theta)^{-1}$, where E is the energy input. Note that any other input could have been chosen to be in place of E. Then (3) becomes

$$X_1^{-1}\theta^{-1} = f(Y, \tilde{X}_{-E}; \beta) \exp v, \tag{4}$$

where $\tilde{X}_{-E} = (L/E, K/E, NEM/E)$. Taking logs of both sides of (4) and denoting $u = -\log\theta \geq 0$, we obtain (Model 1)

$$-\log E = \log f(Y, \tilde{X}_{-E}; \beta) + v - u. \tag{5}$$

The stochastic frontier (SF) formulation in (5) is known as the input distance function formulation, where u is input oriented inefficiency, which measures percentage (when multiplied by 100) over-use of all the inputs. For small values of u, $e^{-u} \approx 1 - u$. That is, technical efficiency is 1 minus technical inefficiency. It is important to keep this relationship in mind because we switch from one to the other quite frequently. Technical efficiency in this model refers to the efficiency of all inputs including energy. That is, in this model, inefficiency, u, is interpreted as over-use of all the all inputs, including energy, at the same rate. The other two models focus exclusively on energy-use efficiency. Before we explain how u can be estimated, we introduce two other approaches.

The transformation function can also be written as a factor requirement function (see, e.g., [36]). Since the focus is on energy use, we can express the technology in terms of E, and write it as,

$$E = G(Y, X_{-E}), \tag{6}$$

where $X_{-E} = (L, K, NEM)$. Again, assuming that both positive and negative shocks v' can influence energy requirement and positing that energy is not used 100% efficiently used, we can rewrite (6) as

$$E = g(Y, X_{-E}; \gamma) \exp v' \exp u', \tag{7}$$

where γ is the vector of parameters of the energy requirement function, v' is a symmetric error term and u' is the energy use inefficiency. Taking to logs of both sides of the (7) gives us the energy requirement function with inefficiency, viz., (Model 2)

$$\log E = \log g(Y, X_{-E}; \gamma) + v' + u'. \tag{8}$$

This approach was, for example, applied by [6,29] to plant-level data using the second-generation SF model. Note that (8) has a stochastic cost function type formulation. Any inefficiency in the use of energy will increase cost.

Finally, in our last model we recognize endogeneity of output Y. That is, we assume profit maximizing behavior to derive the energy demand function

$$E = H(w, X_{-E}), \tag{9}$$

where $w = w_E/p$, w_E is the energy price and p is the output price. Similar to the factor requirement function, we can obtain energy use inefficiency from the demand function (Model 3)

$$\log E = \log h(w, \mathbf{X}_{-E}; \delta) + v'' + u'', \tag{10}$$

where δ is the vector of parameters of the energy demand function, v'' is a symmetric error term and u'' is the energy use inefficiency.

The difference between (5) and (10) is that in the latter energy input is chosen optimally by maximizing profit. In (5) energy overuse treats all other inputs as given. That is, inefficiency in this model shows by how much energy is overused to produce a given level of output and all other inputs. On the other hand, inefficiency in (10) comes from excess use of energy when all other inputs and output are chosen optimally instead of taking them as exogenously given. From econometric estimation point of view this means Y and $\mathbf{X}_{-E}$ are exogenous in Model 2, whereas they are endogenous in Model 3.

In the next sub-section we examine all three models in more detail in the light of panel stochastic frontier framework. In particular, we add firm-heterogeneity and decompose inefficiency into persistent and transient components.

3.2. Stochastic Frontier Approach with Panel Data

The stochastic production frontier function approach was introduced for cross-sectional data independently by [37,38]. This is expressed as

$$\log q_i = r(\mathbf{X}_i; \boldsymbol{\omega}) + v_i - u_i, \tag{11}$$

where $r(\cdot)$ is the technology (namely, the production function in logarithmic form), q_i is an output, $\mathbf{X}_i$ is a vector of inputs (in log) for a production unit i, $\boldsymbol{\omega}$ is a vector of parameters that define the technology, v_i is the usual error/noise term, and $u_i \geq 0$ is the inefficiency. In this model, the data are cross-sectional and hence error components v_i and u_i represents cross-sectional shocks to the production and production unit-specific inefficiency. When panel data are available, shocks and inefficiency can be both time-constant and time-varying. The authors of [22,39,40] were first to recognize this and formulated the following 4-component stochastic frontier model for panel data. We use this framework for our Model 1, and write it as:

$$\log q_{it} = r(\mathbf{X}_{it}, trend; \boldsymbol{\omega}) + v_{0i} - u_{0i} + v_{it} - u_{it}, \tag{12}$$

where t is a time period in which a production unit i is observed. In (12) we have two additional terms compared to (11). More specifically, v_{it} is the usual symmetric error term, v_{0i} is an individual (production unit) effect also known to represent individual production shock (or heterogeneity), $u_{0i} \geq 0$ is the persistent or structural time-invariant inefficiency, and finally $u_{it} \geq 0$ is the transient or short-term time-varying inefficiency. Thus, the overall inefficiency is the sum of persistent and transient inefficiency and overall efficiency $TE^{overall}$ is decomposed into persistent $TE^{persistent}$ and transient $TE^{transient}$, i.e.,

$$TE^{overall} = TE^{persistent} \times TE^{transient} \tag{13}$$

Note that persistent and transient efficiency ($TE^{persistent}$ and $TE^{transient}$) are defined as $e^{-u_{0i}}$ and $e^{-u_{it}}$, respectively. The originally proposed model assumed all 4 components to be random and homoskedastic. This model did not include the determinants of inefficiency. In our analysis, we will use the [11] model that introduces determinants of both types of inefficiency in (12).

To estimate parameters $\boldsymbol{\omega}$ in (12), we assume that $v_{it} \sim \mathcal{N}(0, \sigma_{v_{it}})$, $v_{0i} \sim \mathcal{N}(0, \sigma_{v_{0i}})$, $u_{it} \sim \mathcal{N}^+(0, \sigma_{u_{it}})$, and $u_{0i} \sim \mathcal{N}^+(0, \sigma_{u_{0i}})$, where $\mathcal{N}^+$ means the positive part of the zero mean normal distribution, making u_{it} and u_{0i} half-normally distributed. We assume that both noise v_{it} and individual effects v_{0i} are homoskedastic, so that $\sigma_{v_{it}} = \sigma_v$ and $\sigma_{v_{0i}} = \sigma_{v_0}$. We introduce determinants of time-varying inefficiency via the pre-truncated variance of u_{it}. More specifically, we assume

$$\sigma_{u_{it}}^2 = \exp\left(z_{u_{it}} \boldsymbol{\psi}_u\right), \quad i = 1, \cdots, n, \quad t = 1, \cdots, T_i, \tag{14}$$

where $z_{u_{it}}$ denotes the vector of covariates that explain time-varying inefficiency. Since u_{it} is half-normal, $E(u_{it}) = \sqrt{(2/\pi)}\, \sigma_{u_{it}} = \sqrt{(2/\pi)} \exp\left(\frac{1}{2} z_{u_{it}} \boldsymbol{\psi}_u\right)$, and therefore, anything that affects $\sigma_{u_{it}}$ also affects time-varying inefficiency. The determinants of persistent inefficiency can be modeled similarly. However, because the data-set does not provide natural determinants of the persistent inefficiency, we leave it homoskedastic, i.e., $\sigma_{u_{0i}} = \sigma_{u_0}$.

The parameters ω, as well as variances of the 4 components and their determinants, can be estimated by the single stage maximum simulated likelihood (MSL) method (see Appendix B and [11] for details of the estimation procedure). We follow [39] to calculate the persistent and transient efficiencies. The overall efficiency is then calculated as the product of the persistent and transient efficiencies.

We add firm-heterogeneity and decompose inefficiency into persistent and transient inefficiency in the same way as in Model 1, for both Models 2 and 3, which are outlined in (8) and (10). After adding these components, the models will look quite similar to (12) mathematically. Because of this, we skip the details and avoid repetitions. However, note that the interpretation of inefficiency in these models are different. In Model 2 inefficiency refers to overuse of energy, given everything else. Consequently, persistent and transient inefficiency in Model 2 decompose energy overuse into a time-invariant and a time-varying components, *ceteris paribus*. Similar to Model 2, inefficiency in Model 3 described in (10) after adding firm heterogeneity and persistent inefficiency is specifically related to energy overuse. But it does not take other inputs as given, which is what Model 2 does. In Model 3 inputs are chosen optimally, and inefficiency in production is transmitted to overuse of inputs via demand for energy. That is, we focus only on energy by examining the energy demand function.

4. Data

The source of the data we use in this paper is NBER-CES Manufacturing Industry Database, which can be accessed at http://www.nber.org/nberces/. It covers 473 six-digit 1997 NAICS manufacturing industries over 1958–2011. We split our analysis into five decades: 1958–1969 (labeled "the 1960s"), 1970–1979 (labeled "the 1970s"), 1980–1989 (labeled "the 1980s"), 1990–1999 (labeled "the 1990s"), and 2000–2011 (labeled "the 2000s").

The output Y of an industry is calculated as the difference between the value of industry shipments, which are based on net sales, after discounts and allowances, and the change in end-of-year inventories. The labor L is calculated as $PRODH * PAY/PRODW$, where $PRODH$ is the number of production worker hours, PAY is the total payroll, and $PRODW$ is production workers' wages. Capital stock K is obtained as the sum of real equipment and real structures. Energy E is the expenditure on purchased fuels and electrical energy. The cost of overall materials $MATCOST$ in the database includes delivered cost of raw materials, parts, and supplies put into production or used for repair and maintenance and purchased electric energy and fuels consumed for heat and power and contract work done by others for the plant. The cost excludes the costs of services used, overhead costs, or expenditures related to plant expansion. Because the overall cost of materials includes energy, the non-energy materials, NEM are determined as the difference between overall materials and E. See [9] for more details.

The paper analyzes the differences in energy use efficiency between industries that use relatively little and a lot of energy in their production. We define energy intensity $EN\_INTENSITY$ as the ratio of the expenditures on purchased fuels and electrical energy E and the value of industry shipments $VSHIP$, which is the energy cost per unit of sales. The authors of [10], for example, define energy intensity as energy consumption divided by a measure of economic activity. Alternatively, one can define energy intensity as the cost of energy in total costs. We have tried this approach and the correlation coefficient between these two measures of energy intensity was 0.98. So either of them could be used.

Table A1 shows the summary statistics for output and four inputs for 10 percent of the top and bottom energy-intensive manufacturing industries in the respective decade. The criterion to include an industry is that data on it is available for at least 4 years in a decade.

5. Empirical Results

5.1. Change in Energy Intensity of Industries

First, we analyze how energy intensity has evolved in US manufacturing for over the five decades. We concentrate on the top and bottom 10% of the industries in terms of their energy intensity. More specifically, we calculate the 10th and 90th percentile of energy intensity in the 1960s, then we consider industries whose energy intensity is smaller than the 10th percentile and larger than the 90th percentile in the 1960s. Of these industries, we consider only those for which data are available for a period of at least 4 years. Then we repeat this exercise for the other four decades. Table 1 gives a summary statistics of the energy intensity for all industries for the period 1958–2011 as well as by decade and by energy intensity. As we can see, there are industries for which the energy use is negligible. However, some industries consume quite a lot of energy in the production process. All parts of the distribution were increasing up to the 1990s and then started declining.

Table 1. Descriptive statistics of energy intensity by decade and by the intensity of energy use.

Time Period	Industries	Median	Mean	SD	Min	Max
All	Both most and least energy-intensive	0.0417	0.0489	0.0565	0.00008	0.3414
All	Least energy-intensive	0.0045	0.0046	0.0015	0.00008	0.0082
All	Most energy-intensive	0.0690	0.0873	0.0527	0.03169	0.3414
The 1960s	Least energy-intensive	0.0035	0.0033	0.0011	0.00008	0.0046
The 1960s	Most energy-intensive	0.0489	0.0627	0.0333	0.03169	0.2006
The 1970s	Least energy-intensive	0.0049	0.0048	0.0009	0.00061	0.0061
The 1970s	Most energy-intensive	0.0706	0.0881	0.0513	0.04258	0.3311
The 1980s	Least energy-intensive	0.0068	0.0065	0.0012	0.00304	0.0082
The 1980s	Most energy-intensive	0.0850	0.1128	0.0639	0.05426	0.3414
The 1990s	Least energy-intensive	0.0047	0.0046	0.0010	0.00121	0.0061
The 1990s	Most energy-intensive	0.0645	0.0819	0.0477	0.04202	0.2775
The 2000s	Least energy-intensive	0.0042	0.0042	0.0011	0.00108	0.0058
The 2000s	Most energy-intensive	0.0711	0.0921	0.0513	0.04630	0.2709

It can be seen that in each decade, the 10th and 90th percentiles are specific for the decade. The industries that satisfy the above procedure are shown in Figures 1 and 2. The red decade-specific horizontal lines show 10th and 90th percentiles for low and high energy-intensive industries, respectively. The bold green solid line shows the mean of energy intensity for these industries.

One conclusion that we can draw from Figures 1 and 2 and Table 1 is that the energy intensity has a shape that is closer to a parabola than a flat line. Whether we are looking at the 10th or the 90th percentile, the energy intensity has been increasing from 1960s through the 1980s and then started to fall in the 1990s and then stalled through the 2000s. One possible explanations can be that energy was abundant and relatively cheap up until 1990s when manufacturers started to consider better and more energy-saving technologies.

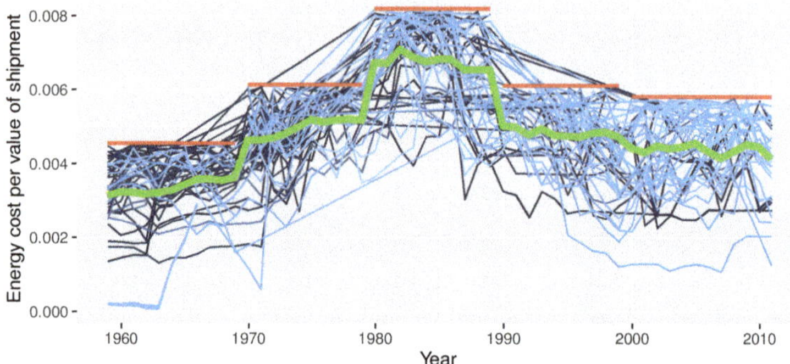

Figure 1. Energy intensity of industries. Shown are industries whose energy intensity are lower than the 10th percentile in a respective decade. Notes: Horizontal red lines show the 10th percentile of energy intensity in a respective decade. The bold green solid line shows the mean of energy intensity for industries whose energy intensity are lower than the 10th percentile in a respective decade.

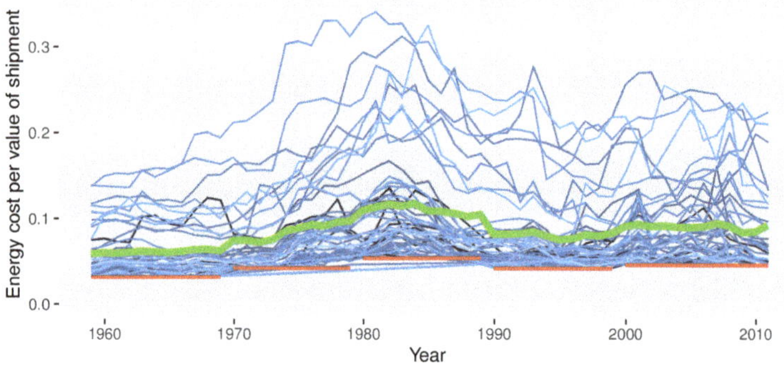

Figure 2. Energy intensity of industries. Shown are industries whose energy intensity exceeds the 90th percentile in a respective decade. Notes: Horizontal red lines show the 90th percentile of energy intensity in a respective decade. The bold green solid line shows the mean of energy intensity for industries whose energy intensity exceeds the 90th percentile in a respective decade.

5.2. Energy Use Efficiency

In this section, we present the results from three models that are presented in Equations (3), (8) and (10). In all three models, the transient inefficiency is modeled to follow either linear or quadratic trend, that is $\sigma_{u_{it}}$ is a function of time in (14). Further, in all three models, we used a translog (log quadratic) specification for the underlying technology. The first model considers energy use inefficiency via an input distance function (IDF). Since inefficiency is radial in the IDF formulation in (3), the energy use efficiency is the same as the efficiency in the use of all other inputs. In the latter two models, inefficiency comes from energy use alone. The difference is that in (10), output can be endogenous, and manufacturing firms are assumed to be profit-maximizing.

The results from models 1, 2, and 3 by decade are presented in Tables 2–4. We observe that in all these models, with an exception of the model 3 for the 1990s, all 4 components are statistically significant and thus use of the [11] model is justified. So, the conclusion about appropriateness of using the 4-component model is in line with [33,34]. This means that models that account for only two components such as [41–43], or three components such as [21] or [18–20] are misspecified and likely to produce wrong results on efficiency. For the 1990s, model 3 could have been estimated using [20] approach.

Table 2. Model 1 as in Equation (3). Dependent variable is $-\log E$. z-values in parentheses.

Parameter	The 1960s		The 1970s		The 1980s		The 1990s		The 2000s	
Intercept	1.430	(24.95)	1.094	(6.44)	1.635	(7.17)	1.109	(6.55)	2.459	(16.77)
$0.5 * \log(K/E)^2$	0.020	(1.51)	0.019	(0.39)	0.028	(0.24)	0.091	(3.14)	-0.031	(-1.37)
$0.5 * \log(L/E)^2$	0.107	(9.48)	-0.041	(-3.25)	0.092	(4.83)	-0.031	(-1.77)	0.072	(2.69)
$0.5 * \log(NEM/E)^2$	0.189	(17.96)	0.124	(6.50)	0.187	(6.92)	0.160	(6.52)	0.071	(2.56)
$0.5 * \log(Y)^2$	0.002	(0.41)	-0.017	(-0.95)	0.044	(14.23)	0.043	(21.12)	0.054	(29.65)
$0.5 * \text{Trend}^2$	0.003	(4.33)	-5.8×10^{-4}	(-0.45)	0.004	(3.40)	0.003	(2.72)	0.003	(4.58)
$\log(K/E)$	-0.082	(-2.68)	0.118	(0.92)	-0.142	(-1.21)	0.026	(0.32)	0.023	(0.32)
$\log(K/E) * \log(L/E)$	0.039	(4.53)	0.061	(6.54)	0.002	(0.08)	-0.031	(-2.72)	-0.052	(-2.78)
$\log(K/E) * \log(NEM/E)$	-0.037	(-3.89)	-0.057	(-2.24)	-0.006	(-0.10)	-0.064	(-2.88)	0.058	(2.45)
$\log(K/E) * \log(Y)$	0.020	(3.53)	0.002	(0.29)	0.042	(2.29)	0.016	(3.78)	0.023	(2.52)
$\log(K/E) * \text{Trend}$	0.005	(2.83)	0.006	(1.12)	0.005	(1.59)	0.004	(1.93)	0.002	(1.32)
$\log(L/E)$	0.493	(17.03)	0.418	(4.42)	0.501	(6.96)	0.862	(11.33)	0.903	(10.50)
$\log(L/E) * \log(NEM/E)$	-0.148	(-10.56)	-0.032	(-2.54)	-0.137	(-4.89)	-0.016	(-1.08)	-0.059	(-2.67)
$\log(L/E) * \log(Y)$	-0.011	(-1.26)	-0.030	(-1.42)	0.024	(1.63)	-0.051	(-6.52)	-0.032	(-7.93)
$\log(L/E) * \text{Trend}$	0.006	(5.75)	-0.002	(-0.42)	-0.001	(-0.84)	0.002	(0.98)	-0.006	(-2.65)
$\log(NEM/E)$	0.431	(23.41)	0.242	(8.68)	0.363	(4.87)	0.331	(5.19)	0.061	(1.06)
$\log(NEM/E) * \log(Y)$	-0.018	(-2.14)	0.018	(1.00)	-0.063	(-3.79)	-0.028	(-6.73)	-0.023	(-2.72)
$\log(NEM/E) * \text{Trend}$	-0.007	(-6.50)	-0.005	(-1.01)	-0.005	(-2.51)	-0.007	(-3.93)	0.003	(1.62)
$\log(Y)$	-0.932	(-76.70)	-0.772	(-10.27)	-1.014	(-37.43)	-0.991	(-65.69)	-1.059	(-60.64)
$\log(Y) * \text{Trend}$	-1.1×10^{-4}	(-0.16)	-0.001	(-0.70)	-0.001	(-0.50)	-0.001	(-1.49)	-0.005	(-4.01)
Trend	-0.017	(-2.39)	0.017	(1.14)	0.006	(0.28)	0.020	(2.40)	0.012	(0.81)
Random effects component: $\log \sigma_{v0i}^2$										
lnVARv0iIntercept	-3.820	(-95.00)	-2.548	(-15.71)	-2.978	(-12.85)	-2.066	(-65.46)	-3.204	(-67.88)
Persistent inefficiency component: $\log \sigma_{u0i}^2$										
lnVARu0iIntercept	-4.570	(-56.15)	-3.936	(-5.12)	-2.197	(-21.74)	-3.943	(-44.12)	-0.764	(-23.66)
Random noise component: $\log \sigma_{vit}^2$										
lnVARvitIntercept	-6.571	(-97.67)	-6.445	(-49.51)	-6.455	(-47.84)	-6.379	(-44.86)	-7.431	(-44.92)
Transient inefficiency component: $\log \sigma_{uit}^2$										
lnVARuitIntercept	-4.548	(-4.04)	-7.251	(-4.62)	-3.859	(-8.03)	-5.909	(-3.74)	-3.658	(-13.88)
lnVARuitTrend	-1.372	(-3.10)	0.462	(0.80)	-1.469	(-5.51)	-0.931	(-2.12)	-0.634	(-6.76)
lnVARuitTrend^2	0.098	(2.91)	-0.067	(-1.17)	0.139	(5.54)	0.104	(3.09)	0.056	(7.84)
Sample Size										
N	105		113		104		98		109	
$\sum_{i=1}^{N} T_i$	968		850		835		817		1005	
Sim. logL	1497.73		1144.89		1055.05		1021.78		1067.50	

Table 3. Model 2 as in Equation (8). Dependent variable is log E. z-values in parentheses.

Parameter	The 1960s		The 1970s		The 1980s		The 1990s		The 2000s	
Intercept	−8.488	(−21.97)	−4.999	(−8.78)	−4.001	(−42.61)	−7.573	(−33.18)	−7.973	(−26.42)
$0.5 * \log(K)^2$	0.019	(0.70)	0.131	(1.90)	0.057	(2.29)	−0.038	(−1.79)	−0.130	(−5.22)
$0.5 * \log(L)^2$	−0.782	(−15.67)	−0.087	(−3.52)	0.106	(6.43)	0.047	(1.26)	−0.049	(−1.44)
$0.5 * \log(NEM)^2$	−0.486	(−5.73)	0.258	(6.97)	0.127	(4.43)	0.791	(10.17)	−0.256	(−15.50)
$0.5 * \log(Y)^2$	0.849	(6.90)	0.353	(7.99)	−0.310	(−6.63)	0.300	(5.12)	−0.678	(−24.19)
$0.5 * Trend^2$	-7.2×10^{-4}	(−0.30)	−0.009	(−3.52)	−0.027	(−15.93)	−0.006	(−3.63)	−0.017	(−11.00)
$\log(K)$	1.504	(26.62)	−0.552	(−2.51)	−0.270	(−3.07)	−0.014	(−0.20)	0.129	(1.03)
$\log(K) * \log(L)$	0.349	(9.52)	−0.038	(−0.97)	−0.130	(−4.60)	−0.055	(−1.17)	−0.007	(−0.16)
$\log(K) * \log(NEM)$	−0.344	(−10.11)	0.198	(2.41)	−0.149	(−10.31)	0.068	(1.75)	−0.124	(−3.52)
$\log(K) * \log(Y)$	−0.158	(−9.43)	−0.150	(−4.94)	0.257	(21.41)	0.056	(2.08)	0.275	(16.63)
$\log(K) * Trend$	0.023	(7.19)	0.008	(1.50)	−0.020	(−6.35)	0.004	(1.15)	0.006	(1.55)
$\log(L)$	1.022	(11.95)	−0.138	(−1.37)	0.464	(7.54)	0.448	(4.34)	−0.114	(−2.11)
$\log(L) * \log(NEM)$	0.982	(9.05)	−0.080	(−1.22)	0.007	(0.19)	−0.413	(−4.51)	-8.7×10^{-5}	(-3.2×10^{-3})
$\log(L) * \log(Y)$	−0.758	(−7.77)	0.167	(1.90)	0.017	(0.36)	0.400	(4.23)	0.065	(1.25)
$\log(L) * Trend$	−0.023	(−2.89)	−0.006	(−0.73)	0.003	(0.76)	−0.007	(−1.45)	-2.1×10^{-4}	(−0.04)
$\log(NEM)$	0.763	(5.20)	0.148	(0.64)	0.527	(4.98)	0.792	(8.17)	0.005	(0.04)
$\log(NEM) * \log(Y)$	−0.052	(−1.69)	−0.405	(−8.70)	0.007	(1.01)	−0.684	(−66.10)	0.339	(29.12)
$\log(NEM) * Trend$	0.039	(3.99)	0.001	(0.11)	−0.019	(−2.58)	0.002	(0.26)	−0.005	(−0.94)
$\log(Y)$	−0.432	(−1.28)	1.182	(5.33)	0.393	(2.43)	0.954	(12.19)	1.475	(6.87)
$\log(Y) * Trend$	−0.030	(−1.75)	0.004	(0.20)	0.036	(2.78)	−0.002	(−0.13)	0.005	(0.43)
Trend	−0.044	(−2.14)	0.061	(2.18)	0.105	(4.91)	0.018	(0.78)	0.085	(3.62)
Random effects component: $\log \sigma^2_{v0i}$										
lnVARv0iIntercept	0.567	(25.03)	−2.153	(−24.98)	−1.542	(−37.01)	−0.367	(−5.24)	−0.643	(−26.28)
Persistent inefficiency component: $\log \sigma^2_{u0i}$										
lnVARu0iIntercept	−1.363	(−23.33)	0.459	(10.69)	−0.469	(−15.46)	−0.730	(−9.23)	−1.022	(−19.82)
Random noise component: $\log \sigma^2_{vit}$										
lnVARvitIntercept	−6.180	(−25.73)	−4.043	(−66.60)	−4.723	(−63.28)	−4.185	(−57.02)	−4.079	(−74.23)
Transient inefficiency component: $\log \sigma^2_{uit}$										
lnVARuitIntercept	−3.195	(−14.71)	−1.013	(−1.92)	−17.732	(−5.89)	−9.850	(−8.06)	−0.320	(−0.88)
lnVARuitTrend	0.214	(7.36)	−2.225	(−4.11)	1.531	(5.13)	0.673	(5.07)	−1.666	(−8.42)
Sample Size										
N	105		113		104		98		109	
$\sum_{i=1}^{N} T_i$	968		850		835		817		1005	
Sim. logL	−274.84		155.72		328.12		163.25		136.60	

Table 4. Model 3 as in Equation (10). Dependent variable is $- \log E$. z-values in parentheses.

Parameter	The 1960s		The 1970s		The 1980s		The 1990s		The 2000s	
Intercept	−5.788	(−27.29)	−4.111	(−13.84)	−4.328	(−19.92)	−6.833	(−27.68)	−6.349	(−19.86)
$0.5 * \log(K)^2$	−0.055	(−1.54)	0.091	(3.74)	0.011	(0.67)	−0.046	(−1.66)	−0.109	(−5.19)
$0.5 * \log(L)^2$	−0.480	(−13.99)	−0.036	(−1.30)	0.246	(19.91)	0.183	(8.27)	0.186	(7.85)
$0.5 * \log(NEM)^2$	−0.203	(−1.30)	−0.010	(−0.50)	0.047	(1.19)	0.176	(5.53)	0.069	(3.56)
$0.5 * \log(wE/wY)^2$	−0.340	(−7.45)	−0.005	(−0.24)	−0.071	(−1.55)	−0.067	(−0.23)	−0.105	(−2.50)
$0.5 * \text{Trend}^2$	3.5×10^{-4}	(0.13)	−0.011	(−4.85)	−0.021	(−12.95)	−0.007	(−4.07)	−0.018	(−12.46)
$\log(K)$	1.180	(8.92)	−0.377	(−3.85)	0.275	(5.94)	0.852	(18.24)	1.061	(14.19)
$\log(K) * \log(L)$	0.207	(6.18)	0.020	(0.64)	0.029	(1.27)	0.060	(1.30)	0.065	(2.07)
$\log(K) * \log(NEM)$	−0.264	(−108.79)	−0.005	(−0.72)	−0.003	(−1.84)	−0.035	(−12.27)	−0.012	(−3.90)
$\log(K) * \log(wE/wY)$	0.088	(0.93)	−0.003	(−0.10)	0.113	(3.90)	−0.075	(−0.87)	−0.070	(−1.43)
$\log(K) * \text{Trend}$	0.016	(5.66)	0.019	(5.67)	−0.006	(−3.55)	0.009	(3.87)	0.006	(2.00)
$\log(L)$	−0.426	(−1.15)	0.675	(8.24)	0.095	(1.04)	1.736	(26.02)	0.452	(5.57)
$\log(L) * \log(NEM)$	0.336	(3.92)	−0.057	(−1.93)	−0.146	(−4.47)	−0.311	(−6.62)	−0.163	(−4.85)
$\log(L) * \log(wE/wY)$	0.050	(0.47)	0.151	(3.91)	−0.031	(−0.71)	0.034	(0.28)	0.055	(1.06)
$\log(L) * \text{Trend}$	−0.027	(−6.74)	−0.010	(−1.85)	0.019	(7.95)	0.001	(0.40)	0.010	(2.29)
$\log(NEM)$	1.128	(2.28)	0.706	(11.62)	0.636	(4.38)	0.292	(3.91)	0.561	(14.16)
$\log(NEM) * \log(wE/wY)$	−0.053	(−0.39)	−0.090	(−2.58)	−0.079	(−1.62)	−0.111	(−1.01)	−0.019	(−0.49)
$\log(NEM) * \text{Trend}$	0.023	(4.45)	−0.004	(−0.83)	−0.006	(−2.29)	−0.013	(−3.77)	−0.006	(−1.55)
$\log(wE/wY)$	−0.542	(−2.10)	0.114	(0.76)	−0.094	(−0.54)	1.371	(2.97)	0.533	(1.91)
$\log(wE/wY) * \text{Trend}$	−0.005	(−0.62)	0.024	(5.10)	0.021	(4.38)	0.013	(0.65)	0.011	(1.43)
Trend	−0.093	(−2.22)	0.092	(4.71)	0.113	(8.89)	0.056	(3.85)	0.111	(6.40)
Random effects component: $\log \sigma_{v0i}^2$										
lnVARv0iIntercept	0.769	(20.77)	−3.036	(−20.12)	−3.841	(−37.08)	−0.039	(−0.93)	−0.071	(−3.18)
Persistent inefficiency component: $\log \sigma_{u0i}^2$										
lnVARu0iIntercept	−4.199	(−9.11)	1.390	(29.23)	0.433	(15.01)	−0.981	(−13.94)	−2.428	(−18.20)
Random noise component: $\log \sigma_{vit}^2$										
lnVARvitIntercept	−5.714	(−8.29)	−4.004	(−70.89)	−4.749	(−71.78)	−4.218	(−57.99)	−3.912	(−74.27)
Transient inefficiency component: $\log \sigma_{uit}^2$										
lnVARuitIntercept	−3.459	(−13.87)	−1.055	(−1.88)	−18.646	(−6.10)	−9.658	(−9.10)	−0.409	(−1.06)
lnVARuitTrend	0.248	(7.63)	−2.300	(−4.16)	1.572	(5.21)	0.688	(5.86)	−1.765	(−7.76)
Sample Size										
N	105		113		104		98		109	
$\sum_{i=1}^{N} T_i$	968		850		835		817		1005	
Sim. logL	−290.77		144.77		376.35		151.07		84.05	

Figure 3 shows the evolution of average efficiency over time by the type of efficiency. Figures 4–6 show densities of three types of efficiencies using the formula in (13) for models 1, 2, and 3, respectively by decade. The three columns in each of the there figures present overall, transient, and persistent efficiencies. Recall that the overall efficiency is the product of transient and persistent efficiencies. The rows from 1 to 5 show the decades from the 1960s through the 2000s.

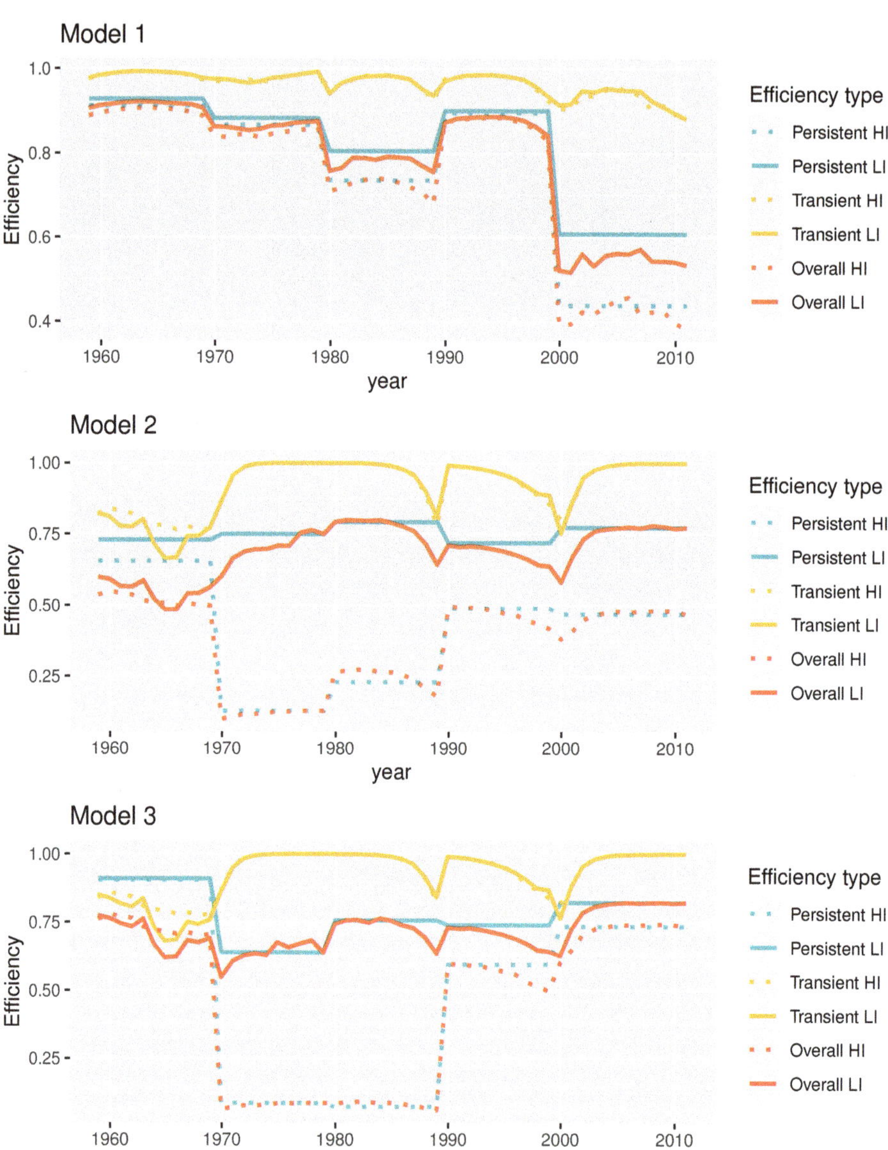

Figure 3. Average efficiency by year, energy intensity, and type of efficiency. The abbreviation HI stands for high intensity and LI means low intensity. Notes: The dotted time-series lines are for the high and solid lines are for low intensity industries.

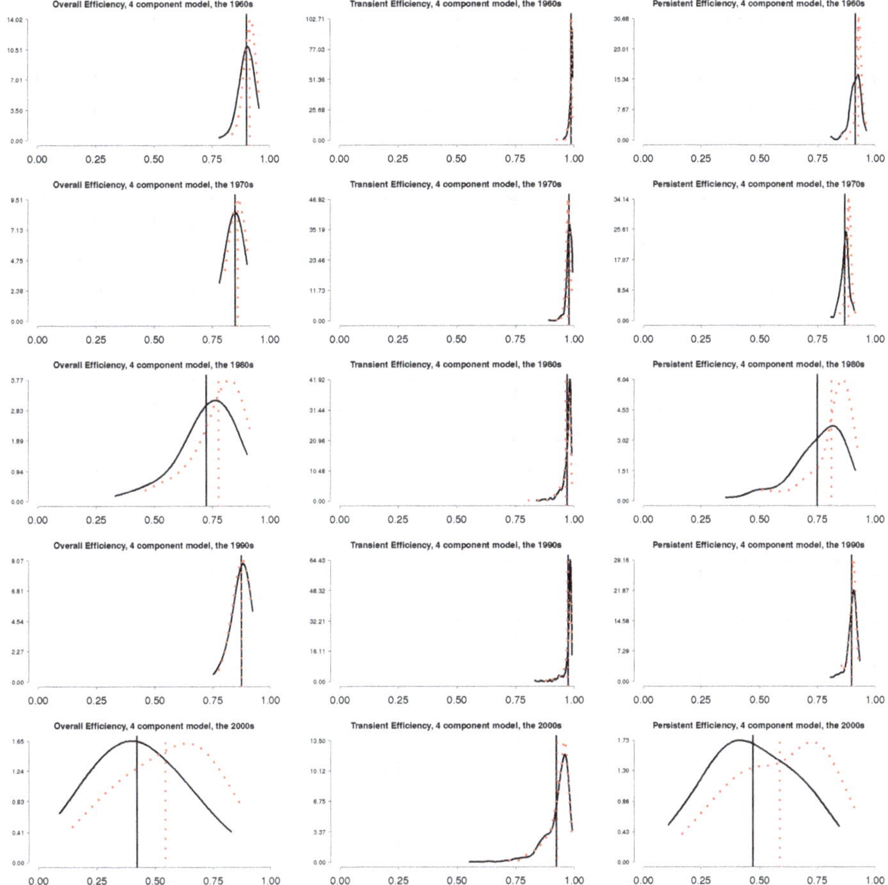

Figure 4. Overall, transient and persistent energy efficiency, Model 1. Notes: Solid black curves are high energy-intensive industries, dotted red curves are low energy-intensive industries. Vertical lines are respective mean values.

Models 2 and 3 measure energy use efficiency directly. Since we are applying the 4-component model, the overall energy use efficiency is decomposed into the persistent and transient components. The left column of Figure 5 reveals that the overall energy use efficiency has deteriorated over time. This is also confirmed by the middle and lower panels in Figure 3. We note again that the efficiencies are not comparable as they are measured relative to decade-specific frontiers, however, we can gauge how industries performed within decades. The energy use efficiency was very low in the 1970s, which could be the result of the oil crisis, which hit all industries of the economy. The overall efficiency figure in the 1970s, however, additionally reveals that the high energy-intensive industries were hit much harder. We have seen in Table 1 that some industries consume energy up to about a third of their actual sales. The lower panels in the left column of Figure 5 and middle panel in Figure 3 indicate that high energy-intensive industries were rebounding from the oil crisis and were only short of reaching the level of overall energy use efficiency only in the 2000s. The levels of overall energy use efficiency are still very low by any standard for energy-intensive industries. The energy use efficiency of the low energy-intensive industries is quite stable relative to the decade specific frontiers. Clearly, if the share of energy costs in production is very low as Table 1 suggests, the shocks to energy use are not that profound.

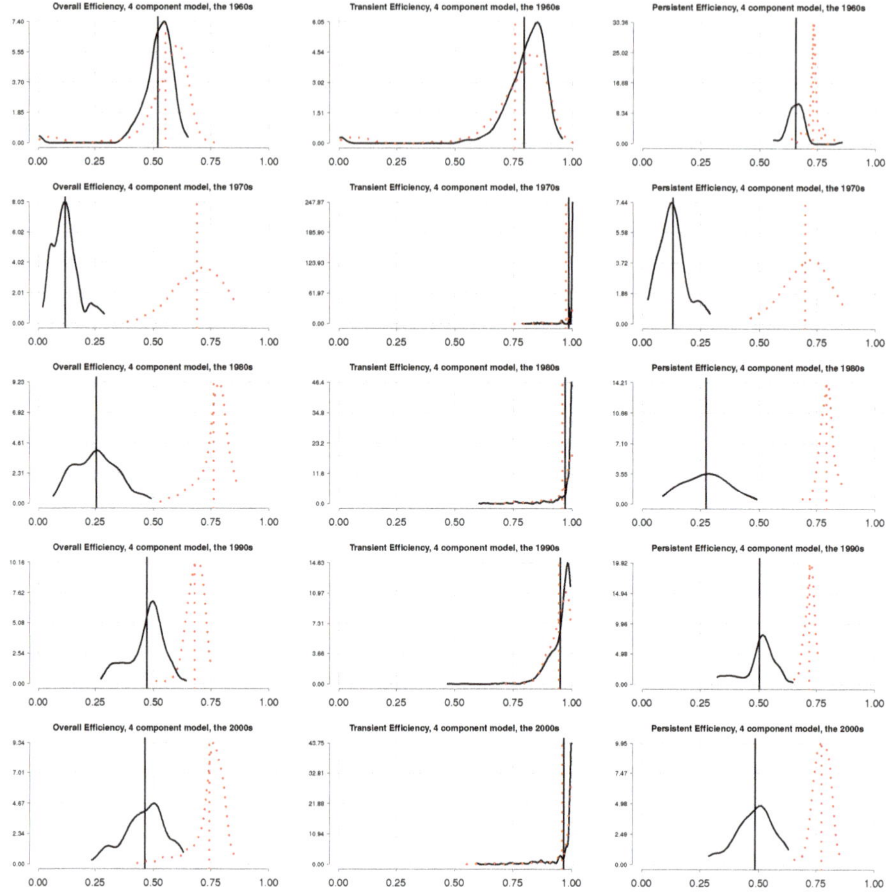

Figure 5. Overall, transient and persistent energy efficiency, Model 2. Notes: Solid black curves are high energy-intensive industries, dotted red curves are low energy-intensive industries. Vertical lines are respective mean values.

Looking at the components of the overall efficiency, we again observe that the overall inefficiency is mainly rooted in the structural energy use inefficiency. The density of the transient efficiency with an exception of the 1960s is concentrated around unity. The structural efficiency is shown in the third column of Figure 5 and as persistent efficiency in Figure 3. For low energy-intensive industries, it remains virtually unchanged, albeit relative to the decade-specific frontier. As is expected after discussion of the overall efficiency, the persistent efficiency of the high energy-intensive industries plummeted in the 1970s and increased gradually only in the 2000s.

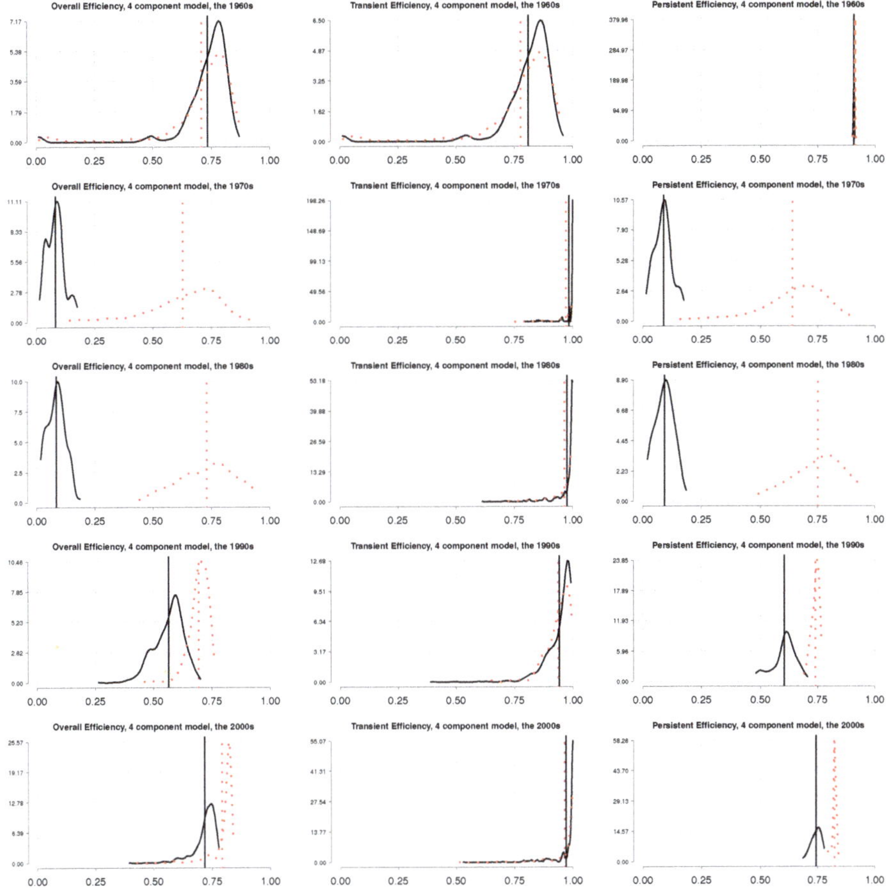

Figure 6. Overall, transient and persistent energy efficiency, Model 3. Notes: Solid black curves are high energy-intensive industries, dotted red curves are low energy-intensive industries. Vertical lines are respective mean values.

5.3. Discussion

It is worth noting that because we have estimated decade-specific frontiers, the efficiencies across decades are not directly comparable. Thus, we discuss the differences in efficiencies that are estimated relative to their frontiers. Overall, the level of efficiency is close to that reported by [44]. Based on Model 1, one result the becomes evident is that the industries move further away from the frontier over time. We cannot say whether this is because they were lagging behind technological progress or whether they were becoming less efficient. The second feature is that transient inefficiency is almost non-existent and input inefficiency almost completely stems from structural inefficiency. Third, we see a drop in efficiency in the 2000s, which can be attributed to downturns at the beginning of the 2000s as well as the financial crisis at the end of the decade. Finally, in terms of overall input inefficiency, both high and low energy-intensive industries perform similarly. Only in the 2000s, low energy-intensive industries seem to slightly over-perform high energy-intensive industries. We find confirmation for average levels in Figure 3.

Figure 6 summarizes the energy use efficiency for the third model, which is only slightly different from Model 2. The change that we observe in Figure 6 relative to Figure 5 is only quantitative.

Conclusions that we drew from Figure 5 can be repeated for Model 3, so that the results of the third model can be seen as a robustness check.

It is difficult to say why we observe the so-called "energy paradox". The US is known to promote energy efficiency policy (see, e.g., [33]). However, such policies lead to different outcomes. In Sweden for example, the adoption rate of energy efficiency measures is over 40% ([45]). Although financial intensive may be an important one in some industries and countries ([46]), Refernece [4] document lack of adoption, which constitutes the above paradox. The authors of [47] find that the most important barriers to more energy-efficient organization are internal economic and behavioral barriers. The authors of [48] name additional barriers including lack of interest in energy efficient technologies. Further, their findings suggest that adopting sound energy management practices is the most important driver of increased energy efficiency. Adopting cost-effective technologies is also important, but less so than the above-mentioned practices.

6. Conclusions

Energy is one of the most important inputs in manufacturing industries. It is a scarce input that is expensive in both monetary and environmental terms. Hence, both policymakers and businesses should consider the efficient use of this input in the long-term.

This study uses the stochastic frontier approach to measure energy use efficiency in the US manufacturing during the time period 1958–2011 using the NBER-CES Manufacturing Industry Database. When panel data are available as in our case, we advocate using the latest or the 4-component SF model. We concentrate on the most and least energy-intensive manufacturing industries. More specifically, we first define energy intensity as the costs of energy in total economic activity. Then for each of five decades, we identify the top 10% and bottom 10% energy-intensive industries. We apply the 4-component stochastic frontier model that decomposes overall efficiency into the long-term or persistent and short-term efficiencies. Our main findings suggest that energy use efficiency in US manufacturing hit hard by the oil shock in the 1970s and it did not rebound until the 2000s. The major culprit of the low overall energy use efficiency was structural inefficiency, a finding that goes hand in hand with the "energy paradox" (see, e.g., [4]). It seems that one of the ways to mitigate low levels of energy use efficiency should be to do more research along the lines of [5,47,48] to promote, adopt, and establish energy-efficient technologies as the new benchmark.

Author Contributions: Conceptualization, O.B. and S.C.K.; Methodology, O.B. and S.C.K.; Data collection S.C.K, formal analysis, O.B.; Writing—original draft preparation, O.B., Writing—Review and Editing, O.B. and S.C.K. All authors have read and agreed to the published version of the manuscript.

Funding: This research received no external funding.

Conflicts of Interest: All authors declare that they have no conflict of interest.

Appendix A

Table A1. Descriptive statistics by decade and by the intensity of energy use.

Time Period	Industries	Median	Mean	SD	Min	Max
		Y				
The 1960s	Least energy-intensive	832.95	1509.60	2214.52	27.10	17,931.30
The 1960s	Most energy-intensive	546.85	1251.66	2780.79	79.70	21,983.20
The 1970s	Least energy-intensive	1438.45	3796.49	7199.52	150.30	43,128.30
The 1970s	Most energy-intensive	1128.70	3024.25	6117.26	111.60	54,446.30
The 1980s	Least energy-intensive	3639.00	8997.84	14,000.49	286.80	73,925.10
The 1980s	Most energy-intensive	2222.15	4855.71	8216.35	136.60	56,738.40
The 1990s	Least energy-intensive	6168.60	16,492.67	23,594.40	219.30	145,256.20
The 1990s	Most energy-intensive	3312.50	6678.93	10,571.67	116.30	56,895.00
The 2000s	Least energy-intensive	7992.90	19,773.51	30,016.89	149.10	16,2181.80
The 2000s	Most energy-intensive	4243.00	9790.06	17,328.47	130.80	123,129.80

Table A1. *Cont.*

Time Period	Industries	Median	Mean	SD	Min	Max
		K				
The 1960s	Least energy-intensive	400.85	1131.83	1841.05	6.20	9507.70
The 1960s	Most energy-intensive	1548.30	4670.45	12,180.26	111.10	90,867.10
The 1970s	Least energy-intensive	631.45	2049.62	3965.95	27.60	20,595.00
The 1970s	Most energy-intensive	2646.90	6482.31	14,042.12	257.70	92,312.80
The 1980s	Least energy-intensive	960.60	3075.50	5135.97	52.80	25,549.50
The 1980s	Most energy-intensive	2475.65	6780.63	13,499.10	236.70	91,786.60
The 1990s	Least energy-intensive	1338.90	4242.53	6078.23	39.40	27,170.30
The 1990s	Most energy-intensive	2933.30	6867.77	12,257.56	172.20	70,528.90
The 2000s	Least energy-intensive	1605.60	5526.12	8706.73	57.30	53,612.40
The 2000s	Most energy-intensive	3162.85	7044.35	11,075.66	165.70	58,371.10
		L				
The 1960s	Least energy-intensive	72.40	127.50	147.79	1.35	1018.47
The 1960s	Most energy-intensive	31.66	83.17	164.73	3.75	1145.57
The 1970s	Least energy-intensive	76.74	121.04	122.07	9.94	727.97
The 1970s	Most energy-intensive	30.70	80.51	150.27	3.90	1070.72
The 1980s	Least energy-intensive	69.93	116.73	125.13	6.91	686.44
The 1980s	Most energy-intensive	28.38	59.72	92.65	2.98	762.71
The 1990s	Least energy-intensive	64.72	113.52	117.39	4.49	659.60
The 1990s	Most energy-intensive	26.66	55.92	74.33	2.09	423.21
The 2000s	Least energy-intensive	55.18	92.35	96.65	2.29	487.05
The 2000s	Most energy-intensive	20.63	39.73	53.02	2.48	380.14
		NEM				
The 1960s	Least energy-intensive	432.60	917.37	1687.38	16.90	15,436.70
The 1960s	Most energy-intensive	193.25	555.57	1414.46	6.50	11,553.60
The 1970s	Least energy-intensive	787.60	2501.26	5759.85	65.80	37,855.40
The 1970s	Most energy-intensive	485.10	1460.07	3318.96	14.70	29,737.30
The 1980s	Least energy-intensive	1832.10	5568.41	10,423.51	129.60	50,316.50
The 1980s	Most energy-intensive	873.20	2306.36	4384.77	27.70	31,836.00
The 1990s	Least energy-intensive	2950.90	9604.14	16,286.32	80.50	102,924.30
The 1990s	Most energy-intensive	1276.85	3034.87	5404.51	25.30	29,446.20
The 2000s	Least energy-intensive	3954.40	10,175.94	18,271.44	63.80	110,074.70
The 2000s	Most energy-intensive	1602.30	4847.67	10,339.06	24.80	75,089.60
		E				
The 1960s	Least energy-intensive	2.50	4.83	8.95	0.10	75.70
The 1960s	Most energy-intensive	29.00	70.23	135.82	2.60	1056.00
The 1970s	Least energy-intensive	7.00	18.22	35.64	0.40	225.20
The 1970s	Most energy-intensive	96.30	248.52	513.12	7.10	5325.20
The 1980s	Least energy-intensive	23.70	56.56	84.23	1.50	348.40
The 1980s	Most energy-intensive	202.60	494.83	782.74	20.30	5858.60
The 1990s	Least energy-intensive	29.30	66.00	86.92	1.00	350.40
The 1990s	Most energy-intensive	219.90	486.65	689.62	8.90	3570.80
The 2000s	Least energy-intensive	35.00	78.93	126.08	0.70	858.00
The 2000s	Most energy-intensive	355.60	777.17	1158.60	12.00	6775.10

Appendix B

Here we describe how to estimate the model in (11). To facilitate the discussion, rewrite

$$\log q_{it} = r(X_{it}, trend; \omega) + v_{0i} - u_{0i} + v_{it} - u_{it} \tag{A1}$$

as

$$\log q_{it} = r(X_{it}, trend; \omega) + \epsilon_{0i} + \epsilon_{it},$$

where $\epsilon_{it} = v_{it} - u_{it}$ and $\epsilon_{0i} = v_{0i} - u_{0i}$ decompose the error term into two 'composed error' terms (both of which contain a two-sided and a one-sided error terms). Assume the most general case where all four components are heteroskedastic

$$\sigma_{u_{it}}^2 = \exp\left(z_{u_{it}}\boldsymbol{\psi}_u\right), \quad i = 1, \cdots, n, \quad t = 1, \cdots, T_i, \tag{A2}$$

$$\sigma_{u_{0i}}^2 = \exp\left(z_{u_{0i}}\boldsymbol{\psi}_{u0}\right), \quad i = 1, \cdots, n, \tag{A3}$$

$$\sigma_{v_{it}}^2 = \exp\left(z_{v_{it}}\boldsymbol{\psi}_v\right), \quad i = 1, \cdots, n, \quad t = 1, \cdots, T_i, \tag{A4}$$

$$\sigma_{v_{0i}}^2 = \exp\left(z_{v_{0i}}\boldsymbol{\psi}_{v0}\right), \quad i = 1, \cdots, n, \tag{A5}$$

where $z_{u_{it}}$ are the determinants of transient inefficiency, $z_{u_{0i}}$ are the determinants of persistent inefficiency, and $z_{v_{it}}$ and $z_{v_{0i}}$ define the heteroskedasticity functions of the noise and random effects. The homoskedastic error component is easily derived from (A2–A5) by setting the vector of determinants to a constant. For example if v_{it} is homoskedastic, $z_{v_{it}}$ is a vector of ones of length $\sum_{i=1}^n T_i$. The conditional density of $\boldsymbol{\epsilon}_i = (\epsilon_{i1}, \dots, \epsilon_{iT_i})$ is given by

$$f\left(\boldsymbol{\epsilon}_i | \epsilon_{0i}\right) = \prod_{t=1}^{T_i} \frac{2}{\sigma_{it}} \phi\left(\frac{\epsilon_{it}}{\sigma_{it}}\right) \Phi\left(\frac{\epsilon_{it}\lambda_{it}}{\sigma_{it}}\right),$$

where $\sigma_{it} = [\exp\left(z_{u_{it}}\boldsymbol{\psi}_u\right) + \exp\left(z_{v_{it}}\boldsymbol{\psi}_v\right)]^{1/2}$ and $\lambda_{it} = [\exp\left(z_{u_{it}}\boldsymbol{\psi}_u\right) / \exp\left(z_{v_{it}}\boldsymbol{\psi}_v\right)]^{1/2}$.

Integrate ϵ_{0i} (the distribution of which we know) out to get the unconditional density of $\boldsymbol{\epsilon}_i$

$$f\left(\boldsymbol{\epsilon}_i\right) = \int_{-\infty}^{\infty} \left[\prod_{t=1}^{T_i} \frac{2}{\sigma_{it}} \phi\left(\frac{\epsilon_{it}}{\sigma_{it}}\right) \Phi\left(\frac{\epsilon_{it}\lambda_{it}}{\sigma_{it}}\right)\right] \times \frac{2}{\sigma_{0i}} \phi\left(\frac{\epsilon_{0i}}{\sigma_{0i}}\right) \Phi\left(\frac{\epsilon_{0i}\lambda_{0i}}{\sigma_{0i}}\right) d\epsilon_{0i},$$

where $\sigma_{0i} = [\exp\left(z_{u_{0i}}\boldsymbol{\psi}_{u0}\right) + \exp\left(z_{v_{0i}}\boldsymbol{\psi}_{v0}\right)]^{1/2}$ and $\lambda_{0i} = [\exp\left(z_{u_{0i}}\boldsymbol{\psi}_{u0}\right) / \exp\left(z_{v_{0i}}\boldsymbol{\psi}_{v0}\right)]^{1/2}$. The log-likelihood function for the i-th observation of model (A1) is therefore given by

$$\log L_i\left(\boldsymbol{\beta}, \boldsymbol{\psi}_{u0}, \boldsymbol{\psi}_{v0}, \boldsymbol{\psi}_u, \boldsymbol{\psi}_v\right)$$

$$= \log\left[\int_{-\infty}^{+\infty} \left(\prod_{t=1}^{T_i}\left\{\frac{2}{\sigma_{it}}\phi\left(\frac{r_{it} - \epsilon_{0i}}{\sigma_{it}}\right)\Phi\left(\frac{(r_{it} - \epsilon_{0i})\lambda_{it}}{\sigma_{it}}\right)\right\}\right) \frac{2}{\sigma_{0i}}\phi\left(\frac{\epsilon_{0i}}{\sigma_{0i}}\right)\Phi\left(\frac{\epsilon_{0i}\lambda_{0i}}{\sigma_{0i}}\right) d\epsilon_{0i}\right]$$

$$= \log\left[\int_{-\infty}^{+\infty} \left(\prod_{t=1}^{T_i}\left\{\frac{2}{\sigma_{it}}\phi\left(\frac{\epsilon_{it}}{\sigma_{it}}\right)\Phi\left(\frac{\epsilon_{it}\lambda_{it}}{\sigma_{it}}\right)\right\}\right) \times \frac{2}{\sigma_{0i}}\phi\left(\frac{\epsilon_{0i}}{\sigma_{0i}}\right)\Phi\left(\frac{\epsilon_{0i}\lambda_{0i}}{\sigma_{0i}}\right) d\epsilon_{0i}\right], \tag{A6}$$

where $\epsilon_{it} = r_{it} - (v_{0i} + u_{0i})$ and $r_{it} = \log q_{it} - r(X_{it}, trend; \omega)$. We rely on the Monte-Carlo integration as a method to approximate the integral in (A6). For estimation purposes, we write $\epsilon_{0i} = [\exp\left(z_{u_{0i}}\boldsymbol{\psi}_{u0}\right)]^{1/2}V_i + [\exp\left(z_{v_{0i}}\boldsymbol{\psi}_{v0}\right)]^{1/2}|U_i|$, where both V_i and U_i are standard normal random variables. The resulting simulated log-likelihood function for the i-th observation is

$$\log L_i^S\left(\boldsymbol{\beta}, \boldsymbol{\psi}_{u0}, \boldsymbol{\psi}_{v0}, \boldsymbol{\psi}_u, \boldsymbol{\psi}_v\right)$$

$$= \log\left[\frac{1}{R}\sum_{r=1}^{R}\left(\prod_{t=1}^{T_i}\left\{\frac{2}{\sigma_{it}}\phi\left(\frac{r_{it} - \left([\exp\left(z_{u_{0i}}\boldsymbol{\psi}_{u0}\right)]^{1/2}V_{ir} + [\exp\left(z_{v_{0i}}\boldsymbol{\psi}_{v0}\right)]^{1/2}|U_{ir}|\right)}{\sigma_{it}}\right) \times \Phi\left(\frac{[r_{it} - \left([\exp\left(z_{u_{0i}}\boldsymbol{\psi}_{u0}\right)]^{1/2}V_{ir} + [\exp\left(z_{v_{0i}}\boldsymbol{\psi}_{v0}\right)]^{1/2}|U_{ir}|\right)]\lambda}{\sigma_{it}}\right)\right\}\right)\right]$$

$$= \log\left[\frac{1}{R}\sum_{r=1}^{R}\left(\prod_{t=1}^{T_i}\left\{\frac{2}{\sigma}\phi\left(\frac{\epsilon_{itr}}{\sigma}\right)\Phi\left(\frac{\epsilon_{itr}\lambda}{\sigma}\right)\right\}\right)\right], \tag{A7}$$

where V_{ir} and U_{ir} are R random deviates from the standard normal distribution, and $\epsilon_{itr} = r_{it} - ([\exp\left(z_{u_{0i}}\boldsymbol{\psi}_{u0}\right)]^{1/2}V_{ir} + [\exp\left(z_{v_{0i}}\boldsymbol{\psi}_{v0}\right)]^{1/2}|U_{ir}|)$. R is the number of draws for approximating the log-likelihood function. The full log-likelihood is the sum of panel-i specific log-likelihoods given in (A7).

We use the results of [39] to calculate persistent and time-varying cost efficiencies. Using the moment generating function of the closed skew normal distribution, the conditional means of $u_{0i}, u_{i1}, \cdots, u_{iT_i}$ are given by:

$$E(\exp\{t'u_i\}|r_i) = \frac{\overline{\Phi}_{T_i+1}\left(R_i r_i + \Lambda_i t, \Lambda_i\right)}{\overline{\Phi}_{T_i+1}\left(R_i r_i, \Lambda_i\right)} \times \exp\left(t'R_i r_i + 0.5 t'\Lambda_i t\right), \tag{A8}$$

where $r_i = \left(r_{i1}, \ldots, r_{iT_i}\right)'$, $A = -[\mathbf{1}_{T_i} I_{T_i}]$, $\mathbf{1}_{T_i}$ is the column vector of length T_i and I_{T_i} is the identity matrix of dimension T_i, the diagonal elements of V_i are $[\exp\left(z_{u_{0i}}\psi_{u0}\right)\exp\left(z_{u_{it}}\psi_u\right)]$, $\Sigma_i = \exp\left(z_{v_{it}}\psi_v\right)I_{T_i} + \exp\left(z_{v_{0i}}\psi_{v0}\right)\mathbf{1}_{T_i}\mathbf{1}'_{T_i}$, $\Lambda_i = V_i - V_i A'\left(\Sigma_i + AV_i A'\right)^{-1}AV_i = \left(V_i^{-1} + A'\Sigma_i^{-1}A\right)^{-1}$, $R_i = V_i A'\left(\Sigma_i + AV_i A'\right)^{-1} = \Lambda_i A'\Sigma_i^{-1}$, $\phi_q\left(x, \mu, \Omega\right)$ is the density function of a q-dimensional normal variable with expected value μ and variance Ω and $\overline{\Phi}_q\left(\mu, \Omega\right)$ is the probability that a q-variate normal variable of expected value μ and variance Ω belongs to the positive orthant., $u_i = \left(u_{0i}, u_{i1}, \ldots, u_{iT_i}\right)'$, and $-t$ is a row of the identity matrix of dimension $(T_i + 1)$. If $-t$ is the τ-th row, Equation (A8) provides the conditional expected value of the τ-th component of the cost efficiency vector $\exp\left(-u_i\right)$. In particular, for $\tau = 1$, we get the conditional expected value of the persistent technical efficiency.

References

1. Menghi, R.; Papetti, A.; Germani, M.; Marconi, M. Energy efficiency of manufacturing systems: A review of energy assessment methods and tools. *J. Clean. Prod.* **2019**, *240*, 118276. [CrossRef]
2. Kanellakis, M.; Martinopoulos, G.; Zachariadis, T. European energy policy-A review. *Energy Policy* **2013**, *62*, 1020–1030. [CrossRef]
3. Wang, C.N.; Ho, H.X.T.; Hsueh, M.H. An integrated approach for estimating the energy efficiency of seventeen countries. *Energies* **2017**, *10*, 1597. [CrossRef]
4. Gerarden, T.D.; Newell, R.G.; Stavins, R.N. Assessing the Energy-Efficiency Gap. *J. Econ. Lit.* **2017**, *55*, 1486–1525. [CrossRef]
5. Blass, V.; Corbett, C.J.; Delmas, M.A.; Muthulingam, S. Top management and the adoption of energy efficiency practices: Evidence from small and medium-sized manufacturing firms in the US. *Energy* **2014**, *65*, 560–571. [CrossRef]
6. Boyd, G.A.; Lee, J.M. Measuring plant level energy efficiency and technical change in the U.S. metal-based durable manufacturing sector using stochastic frontier analysis. *Energy Econ.* **2019**, *81*, 159–174. [CrossRef]
7. Filipović, S.; Verbič, M.; Radovanović, M. Determinants of energy intensity in the European Union: A panel data analysis. *Energy* **2015**, *92*, 547–555. [CrossRef]
8. Bentzen, J. Estimating the rebound effect in US manufacturing energy consumption. *Energy Econ.* **2004**, *26*, 123–134. [CrossRef]
9. Bartelsman, E.J.; Gray, W. *The NBER Manufacturing Productivity Database*; Working Paper 205; National Bureau of Economic Research: Cambridge, MA, USA, 1996. [CrossRef]
10. Parker, S.; Liddle, B. Energy efficiency in the manufacturing sector of the OECD: Analysis of price elasticities. *Energy Econ.* **2016**, *58*, 38–45. [CrossRef]
11. Badunenko, O.; Kumbhakar, S.C. Economies of scale, technical change and persistent and time-varying cost efficiency in Indian banking: Do ownership, regulation and heterogeneity matter? *Eur. J. Oper. Res.* **2017**, *260*, 789–803. [CrossRef]
12. Wei, T. Impact of energy efficiency gains on output and energy use with Cobb-Douglas production function. *Energy Policy* **2007**, *35*, 2023–2030. [CrossRef]
13. Saunders, H.D. Historical evidence for energy efficiency rebound in 30 US sectors and a toolkit for rebound analysts. *Technol. Forecast. Soc. Chang.* **2013**, *80*, 1317–1330. [CrossRef]
14. Allcott, H.; Knittel, C.; Taubinsky, D. Tagging and Targeting of Energy Efficiency Subsidies. *Am. Econ. Rev.* **2015**, *105*, 187–191. [CrossRef]

15. Kumbhakar, S.C.; Lovell, C.A.K. *Stochastic Frontier Analysis*; Cambridge University Press: Cambridge, UK, 2000. [CrossRef]
16. Färe, R.; Grosskopf, S.; Lovell, C.A.K. *Production Frontiers*; Cambridge University Press: Cambridge, UK, 1994. [CrossRef]
17. Kumbhakar, S.C. The Measurement and Decomposition of Cost-Inefficiency: The Translog Cost System. *Oxf. Econ. Pap. New Ser.* **1991**, *43*, 667–683. [CrossRef]
18. Kumbhakar, S.C.; Heshmati, A. Efficiency Measurement in Swedish Dariy Farms: An Application of Rotating Panel Data. *Am. J. Agric. Econ.* **1995**, *77*, 660–674. [CrossRef]
19. Kumbhakar, S.C.; Hjalmarsson, L. Technical Efficiency and Technical Progress in Swedish Dairy Farms. In *The Measurement of Productive Efficiency. Techniques and Applications*; Fried, H.O., Lovell, C.A.K., Schmidt, S.S., Eds.; Oxford University Press: Oxford, UK, 1993; pp. 256–270.
20. Kumbhakar, S.C.; Hjalmarsson, L. Labour-Use Efficiency in Swedish Social Insurance Offices. *J. Appl. Econ.* **1995**, *10*, 33–47. [CrossRef]
21. Greene, H.W. Reconsidering Heterogeneity in Panel Data Estimators of the Stochastic Frontier Model. *J. Econ.* **2005**, *126*, 269–303. [CrossRef]
22. Kumbhakar, S.C.; Lien, G.; Hardaker, J.B. Technical Efficiency in Competing Panel Data Models: A Study of Norwegian Grain Farming. *J. Product. Anal.* **2014**, *41*, 321–337. [CrossRef]
23. Mukherjee, K. Energy use efficiency in U.S. manufacturing: A nonparametric analysis. *Energy Econ.* **2008**, *30*, 76–96. [CrossRef]
24. Mukherjee, K. Energy use efficiency in the Indian manufacturing sector: An interstate analysis. *Energy Policy* **2008**, *36*, 662–672. [CrossRef]
25. Zhou, P.; Ang, B.W. Linear programming models for measuring economy-wide energy efficiency performance. *Energy Policy* **2008**, *36*, 2911–2916. [CrossRef]
26. Mousavi-Avval, S.H.; Rafiee, S.; Jafari, A.; Mohammadi, A. Improving energy use efficiency of canola production using data envelopment analysis (DEA) approach. *Energy* **2011**, *36*, 2765–2772. [CrossRef]
27. Sueyoshi, T.; Yuan, Y.; Goto, M. A literature study for DEA applied to energy and environment. *Energy Econ.* **2017**, *62*, 104–124. [CrossRef]
28. Boyd, G.A. A Method for Measuring the Efficiency Gap between Average and Best Practice Energy Use: The ENERGY STAR Industrial Energy Performance Indicator. *J. Ind. Ecol.* **2005**, *9*, 51–65. [CrossRef]
29. Boyd, G.A. Estimating Plant Level Energy Efficiency with a Stochastic Frontier. *Energy J.* **2008**, *29*, 23–43. [CrossRef]
30. Filippini, M.; Hunt, L.C. US residential energy demand and energy efficiency: A stochastic demand frontier approach. *Energy Econ.* **2012**, *34*, 1484–1491. [CrossRef]
31. Shui, H.; Jin, X.; Ni, J. Manufacturing productivity and energy efficiency: A stochastic efficiency frontier analysis. *Int. J. Energy Res.* **2015**, *39*, 1649–1663. [CrossRef]
32. Farsi, M.; Filippini, M.; Greene, W. Application of Panel Data Models in Benchmarking Analysis of the Electricity Distribution Sector. *Ann. Public Coop. Econ.* **2006**, *77*, 271–290. [CrossRef]
33. Filippini, M.; Hunt, L.C. Measuring persistent and transient energy efficiency in the US. *Energy Effic.* **2016**, *9*, 663–675. [CrossRef]
34. Kumbhakar, S.C.; Lien, G. Yardstick Regulation of Electricity Distribution – Disentangling Short-run and Long-run Inefficiencies. *Energy J.* **2017**, *38*, 1944–9089. [CrossRef]
35. Kumbhakar, S.C. Specification and estimation of multiple output technologies: A primal approach. *Eur. J. Oper. Res.* **2013**, *231*, 465–473. [CrossRef]
36. Diewert, W.E. Functional Forms for Revenue and Factor Requirements Functions. *Int. Econ. Rev.* **1974**, *15*, 119–130. [CrossRef]
37. Aigner, D.; Lovell, C.A.K.; Schmidt, P. Formulation and estimation of stochastic frontier production function models. *J. Econ.* **1977**, *6*, 21–37. [CrossRef]
38. Meeusen, W.; van den Broeck, J. Efficiency estimation from {C}obb-{D}ouglas production functions with composed error. *Int. Econ. Rev.* **1977**, *18*, 435–444. [CrossRef]
39. Colombi, R.; Kumbhakar, S.C.; Martini, G.; Vittadini, G. Closed-skew normality in stochastic frontiers with individual effects and long/short-run efficiency. *J. Product. Anal.* **2014**, *42*, 123–136. [CrossRef]
40. Tsionas, E.G.; Kumbhakar, S.C. Firm Heterogeneity, Persistent And Transient Technical Inefficiency: A Generalized True Random Effects model. *J. Appl. Econ.* **2014**, *29*, 110–132. [CrossRef]

41. Battese, G.E.; Coelli, T.J. Frontier production functions, technical efficiency and panel data: With application to paddy farmers in India. *J. Product. Anal.* **1992**, *3*, 153–169. [CrossRef]
42. Cornwell, C.; Schmidt, P.; Sickles, R.C. Production frontiers with cross-sectional and time-series variation in efficiencylevels. *J. Econ.* **1990**, *46*, 185–200. [CrossRef]
43. Kumbhakar, S.C. Production frontiers, panel data, and time-varying technical inefficiency. *J. Econ.* **1990**, *46*, 201–211. [CrossRef]
44. Hsiao, W.L.; Hu, J.L.; Hsiao, C.; Chang, M.C. Energy Efficiency of the Baltic Sea Countries: An Application of Stochastic Frontier Analysis. *Energies* **2019**, *12*, 104. [CrossRef]
45. Thollander, P.; Danestig, M.; Rohdin, P. Energy policies for increased industrial energy efficiency: Evaluation of a local energy programme for manufacturing SMEs. *Energy Policy* **2007**, *35*, 5774–5783. [CrossRef]
46. Thollander, P.; Backlund, S.; Trianni, A.; Cagno, E. Beyond barriers—A case study on driving forces for improved energy efficiency in the foundry industries in Finland, France, Germany, Italy, Poland, Spain, and Sweden. *Appl. Energy* **2013**, *111*, 636–643. [CrossRef]
47. Brunke, J.C.; Johansson, M.; Thollander, P. Empirical investigation of barriers and drivers to the adoption of energy conservation measures, energy management practices and energy services in the Swedish iron and steel industry. *J. Clean. Prod.* **2014**, *84*, 509–525. [CrossRef]
48. Trianni, A.; Cagno, E.; Worrell, E.; Pugliese, G. Empirical investigation of energy efficiency barriers in Italian manufacturing SMEs. *Energy* **2013**, *49*, 444–458. [CrossRef]

Article

Measurement of Energy Access Using Fuzzy Logic

Diego Seuret-Jimenez [1], Tiare Robles-Bonilla [1,*] and Karla G. Cedano [2]

[1] Centro de Investigación en Ingeniería y Ciencias Aplicadas, Universidad Autónoma del Estado de Morelos, Ave. Universidad 1001, Cuernavaca 62209, Morelos, Mexico; dseuret@uaem.mx

[2] Instituto de Energías Renovables, UNAM, Xochicalco s/n, Azteca, Temixco 62588, Morelos, Mexico; kcedano@ier.unam.mx

* Correspondence: tiare.roblesbno@uaem.edu.mx

Received: 16 March 2020; Accepted: 30 March 2020; Published: 24 June 2020

Abstract: This paper describes an innovative method to evaluate energy access in any of size population by applying fuzzy logic. The obtained results allow ranking regions of Mexico according to their overall energy access. The regions were determined by the country's political division (32 states). The results presented herein are in close correspondence with other studies undertaken. This method is recommended because it is possible to use as an assessment tool due to its representativeness—that is, it poses a heuristic alternative to quantify the level of Energy Access in a particular region through qualitative data. It is also efficient and cost-effective in terms of computer resources. This is extremely important to public policy makers that require more accurate, faster and cheaper methodologies to assess energy access as an indicator of well-being.

Keywords: energy access; energy use; fuzzy logic

1. Introduction

Understanding the way in which people use energy at home is necessary to move forward in the development of public policies which foster more efficient energy usage. Several surveys have been developed to measure energy access and its use. Butera et al. [1] developed a study about Brazil (Rio De Janeiro), in which two cities were analysed on energy access and the level of energy poverty through questionnaires carried out in 400 households. This helped to determine the local living conditions and the availability of basic energy services, as well as explore the actual energy access and energy poverty in the favelas. One of its main findings was that electricity consumption is very high compared to the service provided—as much as Italian or German households, which are much richer—in addition to electricity access being threatened by interruptions and low tension. This method is replicable with small adaptations; however, Butera et al. do not use fuzzy logic. Jimenez et al. [2] performed an analysis of surveys to determine the state of the electrification barriers in Latin America. Taking three variables—household income, household location, and the country's level of economic development—they analyse 12 countries in Latin America (Bolivia, Brazil, Chile, Costa Rica, Dominican Republic, Ecuador, Guatemala, Honduras, Mexico, Peru, Paraguay and El Salvador). The study shows serious inequality in electricity access, a family living in a poor country has a lesser chance of accessing electricity than a family with the same income but living in a richer country. This study does not use fuzzy logic, but it shows the application of a mathematical method—regression analysis.

In Mexico, for example, there is the Household Expenditure and Income Survey, measuring [3], among other things, energy services and expense in Mexican households. Another, of recent implementation, is the National Survey on Energy Consumption in Private Households (ENCEVI, its Spanish acronym) [4]. This was designed to help better understand the existing relationship between people and energy. Nevertheless, these exercises of information gathering do not provide simple and reliable tools for researchers and policy makers to compare and understand energy access and use at a household level. It is necessary to run the data sets produced by these surveys through often costly processing systems that require large and precise data sets to model energy access.

Addressing and measuring energy access is a complex issue. In most indicator sets that are used to measure energy poverty or that are related to basic energy services, the closest indicator is absolute electricity access. For instance, the World Economic Forum reports a 100% with respect to electrification rate in 2018 for Mexico [5]. It also includes another relevant indicator, electricity supply quality, measured by the following question: "In your country, how reliable is the electricity supply (lack of interruptions and lack of voltage fluctuations)?" on a scale from 1 to 7, 1 being "extremely unreliable", 7 being "extremely reliable", for which Mexico scored 4.9. This might be interpreted as "reliable". This indicator was subsequently measured differently in the following edition, also published in 2018 [6], as the percentage of electricity losses (comprising transmission, distribution and non-technical losses). Even though the indicator has the exact same name, it reflects very different things. However, the World Economic Forum's double definition of electricity access shows that the former version of the indicator was far more representative as the impact that the quality of the service had (or the perception from the consumer). This is the main reason behind our decision to measure Energy Access (EA) by asking a question about the availability of energy services to the population.

It is important to highlight that we are not addressing energy poverty (or fuel poverty) in this paper. While both energy poverty and energy access are related, energy poverty goes beyond the availability (or the perception of the availability) of any given energy service, but also energy use and the social behaviours that accompany said use. This is clearly defined by Thomson et al. [7], who state that energy poverty occurs "when a household is incapable of securing a degree of domestic energy services (such as space heating, cooling, cooking) that would allow them to fully participate in the customs and activities that define membership in society".

As part of your desk research, we did a small scientometric analysis using the Web of Science database. Using the phrase "energy access" as search criteria, we found 778 articles with it in either the title, key words or abstract. Simultaneously, we looked for articles with the phrase "fuzzy logic" and found 48,227. Nevertheless, when we intersected the searches, we could not find a paper that talked about both energy access and fuzzy logic. Therefore, this might present a novel methodology for looking at this issue.

Lotfi A. Zadeh [8] defined fuzzy logic in the 1960s for issues regarding language. It has now been applied more broadly to a diversity of knowledge areas, such as the control of automatic electro-mechanical processes [9], human activity control [10], decision making [11], etc. Nobre et al. [12] consider fuzzy logic a computational and mathematical framework suitable to represent approximate reasoning. It takes into consideration everyday life concepts, experiences, observations, etc., with all of them having "fuzzy limits" [13]. Tron y Margaliot [14] describe fuzzy logic as an effective methodology for creating models by considering intuition and agent related behaviour.

Fuzzy logic has been used in economic topics related to energy. Spandagos et al. [15] state that in order to understand the factors that foster consumer energy behaviour and thus enable the development of more efficient polity, it is necessary to create energy consumption models that take economic behaviour into consideration. With this in mind, Spandagos developed a model based on fuzzy logic that includes concepts of bounded rationality, time discounting of gains and pro environmental behaviour. The model is developed from the decision perspective, rules based on human reasoning and behaviour, and also takes into consideration currency, personal comfort and environmental responsibility related variables to generate predictions regarding purchasing decisions and air conditioning use. An important similarity was found between the results generated by the model and historic data on energy usage for the cooling of urban populations. This proved it to be a trustable model. Spandagos showed the feasibility of using fuzzy logic to combine economic, physical quantitative data with qualitative concepts.

Among the several applications of fuzzy logic, there is a model to define and measure sustainability called SAFE, proposed by Phillis et al. [16]. In this model, fuzzy logic is used as well as 75 indicators to classify 128 countries, also considering expert opinions, international agreements and frameworks. This model measures sustainability on a world scale, but it can be adapted to smaller regions since its variables, both input and output, rules and membership functions can be modified. Fuzzy logic has also been used in the evaluation of energy systems in dwellings, as was done by Gamalath et al. [17]. In their paper they propose an assessment framework for the condition of the energy system in multi-unit residential buildings (MURB). Their evaluation method applies fuzzy logic to overcome data uncertainty and imprecision. It also uses the rules to combine different performance categories to obtain a grade on the general condition of a MURB. The application of fuzzy logic can help to account for qualitative data that might be obtained in stakeholders' consultations. Their study demonstrated that fuzzy logic can be used to improve the strategies of asset management and operation of existing buildings.

In a similar manner to Spandagos et al., that achieve a model based on fuzzy logic with concepts both quantitative as qualitative, our work combines qualitative values with energy expenses. Our model is also adaptive, as like that of Phillis et al., because the variables (input and output), the rules and the membership functions can be modified in the light of each country's context. Furthermore, like the Gamalath et al. model, ours can be used to design public policy as well as to improve management strategies.

Fuzzy logic can obtain results from human language, as opposed to other methods, if the variables and their values are presented as quantifiable data. This mathematical tool is helpful for highly complex systems that cannot be represented by differential equations or which cannot be solved through conventional means, as their solution entails a high level of complexity. However, Fuzzy Logic does not require complex mathematical models, but anyone with expertise on a given subject can use the methodology, i.e., it is a heuristic tool. Within the bivalent logic of Charles Boole [18], an element from the whole might be part or not of the whole, using Zadeh logic being a part of a fuzzy whole is neither one nor zero, but gradually varies between one and zero.

When trying to solve a system using fuzzy techniques, there are three things the expert needs to define. Firstly, the properties which will characterize the system (linguistic variables), and the set of values which they will undertake (linguistic values). Having established these, they must define the membership functions linking these properties with the system. The next and last key step is what makes fuzzy logic a great approach for evaluating complex systems, which is the rule definition. These are defined by an expert performing the analyses and include their biases on how the system should behave. To quantify the membership on a given group does not represent either a probability or a percentage, but rather how a given characteristic places us in a group we are referring to a population group. This allows us to implement a frequently used concept in Fuzzy Logic, the membership in a group with specific characteristics. Defining linguistic variables and rules is what brings about a system that can be adjusted from different perspectives. Every fuzzy analysis is unique, as each expert will attach their personal imprint.

For this study, we used data from ENCEVI, the survey conducted by National Institute of Statistic and Geography (INEGI, its Spanish acronym) [4], which gathers data provided by the persons interviewed. It is important to mention that there are no direct measurements involved in this survey. As defined by the fuzzy logic methodology, we needed to select the linguistic variables for the system. These were selected for a population by its energy access. We chose transport, cooking fuel and electricity expenditure, as they all had an associated measure and are of key importance to an individual's well-being. According to Sovacool et al. [19] " . . . for both the rural and urban poor, low mobility—regard less of the technology or mode of transport involved—stifles the attainment of better living standards. It reduces the ability to earn income, strains economic resources, and limits access to education and health services and markets . . . ". It is with this consideration that we include transport as a variable related to energy access. Both cooking fuel and electricity expenditure have been used on previous studies done on energy access [20] and Energy Poverty [21] in Mexico. Furthermore, electricity access at this time of world development is crucial, as it provides numerous benefits in addition to other services being closely related, such as entertainment, education, communication, etc. [22]. Furthermore, as we know, the massive trend towards electrification will make it more and more relevant to individual and community wellbeing. Another important reason for choosing these variables was considering that all of them can be measured in the same unit (Mexican Peso).

2. Theoretical Basis

When doing a fuzzy logic analysis, the first step is to define linguistic variables, which will be the criteria framing the system. For our energy access analysis, we have defined them as: transport, cooking fuel and electricity expenditure. They serve as indicators to evaluate or characterize it, and are made up of values which we call linguistic values. Linguistics values are then subdivided into bands. For the case presented herein, we have set them as follows: low, middle, and high.

The second step is the rule design. For this, we must first calculate the number of rules to be applied in assessing a problem. This is calculated by the expression $A=B^C$. In which A: number of rules. B: number of linguistic variables. C: amount of bands.

Our energy access index will be comprised of 3 linguistic variables, framed within 3 value bands, requiring 27 rules. The number of rules is the first filter to understand if fizzy logic is the right method to solve the problem. It also needs to take into consideration the degree of knowledge the expert might have about the problem, and their capacity to come up with sensible rules. Having a high number of rules will increase the processing time. If we had chosen to evaluate 5 linguistic variables—each one with 5 bands—the system would require 3125 rules (5^5).

The membership function is a very important element in problem solution. It shows the degree to which an element is related or has a characteristic associated to a linguistic value. It defines membership. This function might be of different types, fundamentally it could be triangular, trapezoidal, Z type and S type—we have chosen to use the latter for this analysis [3].

2.1. Description of Experimental Data.

To have homogenized linguistic variables, the defined set of linguistic values can all be expressed in monetary units. Data at state level were obtained from the ENCEVI survey performed in 2018 by INEGI [4]. Figures 1–3 show the averages by state for transport, fuel used in cooking and electricity expenses.

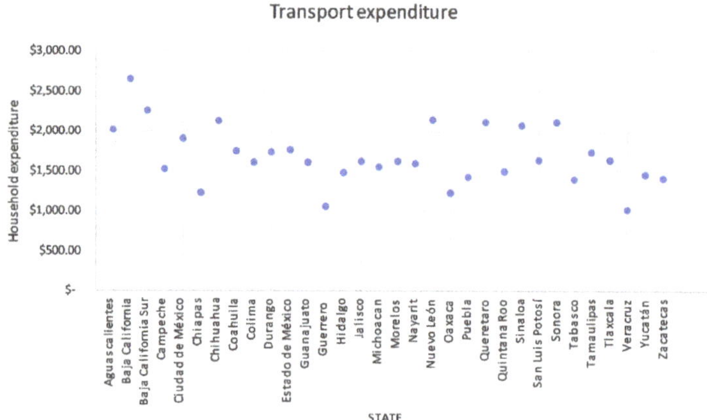

Figure 1. Average monthly transport expenditure per household in pesos, by State. Source ENCEVI, INEGI.

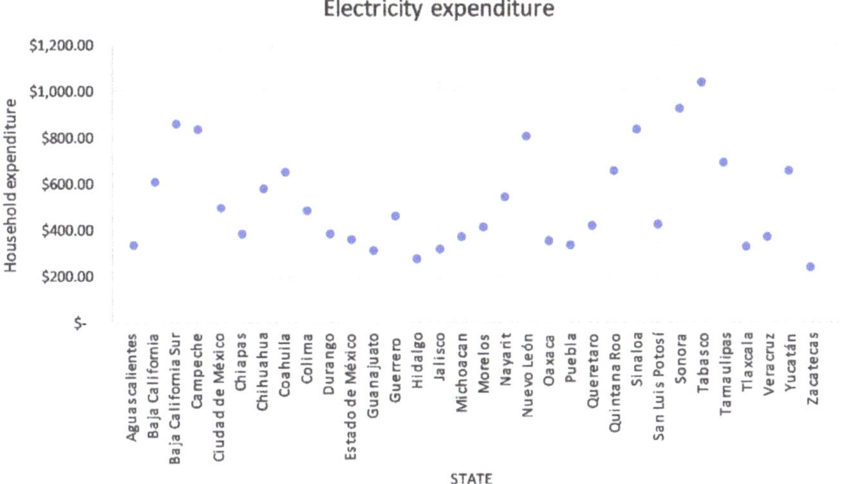

Figure 2. Average bimonthly electricity expenditure per household in pesos, by State. In Mexico, payment for domestic electric service is made every two months. Source ENCEVI, INEGI.

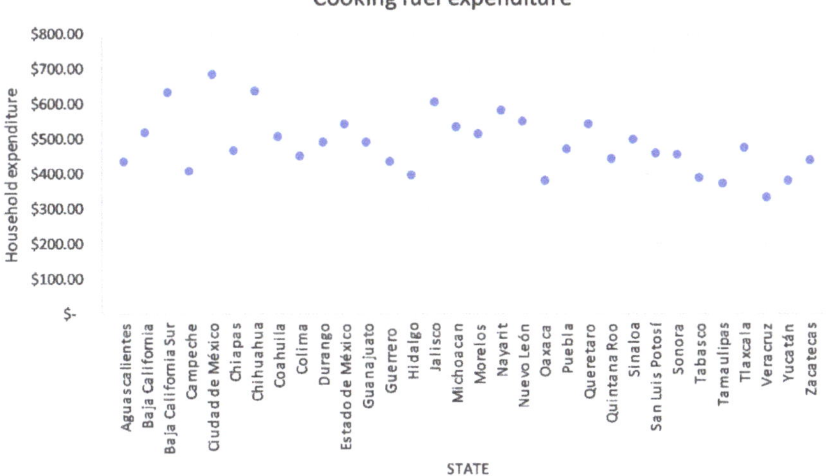

Figure 3. Average monthly cooking fuel expenditure per household in pesos, by State. Source ENCEVI, INEGI.

It is important to mention that these variables, taken from the literature, are coupled to the variables of the survey. Within the survey the expense for household appliances or heating is not included, this is included in the total expenditure of electricity.

In this case, the ENCEVI survey does not consider electricity as a cooking fuel. In Mexico, very few households use electric stoves. The selected linguistic variables show similar behaviour in all states. This could be a problem for the precision of the method. The value-bands need to be adjusted to refine the results (Energy Access) value-range. This procedure will also depend on the experience of the expert in charge of designing the process.

2.2. Calculus

Based on each state's population sample average expenditure, we define the range for the categories each value can fall into. The values can fall into three categories—high, medium and low—and a number is given to each one. For the present case study, the input variables are shown in Figures 4–6 and the Figure 7 represent the output variable.

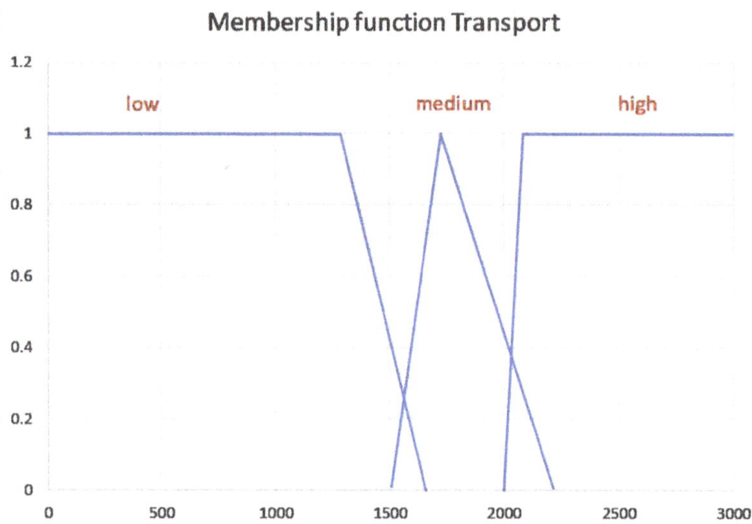

Figure 4. Membership function transport.

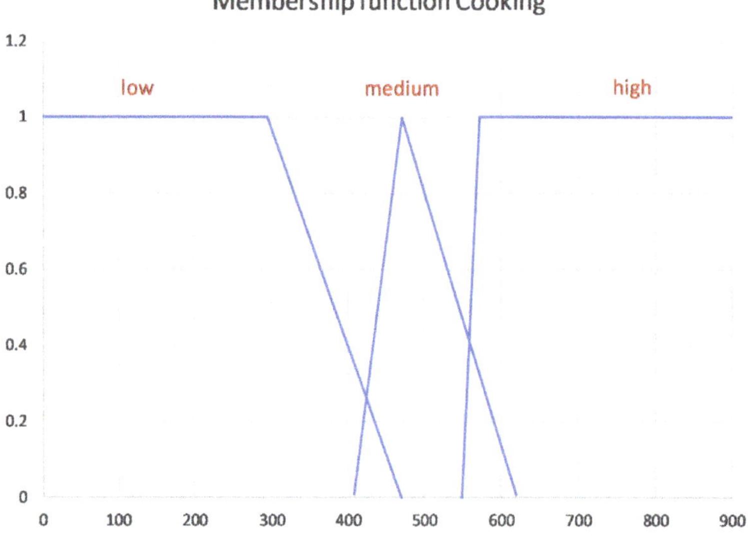

Figure 5. Membership function cooking.

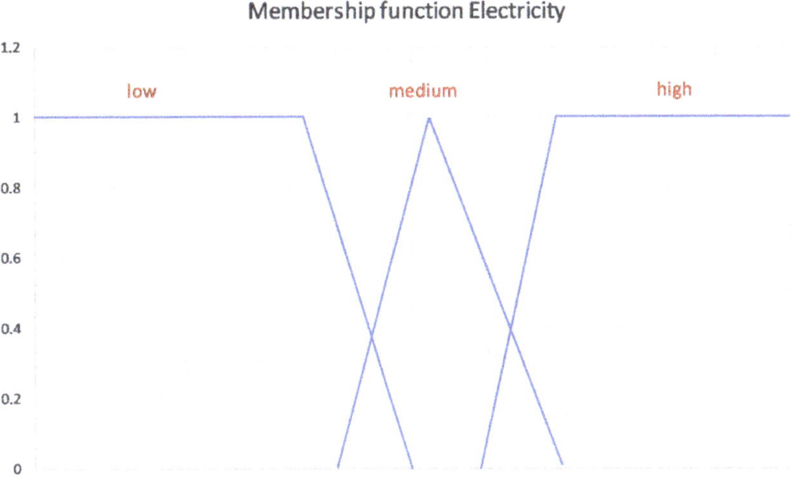

Figure 6. Membership function electricity.

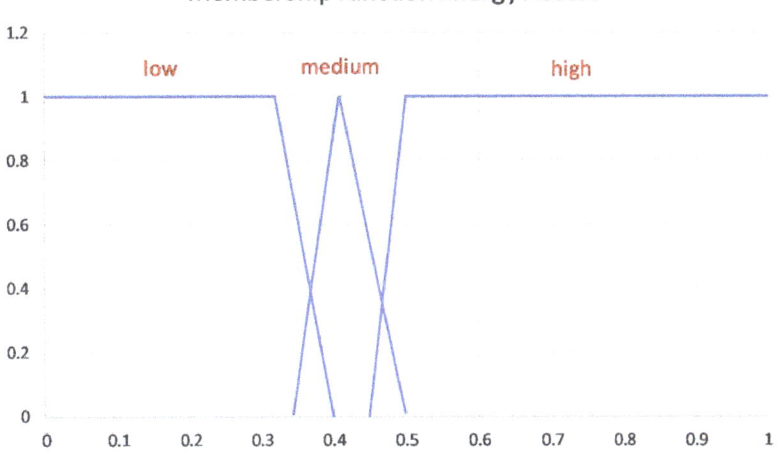

Figure 7. Membership function EA (Energy Access).

In the Figure 7 the values closer to 0 indicate a low energy access, while values closer to 1 indicate a high availability to energy services.

The definition of the values for the membership function "Energy Access" were obtained using a similar method to Nussbaumer et al. [22] to measure energy poverty. They used a threshold to define energy poverty of 0.32, based on experience and through intensive tests. It is a well-known fact that membership function values can be modified according to the context of the specific application. We defined our values due to the similar behaviour that the results from the fuzzy logic algorithm gave when comparing the per capita PIB of each state with the reality of each state. However, those values on energy access in each country should be assigned by experts so to have a model that truthfully represents any given regional context. The experts on the topic are those with the knowledge, the expertise and the access to high quality data, that can determine whether or not a model represents each countries energy landscape.

The next step after defining linguistic variables, values and membership functions is to draft the rules, then to undertake an analysis of each element. This procedure can be performed manually. However, there are several tools that can do this task in a more efficient manner. For this study, we used a tool designed in MATLAB [23].

The same procedure used in defining "EA" output values is used, this output has three variables: high, medium and low, with 0.46, 0.41 and 0.32 values respectively. Below the presentation of a typical rule is shown.

If (transportation is high) and (cooking is low) and (electricity is medium) then (AE is medium).

Using the above rule and the other 26 rules, we designed the system by applying the Mamdani fuzzy inference systems, which closely recreates both human reasoning and the fuzzy if-and-then rules. Moreover, we used the Mamdani method as it generates a fuzzy set as its output. This more complete output is the reason we chose this over more popular methods such as the Sugeno method, whose output is only linear or constant.

It is important to mention that, just as with the linguistic values, rules can be modified depending on the person making the analysis. This presents an important advantage in comparison with other analysis methodologies, as it allows us to assess the system under other conditions.

3. Results

Once the system is ready and the expenditure averages are defined, you can start calculating the value of energy access for each region. Figure 8 shows how the tool works, by inputting the values we need to be analysed and showing the energy access value as a result. For the example below, we introduced Mexico City's values for transportation, cooking and electricity (1901, 682 and 496 respectively). The tool shows us the EA value is of 0.762.

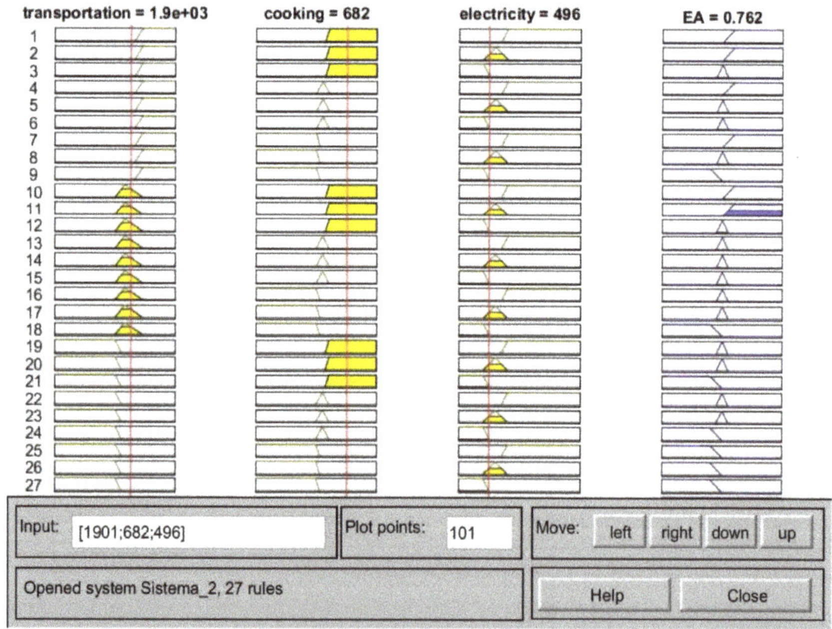

Figure 8. MATLAB screen results. Evaluation of the rules and their results to determine energy access.

Figure 9 shows energy access across all states; as per our previous definition, states over 0.46 are classified as having a high energy accessibility, and those under 0.32 as low accessibility, which means that in general, they will face above average difficulties in gaining access to electricity, cooking fuels and transport compared to the other states.

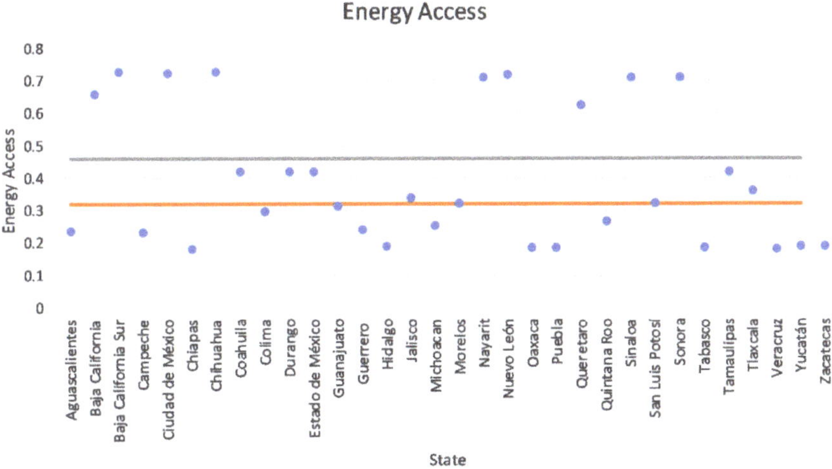

Figure 9. Energy access by state. Less than 0.32 is considered as low energy access, greater than 0.46 is considered as high energy access. Between 0.32 and 0.46 is considered as medium energy access.

4. Discussion

We obtained a distribution of energy access (EA) across Mexico by applying fuzzy logic. As it is a fuzzy set, it can be divided in different ranges or zones. This division varies in response to local information on energy access and the characteristics of the locality. Since it is a heuristic criterion, the most valuable use of the tool is to monitor Energy Access in a locality through time. When comparing the values between regions, other socioeconomic indicators are needed in order to have a better understanding of each region's relationship with energy access. This is not only a strictly theoretical absolute result; rather, it is a methodology that enables comparisons. In this specific case, when we associate a membership function to a number, this element within the whole state reflects the specific property to which we are referring—in this case EA. Each of the EA ranges might be called: Low EA, Medium EA and High EA. The same rule will be used to classify all of them.

Drawing comparisons with research from other contexts is challenging, as the study of social phenomena is complex and involves a different approach. In natural sciences, defining magnitudes and models to estimate a given situation it is a straightforward matter. However, in social sciences, although exact mathematical models are applied to social issues, there are many variables intervening. Furthermore, in most cases, all the different variables and conditions that take place are unknown. So, the application of these type of tools is helpful to grasp the context of various entities, and to start the understanding of social phenomena. Social applications have a complex nature; describing and modelling them is a challenge requiring a complex system approach. For our case study, we wanted to see the relationships between EA and three factors—economic development, geographic characteristics and socio-cultural behaviour. Figure 10 shows our comparison between EA and GDP per capita, which we calculated using 2017 data from INEGI [24] and CONAPO [25].

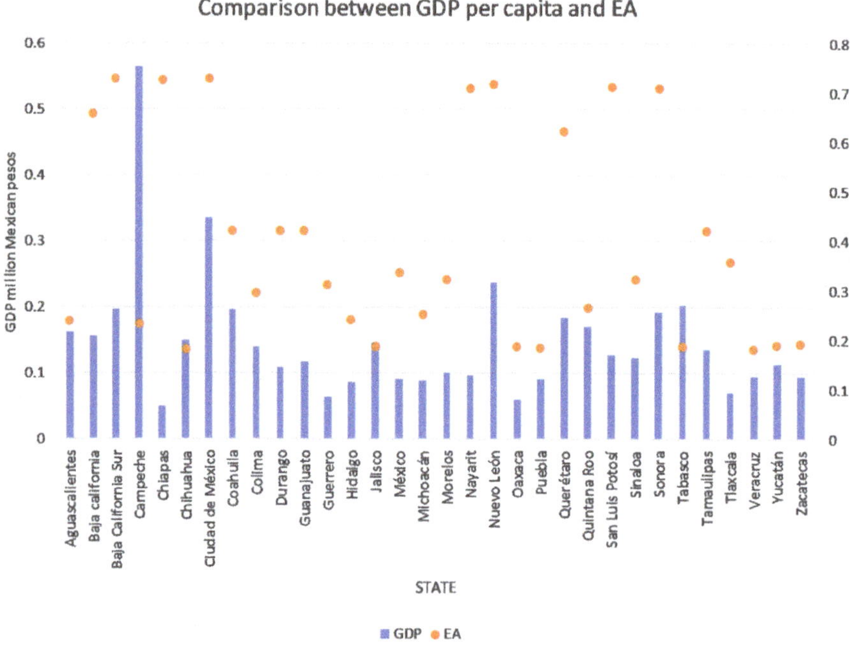

Figure 10. Comparison between GDP per capita and EA by state.

The analysis of this graph takes only one aspect related to energy access into consideration, the overall economic activity per state. Even though the prosperity in each region has a close relationship with the quality of energy access and the socioeconomic level of the population, it is clear that there are several factors that we need to analyse to fully understand each individual case. In general, we can see that the level of EA is relatively equal to or greater than the GDP per capita in almost all of the regions. This is because the largest and most important energy and utilities companies (Pemex and CFE) are public (owned by the state). Therefore, there is a natural bias of those public companies to promote social well-being. However, both Campeche and Tabasco pose an exception to this trend. Campeche is the "richest" state because most of the oil production is based there. However, that wealth is not part of the economic activity of most of the population. Since Pemex, the public oil company, is responsible for that income, GDP per capita does not reflect the economic behaviour of the inhabitants of Campeche. The EA indicator shows an average value, not related to GDP per capita. In Tabasco, something similar happens; it is the second most important oil production location. On the other hand, we have the case of Chiapas, an extremely poor region that has a very important indigenous population and is well known for having a strong political agenda. Its inhabitants have access not only to the services provided by the public energy companies, but they have a long-standing tradition of using biomass, so this analysis portrays the reality of Mexico's landscape regarding EA.

If we were to add geographic characteristics and socio-cultural habits, then this analysis would be even more complex. Many authors prefer to draw their analysis based solely on the "geographic regions" variable, as if the only important factor might be the geography of the place, and they do not take the socioeconomic reality into consideration. As we have already expressed, the solution to social problems includes many unknown variables related to one another. This is a multi-variable problem. When you are dealing with multi-variable problems, the mathematical problems become more challenging. In this case, fuzzy logic plays an important role in solving this, because we would need to design a similar model, only with a greater number of rules.

Within the geographic division performed by INEGI [4], there is a warm region with extreme summers, including the states of Durango and Nuevo Leon. Durango has reported a negative growth of −1.0 while Nuevo Leon has reported a growth of 3.0. On the other hand, if we were to assess a torrid region as defined by INEGI in its regionalization of the climate seasonality, we would find unequal economies such as that of Mexico City and that of the state of Mexico, with other less developed States such as Guerrero. This analysis confirms what we have been suggesting from the beginning—that we need a comprehensive global analysis, including economic, social and climate variables.

Since the objective of this paper is to show the possible application of fuzzy logic, we have decided to simplify the model and shorten the geographic space to states, taking only political division into consideration. What are the advantages and disadvantages of this? The main advantage relies on a more uniform socioeconomic data (per single state). The main disadvantage is the lack of climate-related information, which decreases the model's precision. This relies, on the other hand, on our main purpose of simplifying the model: remembering that increasing the number of linguistic variables increases the number of rules (exponentially). That is why we have decided to draw this analysis by states and not by climate regions.

5. Conclusions

We can reach the conclusion that the method used is pertinent in most of the federate entities when evaluating the EA in each state, especially those with similar bio-climate and/or socioeconomic regions. This method shows that fuzzy logic might be used to measure energy access and to highlight where it is low and deserves special attention.

Nevertheless, to obtain more accurate calculus, we should undertake several possible actions: increase the number of linguistic variables; adjust the values of those linguistic variables; use another survey or even organize our own survey.

We have concluded that the analysis of results by states might be an alternative to the geographic region analysis where the exactitude will depend on the number of variables taken into consideration. Especially when the aim is to implement energy access recovering measures, it is important to precisely define where they will be implemented. It is not the same to define an energy access recovering program for a small city, a municipality or a town as it is for a geographic area in general. That is why it is necessary to include more variables that give a better characterization to the reality of the entities in all-important energy dimensions regarding access.

If we take into consideration all the processing and calculus advantages that fuzzy logic offers us, and combine this with the analysis made, we find out that this approach for evaluating energy access should be taken into consideration by researchers in the field and public policy makers.

The most important result of this paper is that it provides researchers with another tool that has been shown to be useful in the assessment of energy access—fuzzy logic. This technique entails neither high mathematical complexity nor an excessive use of computer time and intensity, while providing a useful model to evaluate and monitor energy access through time.

Author Contributions: Conceptualization, D.S.-J. and K.G.C.; methodology, D.S.-J.; software, T.R.-B.; validation, T.R.-B.and K.G.C.; formal analysis, D.S.-J.; investigation, D.S.-J., T.R.-B. and K.G.C.; data curation, T.R.-B.; writing—original draft preparation, D.S.-J.; writing—review and editing, K.G.C.; visualization, T.R.-B.; supervision, K.G.C. All authors have read and agreed to the published version of the manuscript.

Funding: This research received no external funding.

Acknowledgments: I want to thank the National Science and Technology Council (CONACYT, its Spanish acronym), for the scholarship awarded.

Conflicts of Interest: The authors declare no conflict of interest.

References

1. Butera, F.; Caputo, P.; Adhikari, R.; Mele, R. Energy access in informal settlements. Results of a wide in site survey in Rio De Janeiro. *Energy Policy* **2019**, *134*, 110943. [CrossRef]
2. Jimenez, R. Barriers to electrification in Latin America: Income, location, and economic development. *Energy Strategy Rev.* **2017**, *15*, 9–18. [CrossRef]
3. INEGI (Instituto Nacional de Estadística Geografía e Informática). Encuesta Nacional de Ingresos y Gastos de los Hogares 2016. Available online: https://www.inegi.org.mx/programas/enigh/nc/2016/ (accessed on 22 October 2019).
4. INEGI (Instituto Nacional de Estadística Geografía e Informática). Encuesta Nacional Sobre Consumo de Energéticos en Viviendas Particulares 2018—Información General. 2018. Available online: http://www3.inegi.org.mx/rnm/index.php/catalog/495 (accessed on 22 October 2019).
5. Schwab, K. The Global Competitiveness Report 2017–2018. In World Economic Forum. Available online: http://www3.weforum.org/docs/GCR2017-2018/05FullReport/TheGlobalCompetitivenessReport2017%E2%80%932018.pdf (accessed on 21 February 2020).
6. Schwab, K. The Global Competitiveness Report 2019. In World Economic Forum. Available online: http://www3.weforum.org/docs/WEF_TheGlobalCompetitivenessReport2019.pdf (accessed on 21 February 2020).
7. Thomson, H.; Simcock, N.; Bouzarovski, S.; Petrova, S. Energy & Buildings Energy poverty and indoor cooling: An overlooked issue in Europe. *Energy Build.* **2019**, *196*, 21–29. [CrossRef]
8. Zadeh, L. Fuzzy sets. *Inf. Control* **1965**, *8*, 338–353. [CrossRef]
9. Gutierrez, S.M.C.; Trujillo, O.E. Design and implementation of a speed and position controller for a conveyor belt applying fuzzy logic and digital techniques. *Rev. Colomb. Tecnol. Av.* **2005**, *2*, 27–33.
10. Diaz-Contreras, C.A.; Aguilera-Rojas, A.; Guillén-Barrientos, N. Fuzzy logic vs. multiple regression for selection personnel. *Rev. Chil. Ing.* **2014**, *22*, 547–559.
11. Shang, K.; Hossen, Z. Applying Fuzzy Logic to Risk Assessment and Decision-Making. *Casualty Actuar. Soc. Can. Inst. Actuar. Soc. Actuar.* **2013**, *2*, 1–59.
12. Nobre, F.S.; Tobias, A.M.; Walker, D.S. The impact of cognitive machines on complex decisions and organizational change. *AI Soc.* **2009**, *24*, 365–381. [CrossRef]
13. Tron, E.; Margaliot, M. Modeling observed natural behaviour using fuzzy logic. In Proceedings of the 12th IEEE International Conference on Fuzzy Systems (FUZZ '03), St Louis, MO, USA, 25–28 May 2003; Volume II, pp. 74–78.
14. Bernstein, D.A.; Clarke-Stewart, A.; Roy, E.J.; Wickens, C.D. Boston: Houghton Mifflin Company. *Psychology* **1997**, *4*.
15. Spandagos, C.; Ng, T.L. Fuzzy model of residential energy decision-making considering behavioural economic concepts. *Appl. Energy* **2018**, *213*, 611–625. [CrossRef]
16. Phillis, Y.A.; Grigoroudis, E.; Kouikoglou, V.S. Sustainability ranking and improvement of countries. *Ecol. Econ.* **2011**, *70*, 542–553. [CrossRef]
17. Gamalath, I.; Hewage, K.; Ruparathna, R.; Karunathilake, H.; Prabatha, T.; Sadiq, R. Energy rating system for climate conscious operation of multi-unit residential buildings. *Clean Technol. Environ. Policy* **2018**, *20*, 785–802. [CrossRef]
18. Gossett, E. *Discrete Mathematics with Proof*, 2nd ed.; Bethel University: Ander Hills, MN, USA, 2008; pp. 17–23. ISBN 9780470457931.
19. Sovacool, B.; Cooper, C.; Bazilian, M.; Johnson, K.; Zoppo, D.; Clarke, S.; Raza, H. What moves and works: Broading the consideration of energy poverty. *Energy Policy* **2012**, *42*, 715–719. [CrossRef]
20. De Buen Odón, R.; Navarrete, J.I.; Hernández, P. Análisis del impacto de las Normas Oficiales Mexicanas de Eficiencia Energética en el ingreso-gasto del sector residencial de México a partir de datos de INEGI (1990–2016). *Cuadernos de la Conuee. CONUEE* **2018**, *9*, 1–17.
21. García, R. Spatial Characterization of fuel poverty in Mexico. An analysis at the subnational scale. *Econ. Soc. Territ.* **2016**, *51*, 289–337.
22. Nussbaumer, P.; Bazilian, M.; Modi, V. Measuring energy poverty: Focusing on what matters. *Renew. Sustain. Energy Rev.* **2012**, *16*, 231–243. [CrossRef]
23. Mathworks. Fuzzy Logic ToolboxTM. Available online: https://la.mathworks.com/products/fuzzy-logic.html (accessed on 22 October 2019).

24. INEGI (Instituto Nacional de Estadística Geografía e Informática). Producto Interno Bruto por Entidad Federativa 2017. Available online: https://www.inegi.org.mx/contenidos/saladeprensa/boletines/2018/OtrTemEcon/PIBEntFed2017.pdf (accessed on 22 October 2019).
25. CONAPO (Consejo Nacional de Población). Cuadernillos Estatales de las Proyecciones de la Población de México y de las Entidades Federativas 2016–2050. Available online: https://www.gob.mx/conapo/documentos/cuadernillos -estatales-de-las-proyecciones-de-la-poblacion-de-mexico-y-de-las-entidades-federativas-2016-2050-208243?idiom=es (accessed on 21 February 2020).

Article

Energy Use and Labor Productivity in Ethiopia: The Case of the Manufacturing Industry

Selamawit G. Kebede [1] and Almas Heshmati [2],*

[1]   Department of Economics, Addis Ababa University, Main campus at 6 Kilo, Addis Ababa, Ethiopia; selam3MG@yahoo.com

[2]   Jönköping International Business School, P.O. Box 1026, SE-551 11 Jönköping, Sweden

*   Correspondence: almas.heshmati@ju.se; Tel.: +46-36-101780

Received: 9 April 2020; Accepted: 26 May 2020; Published: 28 May 2020

Abstract: This study investigates the effect of energy use on labor productivity in the Ethiopian manufacturing industry. It uses panel data for the manufacturing industry groups to estimate the coefficients using the dynamic panel estimator. The study's results confirm that energy use increases manufacturing labor productivity. The coefficients for the control variables are in keeping with theoretical predictions. Capital positively augments productivity in the industries. Based on our results, technology induces manufacturing's labor productivity. Likewise, more labor employment induces labor productivity due to the dominance of labor-intensive manufacturing industries in Ethiopia. Alternative model specifications provide evidence of a robust link between energy and labor productivity in the Ethiopian manufacturing industry. Our results imply that there needs to be more focus on the efficient use of energy, labor, capital, and technology to increase the manufacturing industry's labor productivity and to overcome the premature deindustrialization patterns being seen in Ethiopia.

Keywords: manufacturing; labor productivity; energy; Ethiopia

---

1. Introduction

Industrial expansion is essential for socioeconomic development as it generates different opportunities—capital accumulation, structural changes, technological innovations, and productivity—that improve economic performance [1–3]. Industrialization or the shift from agriculture to the manufacturing sector is key to development, making development without industrialization an unthinkable process [1,4]. Industrial development is also the pathway for the structural transformation of an economy and society. High rates of economic growth and capital accumulation are essential but not adequate for structural transformation, unless complemented by industrialization [2]. Industrialization promotes economic diversification, inclusive growth, and the efficient utilization of resources, such as physical, human, and mineral resources, which help eradicate poverty [5].

The productivity advantage of manufacturing over other sectors is a major factor for pursuing sustained industrialization, along with the higher externalities that can arise from manufacturing growth [6]. Unlike agriculture and the service sectors, manufacturing accelerates convergence and, with its huge productivity advantages, will enable developing economies to catch up with their developed counterparts [4]. Different factors are attributed to industrial growth and productivity, including human or physical capital, labor, energy, innovations, and capacity utilization [3,7,8]. Among others, energy is critical for productivity and growth as it enables achieving both industrial development and structural transformation [9]. In fact, the use of energy is a precondition for the development of human society and more energy use is required for sustaining industrial development [9]. Energy use is directly related to growth and economic development and is an essential input for all production and consumption activities [8,10].

The causal relationship between energy consumption and growth has been investigated in different countries but the results remain controversial with diverse outcomes in different countries based on the econometric approaches used and the time spans of the studies [10–14]. Some studies validate the positive effects of energy on growth and productivity [15–17], while others empirically confirm a negative impact of energy on growth and productivity [11,12]. Others find no causal relationship between the two empirically [10,18]. Here we use different econometric estimators for matters of sensitivity analysis of the result to evaluate how energy affects labor productivity in Ethiopian manufacturing.

In Ethiopia, the share of agriculture and services in gross domestic product (GDP) has been more than 60 percent and 20 percent, respectively, for decades, while manufacturing's contribution to GDP has been less than five percent, which, too, is attributable to other industries [19,20]. Currently, the service sector contributes 47 percent, agriculture 43 percent, and industry makes up the rest, leaving a very low share of GDP being contributed by manufacturing [21]. The existing literature confirms that Ethiopian people have been depending on agriculture for their livelihood for decades in terms of production and employment, with significantly small contributions from the manufacturing sector to the economy [22,23]. The dominance of, first, the agriculture sector and, later, the service sector shows premature deindustrialization in Ethiopia, while the low share of manufacturing implies output deindustrialization [21,24].

The low industry performance can be attributed to several factors such as inefficient use of labor, energy, human or physical capital, innovations and capacity utilization [3,8,24]. As established theoretically, energy is a significant factor in determining sustainable industrial production. However, the empirical relationship between energy and growth is mixed [10–14]. Besides, there are very few empirical studies on energy and productivity at the industry level. This motivated us to undertake this study on the empirical relationship between energy and labor productivity in the case of the Ethiopian manufacturing industry. Accordingly, this study addresses the following research question:

How does energy effect labor productivity in the Ethiopian manufacturing industry?

The analysis emphasizes the role of energy use in manufacturing labor productivity in Ethiopia. The study uses panel data for estimating the empirical model using a dynamic generalized method of moments (GMM) estimator. The estimation results confirm that energy use positively effects labor productivity in the manufacturing sector in Ethiopia. This implies that the efficient use of energy is a pillar of labor productivity in the Ethiopian manufacturing industry. Thus, this study adds to the existing literature by empirically confirming the relationship between energy use and labor productivity across different model specifications.

The rest of this research is organized as follows. Section 2 reviews the literature on energy and productivity. The empirical model and estimation approach are presented in Section 3, along with the definitions of the variables used in the model. Data are discussed in Section 4. A descriptive and regression-based analysis of the energy and labor productivity of the manufacturing sector in Ethiopia is discussed in Section 5. The final section gives the conclusion and the implications of the findings.

2. Literature Review on Energy and Productivity Growth

This section presents a general overview of the link between energy and productivity, followed by an empirical review of the relationship between energy and growth. It then discusses existing studies on the determinants of labor productivity. This helps establish the rationale for undertaking this study that links energy with labor productivity at the industry level in Ethiopia.

There are two empirically fundamental questions related to disparities in the level of economic development across nations. Economists inquire why some economies are richer than others, and what accounts for the huge increases in real incomes over time [25,26]. The extensive dispersion of output growth rates across countries is documented economically [27]. A comparison between countries shows that countries that at one time had similar levels of per capita income consequently followed very different patterns, with some seemingly caught in long-term stagnation while others were able to sustain high growth rates [28].

Among others, productivity is a determining factor of growth at the national and industrial levels, with increasing globalization and the expansion of competitive industrial product markets [16,29]. High industrial labor productivity results in lower per unit costs and increases firms' ability to compete in global markets [16]. There are several determinants of labor productivity, including human or physical capital, energy, and technology [29–31]. Energy is an essential input that constrains or induces productivity growth in different firms. It is an essential factor of production that is required in all economic processes [29,31]. This basic production input in economic activities provides a conducive platform for industrial growth and productivity. The efficient use of energy leads to the higher productivity of resources and a more dynamically competitive economy that can respond to the required economic transition from agriculture to industry dominated structure [32].

Energy has countless ways of empowering human beings through increasing productivity, powering industrial and agricultural processes, alleviating poverty, and facilitating sound social and economic development [33]. Limited access to energy cripples economic growth and development, which makes universal access to energy a major emphasis of the sustainable development goals [9]. The increased availability and use of energy increases productivity and enhances economic development [34]. Energy is primarily associated with the provision of power for agricultural or industrial production [35,36]. In fact, sustainable development and modern industry require reliable, affordable, and energy services available for all on a sustainable basis [9,33]. However, access to energy is limited and is accompanied by low quality and poor reliability, affordability, and availability [9]. Energy can be measured in terms of cost or value and can be disaggregated into electricity or other forms of energy based on types. It is possible to measure energy consumption in equivalent kilowatt hours (KWh) [37].

Energy use is a major stimulating factor in industrial productivity [16,32]. Public services and industrial production require access to energy use [12]. Recently, the demand for energy has been increasing, with the world having a population of over 7.2 billion, which is increasing [38]. Access to energy in Africa is low—for every ten people in sub-Saharan Africa (SSA), only four have access to electricity compared to the global access of nine out of ten people having access to energy; 57 percent of the global deficiency in access to electricity energy comes from SSA [9].

There is an increasing interest in identifying energy's role in productivity, as empirical findings on their causal relationship are mixed [13,39]. For instance, Schurr et al. [40] presents the association between energy consumption and growth in the national product (GNP) in the United States over the period 1880-1955. These authors identified two trends in the pattern of the energy share in relation to GNP. The share of energy to GNP was rising, until it declined persistently after the war. This change in the trend is attributed to a compositional change in the national output to light industries, which use less energy compared to heavy manufacturing industries and services and is also due to major improvements in the efficiency of energy conservation in light industries. In a follow-up to the study by Schurr et al. [40], Schurr [41] explored the link between energy use, productive efficiency, and energy efficiency from the 1920s to 1981. His study indicated that energy intensity, defined as energy's share in GNP, declined when multifactor productivity increased during the study period. Unlike the share of energy in output, which is attributed to technological advances that increased overall productive efficiency, energy intensities in terms of factor inputs increased over the study period. This ultimately led to an increase in the final output, which was more than the consumption of energy.

The role of electrification and non-electricity energy in productivity growth for the USA's economy is examined by Jorgenson [42]. His study confirms that electricity energy is related to productivity growth. However, there is also a strong association between non-electricity energy and productivity growth in the US economy. In another related study, Boudreaux [43] examined the impact of electricity energy on manufacturing productivity in the US from 1950 to 1984. This study showed that growth in electricity energy accounted for 79 percent of the value added to the manufacturing sector. Empirically, the study showed that the decline in energy growth accounted for the slowdown in productivity and output growth.

The role of energy in productivity growth in the European Union countries is assessed by Murillo-Zamorano [44] who empirically confirmed that energy is a fundamental input in productivity change. In another related study, the relationship between energy and labor productivity was examined by studying the effect of renewable and non-renewable energy in European countries over the period from 1995 to 2015 using the production frontier approach [45]. This study showed that renewable and non-renewable energy had an effect on the growth of the countries in the European Union. Based on his study, the author concluded that non-renewable energy had a positive impact, leading to divergence, while renewable energy had a negative impact, leading to convergence.

Energy and income causality for ten emerging markets, excluding China because of limited data availability and the G-7 countries, is examined in Soytas and Sari [17]. Their results show the bidirectional causality in Argentina, causality running from energy to GDP in France, Germany, Japan, and Turkey and causality running from GDP to energy consumption in Italy and Korea. The nexus between energy and growth for 20 net importer and exporter countries from 1971 to 2002 using the panel vector correction model is investigated by Mahadevan and Asafu-Adjaye [46]. Their findings show that for energy exporter developed countries this causal relationship is bidirectional, while for developing countries energy stimulates growth in the short term.

The effect of energy consumption and human capital on economic growth for 130 oil-exporting and developed countries from 1981 to 2009 is investigated by Alaali et al. [15]. Using GMM, they estimate an augmented neoclassical growth model, including education and health as human capital along with energy consumption. Their results show that energy had a positive and significant effect on growth. The empirical relationship between energy consumption and gross domestic product for six Gulf Cooperation Council (GCC) countries using cointegration and causality methods is investigated by Al-Irani [13]. His results show a unidirectional causal relationship running from GDP to energy consumption, but not the other way around. Moghaddasi [11] investigated the role of energy consumption in total-factor productivity in Iranian agriculture using the Solow residual model and their results show a negative impact, which they attribute to cheap and inefficient use of energy in this sector.

Kebede et al. [12] investigated energy demand in east, west, central, and south sub-Saharan countries using time series cross-sectional data for 20 countries for a 25-year time span. Their results show that energy demand was positively related to GDP, the population growth rate, and agricultural expansion, while it was negatively correlated with industrial development and the price of petroleum. The causal relationship between energy consumption and economic growth for 11 sub-Saharan African countries is investigated by Skinlo [18] using the ARDL bound test and Granger causality. His results show that there was cointegration between energy use and economic growth in seven countries included in the study: Ghana, Cameron, Senegal, Cote d'Ivoire, Zimbabwe, Gambia, and Sudan. In Sudan and Zimbabwe, the Granger causality ran from economic growth to energy use while in Cameroon and Cote d'Ivoire he found no Granger causality between energy consumption and economic growth.

Wolde-Rufael [39] investigated the causal relationship between energy consumption and economic growth for 17 African countries using the variance decomposition factor and impulse response analysis. The variance decomposition analysis confirmed that labor and capital were important, while energy was not as important as these factors. A meta-analysis using a multinomial logit model for 174 samples was conducted by Chen et al. [10] to explore the relationship between energy and GDP, with controversial results that show that the time span, econometric model, and selection characteristics affected the debatable outcomes of the casual relationship significantly.

The second part of this section explores labor productivity and its determinants, as studied by different researchers. Su and Heshmati [30] studied the development and source of labor productivity in 31 provinces of China during 2000-09. They used a fixed effects model adjusted for heteroscedasticity to estimate the coefficients' fixed assets, average labor wage, total volume of business, post and telecommunications, and profits, which had a positive effect on labor productivity. Accounting for heterogeneity, Velucchi and Viviani [47] examined the determinants of labor productivity in Italian firms using panel data and a quantile regression. Their results show that human capital and assets had a strong positive impact on fostering the productivity of low productive firms compared to high productive ones. Islam and Syed-Shazali [48] studied the impact of the degree of skills, research and development (R&D), and a favorable work environment on the productivity of labor-intensive manufacturing industries in Bangladesh. Their results confirmed a positive correlation between productivity and the degree of skills and the work environment, though it was a weak correlation; R&D had a strong positive correlation with productivity in Bangladesh.

Recently, Heshmati and Rashidghalam [49] studied the determinants of labor productivity in manufacturing and service sectors in Kenya using the World Bank Enterprise Survey database for 2013. Their findings confirm a positive effect of capital intensity and wages on labor productivity while female participation reduced productivity in these sectors. In a comparative study, Nagler and Naudé [50] examined the factors determining the labor productivity of non-farm enterprises in rural sub-Saharan Africa in Ethiopia, Nigeria, Uganda, and Malawi using the World Bank's Living Standards Measurement Study – Integrated Surveys on Agriculture (LSMS-ISA) database. They found that rural enterprises were less productive than urban enterprises. By estimating Heckman selection and using panel data models, their study confirmed that education and credit availability induced enterprises' labor productivity.

Samuel and Aram [51] studied the main factors that helped or hindered the realization of industrial productivity in Africa. The study concluded that financial development, economic development, the labor market's flexibility, and the real effective exchange rate were clear determinants of industrialization in the entire region. In a time-series analysis Otalu and Anderu [7], the determinants of industrial sector growth in Nigeria were examined using the cointegration and error correction model (ECM). Their results show that both labor and capital had significant effects on economic growth. The exchange rate showed a positive and significant impact, signifying that currency appreciation might be detrimental to the growth of the industrial sector. In addition, the authors also found that these factors had a more permanent and not a transitory effect on industrial output.

In the energy literature, the contribution of energy use to productivity in practice is controversial, with some studies claiming that energy use is a fundamental pillar of productivity growth, while others argue that energy has little effect on productivity growth [10,44]. In studies on labor productivity, energy seems to be missing as a major determinant factor in explaining labor productivity [47–51]. Furthermore, there is little focus on investigating the explicit role of energy in labor productivity from the manufacturing industry's perspective [47–49]. Most growth theories fail to include energy use as a pillar of productivity or as one argument for the growth differences between nations [45].

Thus, this study adds to the existing literature by addressing the controversial nature of previous studies' results by empirically investigating the association between energy and productivity in the Ethiopian manufacturing industry. It also considers energy as a major variable of interest for explaining labor productivity in addition to capital and technical changes. This link is investigated from the manufacturing industry's perspective. Moreover, this study uses different model specifications to confirm the consistency of this relationship by using both static and dynamic panel data estimators for the manufacturing industry groups.

3. Model Specification and Estimation

3.1. Model Specification

Productivity is a fundamental indicator for assessing economic performance [52]. In general terms, productivity can be defined as the ratio of total output produced to the inputs used. There are different measures of productivity, which can be classified as multifactor productivity measures and single factor measures of productivity [53]. The former relate output to a bundle of inputs, while the latter measure the ratio of output to a single input [52]. For instance, labor productivity is defined as the ratio of the quantity index of gross output to the quantity index of labor input [53]. Among other factors, energy is a key driver of economic growth and industrialization as it enhances the productivity of labor, capital, and other factors of production. In fact, energy use has received considerable attention as a pillar of productivity in the literature on energy economics, but with mixed empirical results for different countries on the causal relationship between the two [13,15,46].

This study empirically investigates the relationship between energy use and labor productivity in Ethiopian manufacturing industries. Like labor and capital production factors, energy is seen as an essential factor for economic development [15]. The production function is a useful tool for analyzing the technological relationship between labor, capital, other inputs, and the output produced [54]. The production function which relates output to the vector of inputs is mostly used for analyzing productivity [55,56]. Accordingly, in this study, the production function developed by Cobb and Douglas [57] is used for estimating the productivity of labor in the manufacturing sector in Ethiopia. The Cobb–Douglas production function, with two inputs in its basic form [58,59], is represented as:

$$Y = AL^{\alpha}K^{\beta} \tag{1}$$

where Y denotes the quantity of production or output or its value, L represents labor or its value, and K stands for the value of capital. α and β are parameters of inputs labor and capital respectively and A is technology. This standard production function can be generalized to include more inputs such as energy and other material inputs:

$$Y = AL^{\alpha}K^{\beta}E^{\gamma} \tag{2}$$

where the other variables are defined in the same manner as in Equation (1). E stands for energy inputs in the production process and γ denotes a parameter to be estimated as a coefficient for energy input. We can linearize the production function by log transformation as:

$$LogY = logA + \alpha logL + \beta logK + \gamma logE + U \tag{3}$$

$$\begin{aligned} &\text{if } \alpha + \beta + \gamma > 1, \text{IRS} \\ &\text{if } \alpha + \beta + \gamma < 1, \text{DRS} \\ &\text{if } \alpha + \beta + \gamma = 1, \text{CRS} \end{aligned} \tag{4}$$

where α, β, and γ stand for elasticities of production with respect to labor, capital, and energy respectively. Equation (3) is the first model to be estimated to decide production's returns to scale in the manufacturing industry in Ethiopia. The sum of the parameters will give us a measure of the returns to scale from a proportional increase in inputs. If the sum of the parameters is greater than one we have increasing returns to scale (IRS); if the sum is less than one, we get decreasing returns to scale (DRS); if the sum is one then the returns to scale are constant (CRT).

As labor productivity shows how effectively labor inputs are converted into outputs [60], we take production or output per employee to measure labor productivity. There are two ways of doing this. First, if one is interested in the scale effects of energy and capital use on labor productivity, then the right-hand side of the equation to include all inputs in the original form per labor, while the left side is measured as productivity—that is, output is divided by labor. In this case, labor, on the right-hand side, represents the scale of production as:

$$\frac{Y}{L} = \frac{AL^{\alpha}K^{\beta}E^{\gamma}}{L} \tag{5a}$$

$$Y/L = AL^{\alpha-1}K^{\beta}E^{\gamma} \tag{5b}$$

$$LogY/L_{it} = logA_{it} + (\alpha - 1)logL_{it} + \beta logK_{it} + \gamma logE_{it} + U_{it} \tag{5c}$$

$$\rho = \alpha - 1; then, \alpha = \rho + 1 \tag{5d}$$

$$LogY/L_{it} = \lambda + \rho logL_{it} + \beta logK_{it} + \gamma logE_{it} + t_{it} + U_{it} \tag{5e}$$

where the dependent variable is labor productivity, which measures the scale effect of the factors on labor productivity. Value of energy is used for the manufacturing industry as a major variable of interest. Labor is a control variable that represents the scale of production and is defined as the number of employees in the industry group. The second key control variable is capital, which is defined as the value of the industry groups' fixed assets. All variables are in logarithm form, so that the coefficients are defined elasticities. T represents the trend, which is included for capturing the technical change effect. U represents the error term of the panel model and subscripts i and t represent the industry sector and time period respectively. U contains unobservable sector- and time-specific effects. βs are unknown coefficients of the explanatory variables, where λ is the constant term.

Equations (6a) and (6b) represent the third model, which measures the intensity effect of factors on labor productivity. The other way of specifying the model is by dividing the right-hand side variables (L, K, E) with labor to express energy and capital in the form of capital intensity and energy intensity, respectively, while the L ratio will end up in the intercept. Thus, the third model to be estimated is written as:

$$\frac{Y}{L} = \left(\frac{A}{L}\right)\left(\frac{L}{L}\right)^{\alpha}\left(\frac{K}{L}\right)^{\beta}\left(\frac{E}{L}\right)^{\gamma} \tag{6a}$$

$$\frac{Y}{L} = \left(\frac{A}{L}\right)\left(\frac{K}{L}\right)^{\beta}\left(\frac{E}{L}\right)^{\gamma}; \left(\frac{L}{L}\right)^{\alpha} = 1^{\alpha} = 1 \tag{6b}$$

For all the three models to be estimated, an error term is included and the models are linearized and transformed into logarithm forms before estimation. The third model to be estimated (7a) measures energy and capital intensity and their effect on labor productivity in manufacturing industrial groups in Ethiopia:

$$Log(Y/L)_{it} = \mu + \beta logk_{it} + \gamma loge_{it} + t_{it} + u_{it} \tag{7a}$$

$$LogMLP_{it} = \alpha + \beta logCapital\ Intensity_{it} + \gamma logEnergy\ Intensity_{it} + \lambda trend_{it} + U_{it} \tag{7b}$$

where manufacturing labor productivity is the dependent variable defined as the manufacturing output of an industry group per employee. μ is the intercept, β is a slope coefficient for capital intensity, γ is a slope coefficient for energy intensity, while t stands for time trend to represent a shift in the production function over time and thus λ is the rate of technological change. U is the error term in the model with i and t representing industry group and time respectively. It follows an error component structure consisting of industry effects and random error components.

3.2. Model Estimation

Panel data models can be static or dynamic. Static panel data models can be estimated using pooled ordinary least squares (OLS), fixed effects (FE), and random effects (RE) models, but these models do not take the problems of heteroscedasticity, serial correlation, and the endogeneity of the explanatory variables into account [61–63]. The pooled OLS model ignores fixed industry and time effects. In FE, these are fixed effects correlated with the inputs, while it is assumed that they do not correlate with inputs in the RE model. In all the models, the time effects are captured by the trend. In the FE model, we estimate the effects in the form of industry intercepts, while, in RE, we estimate the parameters of the distribution of the industry effects which, are assumed to have means of zero and constant variance [63].

To solve the estimation problems related to a static panel formulation, we use the dynamic panel model of difference GMM and system GMM estimators, as proposed by Arellano Bond [64] and Arellano and Bover [65], respectively. The difference GMM and system GMM are dynamic panel estimators designed for large N and small T, many groups/individuals, a few time periods, a linear functional relationship, one left-hand side that is dynamic depending on its own past realization, and for independent variables that are not strictly exogenous [66]. System GMM contains both level and first difference equation parts, it uses instruments in levels for equations in first difference and uses instruments in first difference for equations in levels [61]. After estimating the dynamic panel data models, tests for the serial correlation of the residuals and overidentification were done using Hausman or Sargan tests and the autoregressive AR (2) test, respectively [64,65].

4. The Data

4.1. Data and Variables

All data used in this study are taken from the Ethiopian Central Statistical Authority (CSA). The period 2005–2016 is chosen for the study since the latest information on all variables is available only up to 2016. The number of industry groups and the study period were determined by data availability. A two-digit industry sector level is the most disaggregated data level available for this specific case. The number of observations for industry groups (industrial sectors) is 15, where, for every industry group, the relevant variables available are included. Table 1 provides a list of the industry groups. The medium and large manufacturing industries in Ethiopia are categorized into 15 industry groups.

Table 1. List of industry groups.

Industry Code	Industry Group (Sector)
1	Food Products and Beverages Industry
2	Tobacco Products Industry
3	Textiles Industry
4	Wearing Apparel, Except Fur Apparel Industry
5	Tanning and Dressing of Leather; Footwear, Luggage, and Handbags Industry
6	Wood and of Products of Wood and Cork, Except Furniture Industry
7	Paper, Paper Products, and Printing Industry
8	Chemicals and Chemical Products Industry
9	Rubber and Plastic Products Industry
10	Other Non-Metallic Mineral Products Industry
11	Basic Iron and Steel Industry
12	Fabricated Metal Products Except Machinery and Equipment Industry
13	Machinery and Equipment Industry
14	Motor Vehicles, Trailers and Semi-Trailer Industry
15	Furniture; Manufacturing Industry

Source: Central Statistical Authority (CSA).

Table 2 gives the list of variables used in this study and their definitions. To define labor productivity, we need information on production and employment. Production, in our case, is defined as the gross value of production by industry group. Employment is defined as the number of employees by industry group. Accordingly, labor productivity is defined as the ratio of production to employment by industry group or per capita employed production, labeled in the literature as labor productivity. Energy is defined as the ratio of the value of energy consumed by the industry groups. Capital is defined as the total value of the fixed assets by industry groups. Table 2 also shows the expected effects of the variables in the model on labor productivity. Labor productivity is the dependent variable and the explanatory variables are energy use, employment, capital, and trend which are expected to be statistically significant in the empirical estimation. The expected sign for employment is positive as industries in Ethiopia are more labor intensive, so adding more labor is expected to increase production. Similarly, the expected signs of the parameters for energy, capital, and technical change are expected to be positive. It is assumed that energy use and capital will increase labor productivity in the manufacturing industries in Ethiopia. Wages and salaries were included as a proxy for human capital but they were excluded from the estimation due to high collinearity problem. An increase in wages and salaries is expected to positively affect labor productivity and higher wages per capita reflect the laborers' skills and education levels.

Table 2. List of variables, expected level of significance, and coefficient signs.

Variables	Variable Definitions	Expected Effect
	Dependent variable:	
Labor Productivity	Ratio of gross value of production to number of employees	-
	Independent variables:	
Production	Gross value of production by industrial group (in 000 Birr)	-
Employment	Number of employees by industrial group	positive
Energy	Ratio of value of energy consumed to total industrial expenditure by industry group	positive
Capital	Total value of fixed assets by industry group (in 000 Birr)	positive
Time trend	Is a proxy for technical change and is included in the model as a control variable	positive

4.2. The Variables' Development Over Time

Figure 1 gives the trends of production for the 15 industries included in this study. The industry classification is standard, as provided by the Statistics Authority of Ethiopia. A list of the 15 industry groups is reported in Table 1. Based on this, the food and beverage industry (industry code 1) shows an increasing trend for 10 years (2005–2016). Similarly, the other non-metallic mineral products industry (industry code 10) and the motor vehicle and trailer industry (industry code 14) show an increase in the recent years of the study period. However, the remaining industries have constant trends in production. Thus, the outcome of policies in the form of industrial development's effects are heterogeneous across industry groups. Figure 2 presents the trends of energy use across the industry groups. With the exception of the wood products industry (industry code 6) and the non-metallic mineral industry (industry code 10), the overall trends in energy use throughout the decade, on average, show steady growth. However, these two industries are relatively more energy intensive and, very recently, a decline in energy use has been witnessed in both these industries.

Figures 3 and 4 give the trends of capital and employment in the 15 industry groups in the study period. The use of capital increased over time for the food and beverage industry (industry code 1) and the non-metallic mineral products industry (industry code 10) compared to the other industry groups. Employment in the food and beverage industry (industry code 1) as well as the textile industry (industry code 3), on average, showed an upward trend throughout while the rubber and plastic industry (industry code 9) and the metallic industry (industry code 12) had huge employment in the second half of the study period but, overall, had a flatter upward trend over time. In the remaining industry groups, the overall employment trend was steady.

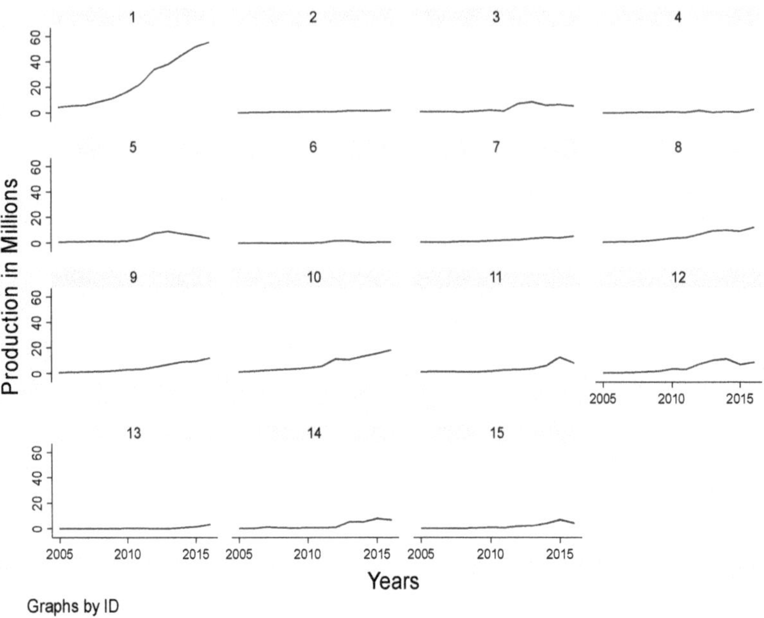

Figure 1. Production trends by industry groups.

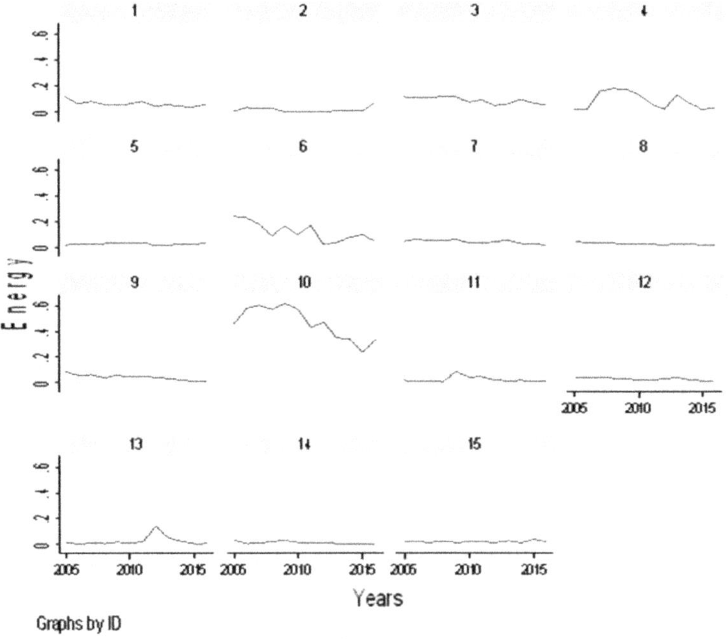

Figure 2. Energy use trends by industry groups.

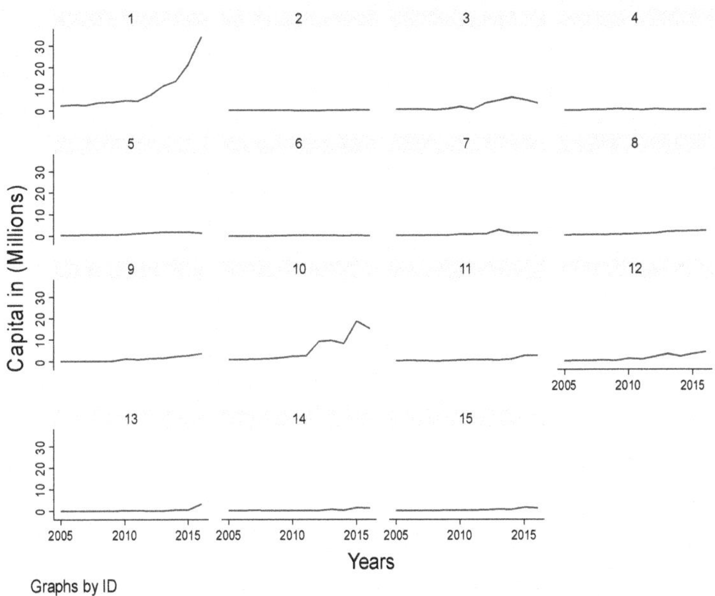

Figure 3. Capital trend by industry groups.

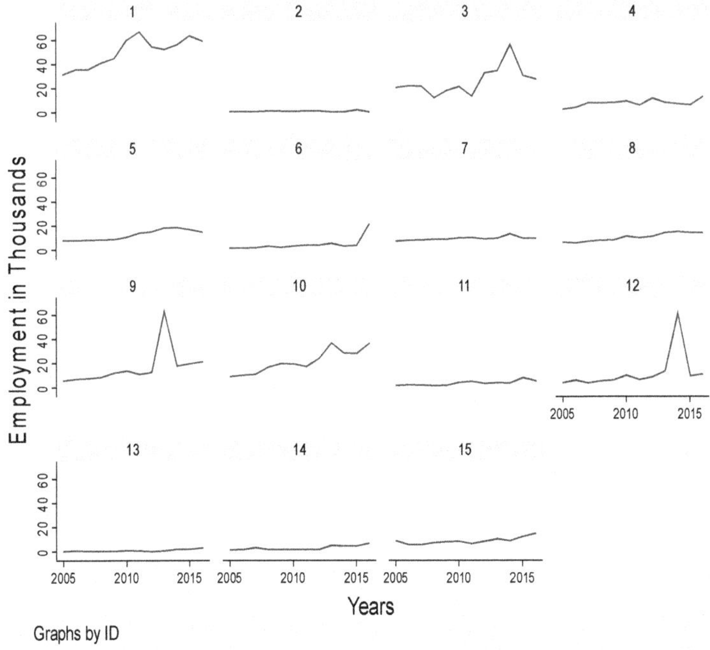

Figure 4. Employment trends by industry groups.

Figures 5 and 6 show the share of production and energy use by the manufacturing industry groups. The food and beverage industry (code 1) had the lion's share in terms of production followed by the non-metallic mineral products industry (code 10). The apparel industry (code 4), wood industry (code 6), and machinery industry (code 13) had the lowest shares compared to the other industry groups. The energy use share was the highest in the metallic industry (code 10), followed by the wood industry (code 6), the apparel industry (code 4), and the textile industry (code 3).

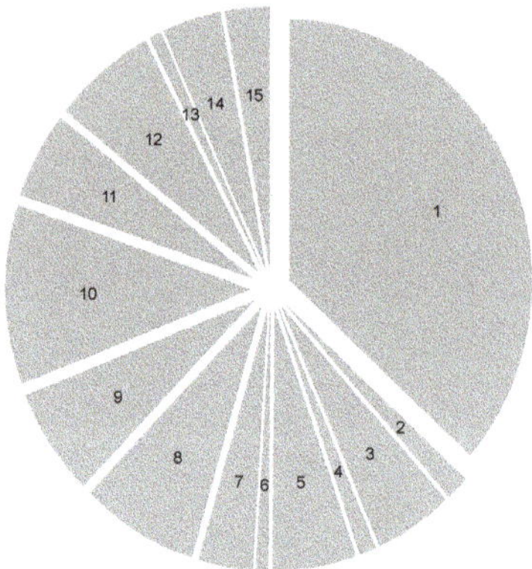

Figure 5. Gross value of production by industry groups.

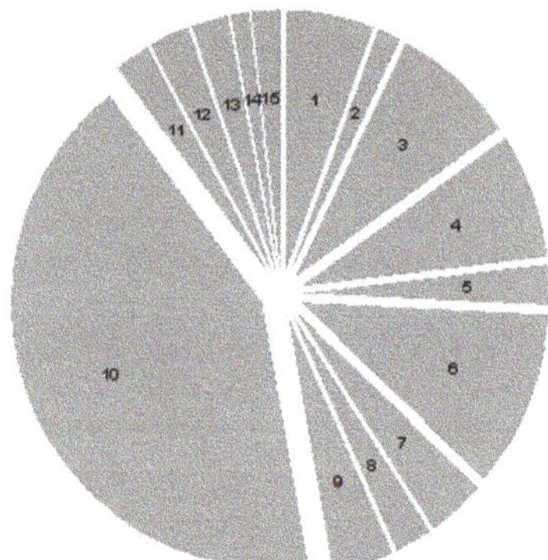

Figure 6. Energy use across industry groups.

5. Empirical Results and Discussion

5.1. Descriptive Statistics

Table 3 gives the summary statistics of the variables of interest. It gives information about the overall, between, and within variations in terms of mean and standard deviations, together with the minimum and maximum values of the variables. The total sample is 180 observations: 15 industry groups and 12 years of data from 2005 to 2016. In the summary, we included variables such as industry production, employment, and labor productivity, defined as the ratio of production per employee in the industry groups, capital proxied by fixed assets, and the value of energy and human capital proxied by wages and salaries. Accordingly, for variables such as production per employee and the value of energy, the within variations are higher than the between variations, while the within variations of labor and capital are higher than the between variations. The minimum and maximum values of each variable are also given in Table 3.

Table 3. Summary statistics of the variables of interest.

Variable	Variations	Mean	Std. Dev.	Minimum	Maximum	Observations
	Overall	8	4.3349	1	15	NT = 180
ID	Between		4.4721	1	15	N = 15
	Within		0	8	8	T = 12
	Overall	2010.5	3.4616	2005	2016	NT = 180
Years	Between		0	2010.5	2010.5	N = 15
	Within		3.4617	2007	2016	T = 12
	Overall	453,240	8,002,421	13,673	5.54×10^7	NT =180
Production	Between		5,985,648	55,1875.1	2.50×10^7	N = 15
	Within		5,514,752	-1.60×10^7	3.50×10^7	T = 12
	Overall	12,512.6	14,497.45	48	67,072	NT = 180
Employment	Between		12699.86	813	50190.67	N = 15
	Within		7668.181	−5985.011	62091.91	T = 12
	Overall	428.838	563.5947	19.6428	4078.363	NT = 180
Productivity	Between		364.3505	85.9931	1470.145	N = 15
	Within		439.3695	−587.8283	3037.057	T = 12
	Overall	175,329	3,860,281	4686	3.42×10^7	NT = 180
Capital	Between		2,541,552	160,494.1	9,332,244	N = 15
	Within		2973086	−5173360	2.66×10^7	T = 12
	Overall	0.0730	0.11774	0.0010	0.6210	NT = 180
Energy	Between		0.11285	0.0132	0.4650	N = 15
	Within		0.04369	−0.1539	0.2290	T = 12
Cost of	Overall	275,439	488,709.7	1329	4,023,882	NT = 180
Labor	Between		351,034.5	30,176.83	1,466,912	N = 15
	Within		350,976.4	−867,761	2,832,410	T = 12

Source: Authors' computations using Stata.

5.2. Regression Results and Analysis

In this section, static and dynamic panel data models are estimated for the industry panel data available from 2005 to 2016. The data contains 15 industry groups listed in Table 1 and all of them are included in the analysis. Thus, the data includes the entire population of the industry groups. The estimated models are pooled OLS, fixed effects (FE), and random effects (RE) models from the static panel estimators, while difference GMM and system GMM are presented as dynamic estimators. Three different model specifications are used in the estimation. In the first model, industry group production is the dependent variable, while energy, labor, and capital are explanatory variables. In this model, the returns in relation to the scale of production are calculated based on the sum of the coefficients for the three input variables. In the second model, manufacturing labor productivity is specified as employment (labor), capital (fixed assets), value of energy, and time trend (technology) as the explanatory variables. In this model, the coefficients measure the scale effect of the explanatory variables on labor productivity of the industry groups and labor represents the scale effect. In the third model, the manufacturing sector's labor productivity is explained by measuring energy and capital intensities respectively. In all the three model specifications, a trend is included to capture a shift in the labor productivity function or rate of technological change. All variables (with the exception of trend) are transformed into logarithmic form so that the coefficients are interpreted as input elasticities.

Accordingly, Table 4 shows the results of the pooled OLS for the three model specifications. In the first model, labor, capital, energy, and technology are found to be statistically significant and positive. These are among the key factors used for explaining the manufacturing industry's production growth. The elasticity of the output with respect to capital is higher than the corresponding figures for labor and energy in these industries. The returns in relation to the scale of the production process are 1.06 implying increasing returns in relation to scale in this specification coinciding with predictions in the literature [1,67]. In the second model, labor is significant and positive at the one percent significance level. However, we do not interpret the coefficient of labor and, instead, based on Equation (5d), we find the value of α by adding one to the estimated coefficient in our model, which is zero. Then α, in our case, will be positive, indicating the positive effect of labor on productivity in the manufacturing industries. This can be attributed to the increasing returns in relation to the scale of production and the type of existing industries, which are dominated by labor-intensive industries. In this model, capital is significant and positive for labor productivity, which is a boost for the industry groups. These results are in accordance with Otalu and Anderu and Velucchi Viviani [7,43]. Energy use also positively effects productivity in line with other empirical studies [40–45]. In the third model, capital and energy intensities are significant and positive and help explain labor productivity in the manufacturing industries in line with other studies [46,50,51]. Our results confirm that labor productivity is high and more elastic for energy intensity than for capital in the Ethiopian manufacturing industries. The models show that adjusted R^2 is high and the probability of F-statistics is significant, confirming the appropriateness of the model's specifications (see Table 4).

Table 4. Pooled OLS estimation results for the three models.

Variables	Model 1		Model 2		Model 3	
	Coef.	Robust Std. Err	Coef.	Robust Std. Err	Coef.	Robust Std. Err
Labor (log)	0.2730 ***	(0.0755)	−0.7269 ***	(0.0755)	-	-
Capital (log)	0.7029 ***	(0.0544)	0.7027 ***	(0.0544)	0.0014 ***	0.0004
Energy (log)	0.0895 ***	(0.0146)	0.0895 ***	(0.0146)	0.1082 ***	0.0127
Time trend	0.0226 ***	(0.0502)	0.0226 ***	(0.0050)	0.0374 ***	0.0088
Constant	0.7930 ***	(0.1996)	0.7930 ***	(0.1996)	1.7272 ***	0.0541
RTS	1.0655					
AdjR2	0.8979		0.8285		0.6074	
F-statistics (*p*-value)	0.0000		0.0000		0.0000	

Notes: ***, **, * denote the statistical significance levels at 1%, 5%, and 10%, respectively. *Model 1: Output is the dependent variable. *Model 2: Labor productivity is the dependent variable (scale effect). *Model 3: Labor productivity is the dependent variable (input intensity effect).

It should be noted that the pooled OLS model ignores industry effects that may generate biased results. However, it serves well to establish the model's specifications. Table 5 presents the static panel data model estimation results. In this section, only the second and third models are estimated using fixed effects (FE) and random effects (RE) estimation methods. The fixed effects model allows the industry effects and inputs to be correlated, while the random effects model assumes that these are not correlated. The fixed effects model is consistent and unbiased regardless of the correlated effects, but the random effects model is valid and efficient. In this case, since the industry groups are made up of the population of industries, the fixed effects model is a better choice. For a comparison, we estimate the models using both estimation methods.

In the fixed effects model, labor is statistically significant and is a positive factor in explaining the variations in manufacturing productivity in Ethiopia. This is expected based on theoretical predictions as more labor employment induces labor productivity. The fixed effects model's estimation results confirm that energy, capital, and technology positively effects labor productivity, and all of them are statistically significant at the one percent level of significance. The input intensity model based on the fixed effects estimation shows that capital intensity and energy intensity are statistically significant factors for explaining labor productivity in the Ethiopian manufacturing industries. However, in this case, productivity is more elastic in relation to capital intensity than energy intensity. In the random effects model, energy, capital, and technology are positive and statistically significant in explaining the industry groups' labor productivity, while the coefficient for labor is negative, but, based on Equation (5d), α is found by adding one to the coefficient, which gives us a positive coefficient with a value of 0.45. For the intensity model, the random effects estimation approach confirms the significance of energy and capital intensities positively effecting labor productivity. Like the fixed effects model's results, productivity is less elastic in relation to energy intensity than it is to capital intensity. In all the models, the coefficients for trends are positive and significant, implying a positive shift in labor productivity because of technological changes in Ethiopian manufacturing industries during the study period.

The models give different results for some of the explanatory variables, so we cannot take into account the results of all the models. Instead, we must select a model that explains the data using different tests and base the analysis on the optimal model's specifications. To choose between pooled and random effects models, we used the Breusch and Pagan lagrange multiplier (LM) tests with the null hypothesis that pooled OLS is an appropriate model that explains the data better relative to the random effects model. The Hausman test compares the random effects model with the fixed effects model and the null hypothesis for the Hausman test shows that the random effects model is not appropriate for representing the data relative to the fixed effects model. Accordingly, in both cases, the *p*-value of chi2 and chibar2 forces us to reject the null hypothesis. Therefore, the fixed effects model is preferred to the pooled OLS model and the fixed effects model is preferred to the random effects model to represent our data. To control for the heteroscedasticity problem, standard errors reported in all the models are robust.

Table 5. Static panel estimation results for Models 2 and 3.

Variables	Fixed Effects		Random Effects	
	Model 2	Model 3	Model 2	Model 3
	Coef.	Coef.	Coef.	Coef.
Log Labor	−0.5541 ***	-	−0.5748 ***	-
	(0.1287)	-	(0.1311)	-
Log Capital	0.3545 ***	0.3807 ***	0.4205 ***	0.4513 ***
	(0.0420)	(0.0475)	(0.0449)	(0.0562)
Log Energy	0.0405 **	0.0335 ***	0.0474 ***	0.0487 ***
	(0.0209)	(0.0113)	(0.0201)	(0.0182)
Time Trend	0.0552 ***	0.0454 ***	0.0486 ***	0.0400 ***
	(0.0075)	(0.0045)	(0.0065)	(0.0046)
Constant	2.0353 ***	1.3045 ***	1.7624 ***	1.1679 ***
	(0.5567)	(0.0822)	(0.4978)	(0.1056)
Test	H_0 & H_1	Appropriate Model	Prob of chi2 & chibar2	Decision
Breusch and Pagan LM Test	H0 H1	Pooled OLS Random Effects	0.000	reject H_0
Hausman test	H0 H1	Random Effects Fixed Effects	0.000	reject H_0

Notes: ***, **, * denote statistical significance levels at the 1%, 5%, and 10%, respectively. *Model 2: Labor productivity as the dependent variable (scale effect). *Model 3: Labor productivity as the dependent variable (intensity effect).

Table 6 gives the dynamic panel model's estimation results for both the difference GMM and system GMM models. Unlike static panel models, these models include the lag of the dependent variable as an explanatory variable in addition to the other variables. In the dynamic models, problems of heteroscedasticity and autocorrelation are considered. In both the scale effects (Model 2) and the input intensity models (Model 3), lagged labor productivity is found to be significant and positive in explaining changes in the manufacturing industry's labor productivity in Ethiopia. This shows that the previous year's productivity increases current productivity, which, in our case, is labor productivity. An increase in employment for the industry groups has a positive and significant effect which is attributed to increasing returns to scale and the labor-intensive nature of manufacturing industries in both the cases. In both the difference GMM and system GMM models, energy induces labor productivity. Comparing our results with those from developing countries suggests that our results are in line with those from some sub-Saharan African countries, such as those reported by Kebede et al. and Akinlo [12,18]. However, the effect of energy on productivity for some African countries shows that it is not as important as labor and capital [10,39], signifying the mixed empirical results of the relationship between energy and growth as one major reason for undertaking this specific study. The empirical validation in our case is at the industry level and not at the aggregate national level and this is one of the contributions of this study to the existing literature, as it is what makes this study different from the existing studies. Unlike other studies, the consistency of our results is empirically confirmed using different model specifications and alternative estimation strategies. In addition to the role of energy in productivity, the effects of labor, capital, and technological change on manufacturing productivity are also empirically validated in Ethiopia. This provides crucial policy input for the country's industrial policy.

Table 6. Dynamic panel estimation results for Models 2 and 3.

	Difference GMM		System GMM	
Variables	**Model 2**	**Model 3**	**Model 2**	**Model 3**
	Coef.	**Coef.**	**Coef.**	**Coef.**
Productivity_L1	0.1443 (0.1286)	0.1210 (0.1667)	0.1342 ** (0.0592)	0.0990 (0.1004)
Log Labor	−0.6557 *** (0.1155)	- -	−0.5997 *** (0.0549)	- -
Log Capital	0.5438 *** (0.0516)	0.0007 *** (0.0002)	0.5276 *** (0.0442)	0.5391 *** (0.0373)
Log Energy	0.0393 *** (0.0188)	0.0221 (0.0153)	0.0357 *** (0.0098)	0.0311 *** (0.0071)
Time trend	0.0263 *** (0.0091)	0.0451 *** (0.0157)	0.0255 *** (0.0039)	0.0251 *** (0.0086)
Constant	1.1946 *** (0.4445)	1.6935 *** (0.3551)	1.0919 *** (0.2333)	0.8973 *** (0.2006)
AR (2)			0.499	0.520
Test for autocorrelation			0.1958	0.1287
Number of instruments		5	4	
Number of groups		15	15	

Notes: ***, **, * denote the statistical significance levels at 1%, 5%, and 10% levels respectively. *Model 2: Labor productivity as the dependent variable (scale effect) *Model 3: Labor productivity as the dependent variable (input intensity effect).

In the input intensity model (Model 3), the elasticity productivity for energy intensity is higher than capital intensity, while the opposite is the case for the system GMM model. Capital is positive and significant in all the models for increasing labor productivity. The coefficient of the time trend has a positive sign in all the models, indicating technological progress with an expected positive effect on the productivity of the industries (see Table 6).

Table 7 discusses the results of the system GMM dynamic estimator, including dummies for trends. Our results show that, in both the models, energy magnitude and energy intensity are statistically significant and positive factors in increasing labor productivity in the manufacturing industry groups; this finding coincides with other findings in the literature [34–36]. Besides, the magnitude of capital and capital intensity are positive factors for labor productivity. In both the models, time dummies are positive throughout. The results show that there is no cyclical effect and, instead, labor productivity increases in both cases over time, which can be attributed to technical changes, increasing labor productivity.

Table 7. System GMM dynamic panel with time dummies for Models 2 and 3.

	System GMM Dynamic Panel (With Time Dummies)			
	Scale Effect Model (Model 2)		Input Intensity Effect Model (Model 3)	
	Coff.	Std. Err	Coff.	Std. Err
Productivity_L1	0.1653 *	(0.0877)	0.1660 *	(0.0874)
Log Labor	−0.6872 ***	(0.0607)	-	-
Log Capital	0.6784 ***	(0.0545)	0.7145 ***	(0.0481)
Log Energy	0.0954 ***	(0.0171)	0.0947 ***	(0.0108)
D.trend(2)	0.7977 ***	(0.1921)	0.0709	(0.0805)
D.trend(3)	0.8422 ***	(0.1929)	0.1068 **	(0.0805)
D.trend(4)	0.8978 ***	(0.1933)	0.1632	(0.0805)
D.trend(5)	0.8448 ***	(0.1982)	0.1036	(0.0807)
D.trend(6)	0.8115 ***	(0.2085)	0.0744	(0.0817)
D.trend(7)	0.9268 ***	(0.2058)	0.1885 **	(0.0814)
D.trend(8)	0.9867 ***	(0.2121)	0.2669 ***	(0.0832)
D.trend(9)	1.0325 ***	(0.2178)	0.2701 ***	(0.0835)
D.trend(10)	1.0505 ***	(0.2191)	0.2948 ***	(0.0832)
D.trend(11)	0.9460 ***	(0.2271)	0.1892 **	(0.0858)
D.trend(12)	0.9808 ***	(0.2277)	0.2142 **	(0.0857)
AR (2)	0.853		0.779	
Test for Autocorrelation	0.1200		0.1287	
Number of Instruments	14		14	
Number of groups	15		15	

Notes: ***, **, * denote the statistical significance levels at 1%, 5%, and 10% levels respectively. *Model 2: Labor productivity as the dependent variable (scale effect). *Model 3: Labor productivity as the dependent variable (input intensity effect).

One major objective of this study was to ascertain whether an empirical relationship existed between energy and labor productivity in Ethiopian industries, along with investigating whether it positively affected productivity or limited it. The results of all the models confirm that the energy-related parameter is significant and positive, showing that an increase in energy consumption enhances labor productivity in Ethiopian manufacturing industry groups. This result coincides with other empirical studies [15–17,46]. However, in Ethiopia, agriculture was previously a major source of livelihood for the population. Agriculture was a dominant sector in terms of the employment share up until recently, when traditional services emerged to dominate the economy [21,23]. The share of manufacturing in Ethiopian GDP was very low, indicating output and premature deindustrialization [20,21,24]. This requires serious engagement for identifying and prioritizing the major explanatory factors for the manufacturing industry. Furthermore, manufacturing is more energy intensive relative to other sectors and the interdependence between energy and industries is a crucial tool for sustainable economic development [6,32]. Accordingly, empirically identifying the role of energy in the manufacturing productivity of Ethiopia can contribute to industrial policy input. The labor input is significant and positive in the scale effects model (Model 2). This means that an increase in labor employment will increase labor productivity due to increasing returns and the labor-intensive nature of the industries [7,30]. Finally, we reported the diagnostic tests for serial correlation and heteroscedasticity. The AR (2) test validated the model, free from the serial correlation problem. The number of instruments used were less than the groups in both the dynamic panel estimation approaches.

6. Conclusions and Policy Implications

This study investigated the effect of energy on manufacturing labor productivity in Ethiopia using panel data for manufacturing industry groups. Fifteen industries were included in the study covering 12 years of data from 2005 to 2016. The number of industry groups and the period was determined by data availability. Data were obtained from the Central Statistical Authority (CSA) in Ethiopia. We used both descriptive and econometric approaches for examining the empirical relationships between the variables of interest conditional on some other variables and characteristics. This study had two specific objectives: examining the existence of an empirical relationship between energy and labor productivity in the manufacturing industry and estimating the elasticity effect of energy on labor productivity.

Three models were estimated. The first model is a conventional production function with labor, capital, and energy as the explanatory variables along with a time trend to proxy for capturing technological change. The second model measures the scale effect of energy with the control variables labor, capital, and technology. The third model measures the intensity effect of energy and capital on labor productivity in Ethiopian manufacturing industries. Accordingly, static and dynamic panel data models were estimated—pooled OLS, fixed effects, and random effects static panel estimators, along with difference and system GMM dynamic panel models.

The data for industrial group production showed that the overall trends in production were steady and constant over the study period, except for the food and beverage industry, which rapidly increased (industry code 1). On average, the energy use trend increased in the food and beverage industry (industry code 1) as well as the textile industry (industry code 3). The share of production across the 15 industry groups was dominated by the food and beverage industry (industry code 1), followed by the non-metallic mineral products industry (industry code 10). The non-metallic mineral products industry was found to be more energy intensive than the others.

In the first model, the manufacturing production function was estimated with labor, capital, and energy as the inputs in the production process. The time trend was included to capture technological change. In this model, energy, capital, and labor were statically significant and positive in augmenting manufacturing production in Ethiopia; this result is similar to that of other empirical studies [15–17]. Technology was also significant and a positive factor in industrial growth in Ethiopia. In this model, the sample average returns in relation to the scale of production were 1.07, implying increasing returns in relation to the scale of the manufacturing industries. Labor and capital were statistically significant in all the models at the one percent level of significance.

Across the models, some variables had different significance levels, which led us to select an appropriate model that fit the data best. Both static and dynamic model estimation methods were considered, and we got different estimated coefficient results. For the static models, limitations in considering endogeneity, omitted variable bias, autocorrelation, and heteroscedasticity led to the dynamic panel model estimator being selected over the static panel estimator. The system GMM estimator was chosen over the difference GMM model based on the diagnostic tests and to overcome the limitations of missing observations in the difference GMM model.

In all the models, an increase in employment induced labor productivity due to increasing returns to scale and the labor-intensive nature of the industries. Energy positively explains labor productivity in manufacturing industries in Ethiopia. This means an increase in the use of energy-enhanced labor productivity in the industry groups. Capital intensity use gave a boost to labor productivity, which is consistent with theoretical predictions. In addition, a system GMM model was estimated, including time dummies for the scale effect and input intensity models. In both the cases, labor productivity increased over time, signifying the positive effect of technical change on manufacturing labor productivity in Ethiopia. Across the different approaches used, the role of energy use and energy intensity was consistently significant and positive in explaining labor productivity changes in Ethiopian manufacturing industries.

This study showed that energy induces labor productivity in the manufacturing industry groups in Ethiopia, showing that the efficient use of energy increases industrial growth. It also empirically identified labor and capital as essential determinant factors of productivity in the manufacturing industries in Ethiopia, complemented by technological change effects. This indicates a need to organize resources in a way that boosts the growth of the industries. Energy and capital should also be efficiently used, as the results show that productivity is elastic in relation to a change in energy and capital input intensities in the manufacturing industries in Ethiopia.

A review of the existing literature showed that the role of energy in productivity is controversial across countries [10,17,44,57]. This study adds to the literature by empirically validating the positive role of energy in productivity, applying different model specifications and estimation methods to Ethiopia's manufacturing industries. This implies that industrial policies in Ethiopia should focus on the efficient use of energy along with labor, capital, and technical changes to overcome the premature deindustrialization pattern over time. Research on the energy efficiency and energy productivity of the manufacturing industries in Ethiopia is expected to provide additional policy inputs. This type of research can be extended to cross-country analyses in developing countries, using the manufacturing industry as a case study.

Author Contributions: Conceptualization, S.G.K. and A.H.; methodology, S.G.K. and A.H.; Data collection, S.G.K.; Formal analysis, S.G.K.; writing—original draft preparation, S.G.K. and A.H.; writing—review and editing, S.G.K. and A.H. All authors have read and agreed to the published version of the manuscript.

Funding: This research received no external funding.

Acknowledgments: The authors gratefully acknowledge the comments and suggestions by the editor of the journal and three anonymous referees on an earlier version of this manuscript.

Conflicts of Interest: The authors declare no conflict of interest.

References

1. Kaldor, N. *Causes of the Slow Rate of Economic Growth of the United Kingdom. An Inaugural Lecture*; Cambridge University Press: Cambridge, UK, 1966.
2. Cornwall, J. *Modern Capitalism. Its Growth and Transformation*; St. Martin's Press: New York, NY, USA, 1977.
3. Guadagno, F. The determinants of Industrialization in Developing Countries, 1960–2005. *UNU-MERIT Maastricht Univ.* **2016**, *31*, 26.
4. UNECA. *Dynamic Industrial Policy in Africa: Economic Report on Africa*; UNECA: Addis Ababa, Ethiopia, 2014.
5. UNECA. *Industrializing Through Trade. Economic Report on Africa*; UNECA: Addis Ababa, Ethiopia, 2015.
6. UNIDO. *Demand for Manufacturing: Driving Inclusive and Sustainable Industrial Development. Industrial Development Report*; United Nations Industrial Development Organization UNIDO: Vienna, Austria, 2018.
7. Otalu, J.A.; Anderu, K.S. An Assessment of the Determinants of Industrial Sector Growth in Nigeria. *J. Res. Bus Man* **2015**, *3*, 1–9.
8. Story, D. Time-Series Analysis of Industrial Growth in Latin America: Political and Economic Factors. *Soc. Sci. Quart.* **1980**, *61*, 293–307.
9. UNDP (United Nations Development Program). Transforming Lives Through Renewable Energy Access in Africa UNDP's Contributions. *UNDP Afr. Policy Brief* **2018**, *1*, 1–26.
10. Chen, P.Y.; Chen, S.T.; Chen, C.C. Energy consumption and economic growth—New evidence from meta-analysis. *Energy Policy* **2012**, *44*, 245–255. [CrossRef]
11. Moghaddasi, R.; Pour, A.A. Energy Consumption and Total Factor productivity in Iranian Agriculture. *Energy Rep.* **2016**, *2*, 218–220. [CrossRef]
12. Kebede, E.; Kagochi, J.; Jolly, C.M. Energy consumption and economic development in Sub-Sahara Africa. *Energy Econ.* **2010**, *32*, 532–537.
13. Al-Iriani, M.A. Energy-GDP relationship revisited: An Example from GCC Countries Using Panel Causality. *Energy Policy* **2005**, *34*, 3342–3350. [CrossRef]
14. Cleveland, C.J.; Kaufmann, R.K.; Stern, D.I. Aggregation and the role of energy in the economy. *Ecol. Econ.* **2000**, *32*, 301–317. [CrossRef]
15. Alaali, F.; Roberts, J.; Taylor, K. The Effect of Energy Consumption and Human Capital on Economic Growth: An Exploration of Oil Exporting and Developed Countries. *SERPS (Sheffield Econ. Res. Pap. Ser.)* **2015**, 2015015.
16. Fallahi, F.; Sojood, S.; Aslaninia, N.M. Determinants of Labor Productivity in Iran's Manufacturing Firms: With Emphasis on Labor Education and Training. *MPRA* **2010**, 27447.
17. Soytas, U.; Sari, R. Energy consumption and GDP: Causality relationship in G-7 countries and emerging markets. *Energy Econ.* **2003**, *25*, 33–37. [CrossRef]

18. Akinlo, A.E. Energy consumption and economic growth: Evidence from 11 Sub-Sahara African countries. *Energy Econ.* **2008**, *30*, 2391–2400. [CrossRef]

19. CSA (Central Statistic Authority). *Industry Sector Data Base of Ethiopian Industries*; CSA: Addis Ababa, Ethiopia, 2018.

20. EEA (Ethiopian Economic Association). *Data Base on Macroeconomic Variables*; EEA: Addis Ababa, Ethiopia, 2017.

21. Ejigu, Z.K.; Singh, I.S. Service Sector: The Source of Output and Employment Growth in Ethiopia. *Acad. J Econ.* **2016**, *2*, 139–156.

22. Oqubay, A. *Industrial Policy and Late Industrialization in Ethiopia*; Working paper Series No. 303; African Development Bank: Abidjan, Cote D'Ivoire, 2018.

23. Gebreeyesus, M. *Industrial Policy and Development in Ethiopia: Evolution and Present Experimentation*; UNU.WIDER Working Paper No 6; UNU.WIDER: Helsinki, Finland, 2010.

24. Rodrik, D. Premature deindustrialization. *J Econ. Growth* **2016**, *21*, 1–33. [CrossRef]

25. Ray, D. *Development Economics*; Princeton University Press: Princeton, NJ, USA, 1998.

26. Romer, D. *Advanced Macroeconomics*, 4th ed.; University of California: Berkeley, CA, USA, 2011.

27. Todaro, M.P.; Smith, S.C. *Economic Development*, 12th ed.; Pearson Education Inc.: New Jersey, USA, 2015.

28. Agénor, P.R.; Montiel, P. *Development Macroeconomics*, 3rd ed.; Princeton University Press: Princeton, NJ, USA, 2008.

29. Stern, D.I. *The Role of Energy in Economic Growth CCEP*; Working Paper 3.10; Crawford School, The Australian National University: Canberra, Australia, 2010.

30. Su, B.; Heshmati, A. Development and Sources of Labor Productivity in Chinese Provinces. *China Econ. Policy Rev.* **2014**, *2*, 1–30.

31. Stern, D.I. Limits to Substitution and Irreversibility in Production and Consumption: A Neoclassical Interpretation of Ecological Economics. *Ecol. Econ.* **1997**, *21*, 197–215. [CrossRef]

32. OECD. *Energy*; OECD Green Growth Studies; OECD Publishing: Paris, France, 2012.

33. UNDP (United Nations Development Program). *Energy for Sustainable Development a Policy Agenda*; UNDP: New York, NY, USA, 2002.

34. Toman, M.A.; Jemelkova, B. Energy and economic development: An assessment of the state of knowledge. *Energy J.* **2003**, *24*, 93. [CrossRef]

35. Cabraal, R.A.; Barnes, D.F.; Agarwal, S.G. Productive Uses of Energy for Rural Development. *Ann. Rev. Environ. Res.* **2005**, *30*, 117–144. [CrossRef]

36. Ejaz, Z.; Aman, M.U.; Usman, M.K. Determinants of Industrial Growth in South Asia: Evidence from Panel Analysis. In *Papers and Proceedings*; Unpublished; 2016; pp. 97–110. Available online: https://www.pide.org.pk/psde/pdf/AGM33/papers/M%20Aman%20Ullah.pdf (accessed on 26 May 2020).

37. Heshmati, A. Productivity Growth, Efficiency and Outsourcing in Manufacturing and Service Industries. *J. Econ. Sur.* **2003**, *17*, 79–112. [CrossRef]

38. Rybár, R.; Kudelas, D.; Beer, M. Selected Problems of Classification of Energy Sources: What are Renewable Energy Sources? *Acta Montan. Slovaca* **2015**, *20*, 172–180.

39. Wolde-Rufael, Y. Energy consumption and economic growth: The experience of African countries revisited. *Energy Econ.* **2009**, *31*, 217–224. [CrossRef]

40. Schurr, S.H.; Netschert, B.C.; Eliasberg, V.E.; Lerner, J.; Landsberg, H.H. *Energy in the American Economy*; Johns Hopkins University Press: Baltimore, MD, USA, 1960.

41. Schurr, S.H. Energy use, technological change, and productive efficiency: An economic-historical interpretation. *Ann. Rev. Energy* **1984**, *9*, 409–425. [CrossRef]

42. Jorgenson, D.W. The role of energy in productivity growth. *Energy J.* **1984**, *5*, 26–30. [CrossRef]

43. Boudreaux, B.C. The impact of electric power on productivity: A study of US manufacturing 1950–1984. *Energy Econ.* **1995**, *17*, 231–236. [CrossRef]

44. Murillo-Zamorano, L.R. The role of energy in productivity growth: A controversial issue? *Energy J.* **2005**, *26*, 69–88. [CrossRef]

45. Walheer, B. Labour productivity growth and energy in Europe: A production-frontier approach. *Energy* **2018**, *152*, 129–143. [CrossRef]

46. Mahadevan, R.; Asafu-Adjaye, J. Energy consumption, economic growth and prices: A reassessment using panel VECM for developed and developing countries. *Energy Policy* **2007**, *35*, 2481–2490. [CrossRef]

47. Velucchi, M.; Viviani, A. Determinants of the Italian Labor Productivity: A Quantile Regression Approach. *Statistica* **2011**, *71*, 213–238.
48. Islam, S.; Shazali, S.S. Determinants of Manufacturing Productivity: Pilot Study on Labor-Intensive Industries. *Int. J. Product. Perform. Manag.* **2010**, *60*, 567–582. [CrossRef]
49. Heshmati, A.; Rashidghalam, M. Labour productivity in Kenyan manufacturing and service industries. In *Determinants of Economic Growth in Africa*; IZA: Bonn, Germany, 2016.
50. Nagler, P.; Naudé, W. *Labor Productivity in Rural African Enterprises: Empirical Evidence from the LSMS-ISA*; IZA Discussion Papers 8524; Institute of Labor Economics (IZA): Bonn, Germany, 2014.
51. Samuel, B.; Aram, B. The Determinants of Industrialization: Empirical Evidence for Africa. *Eur. Sci. J.* **2016**, *12*, 219–239.
52. OECD. *Productivity Measurement and Analysis. Proceedings from OECD Workshops*; Swiss Federal Statistical Office (FSO): Neuchâtel, Switzerland, 2008.
53. OECD. *Measuring Productivity—OECD Manual: Measurement of Aggregate and Industry-Level Productivity Growth*; OECD Publishing: Paris, France, 2001.
54. Hajkova, D.; Hurnik, J. Cobb-Douglas production function: The case of a converging economy. *Czech J. Econ. Financ.* **2007**, *57*, 465–476.
55. Van Beven, L. Total Factor Productivity: A Practical View. *J. Econ. Surv.* **2010**, *26*, 98–128. [CrossRef]
56. Del Gatto, M.; Di Liberto, A.; Petraglia, C. Measuring productivity. *J. Econ. Surv.* **2011**, *25*, 952–1008. [CrossRef]
57. Cobb, C.W.; Douglas, P.H. A theory of production. *Am. Econ. Rev.* **1928**, *18*, 139–165.
58. Murthy, K.V. Arguing a case for Cobb-Douglas production function. *Rev. Commer. Stud.* **2002**, *20*, 21.
59. Zellner, A.; Kmenta, J.; Dreze, J. Specification and estimation of Cobb-Douglas production function models. *Econom. J. Econom. Soc.* **1966**, *34*, 784–795. [CrossRef]
60. Eldridge, L.P.; Price, J. Measuring quarterly labor productivity by industry. *Mon. Labor Rev.* **2016**, *139*, 1–27. [CrossRef]
61. Faustino, H.C.; Leitão, N.C. Intra-industry trade: A static and dynamic panel data analysis. *Int. Adv. Econ. Res.* **2007**, *13*, 313–333. [CrossRef]
62. Hummels, D.; Levinsohn, J. Monopolistic competition and international trade: Reconsidering the evidence. *Q. J. Econ.* **1995**, *110*, 799–836. [CrossRef]
63. Zhang, J.; van Witteloostuijn, A.; Zhou, C. Chinese bilateral intra-industry trade: A panel data study for 50 countries in the 1992–2001 period. *Rev. World Econ.* **2005**, *141*, 510–540. [CrossRef]
64. Arellano, M.; Bond, S. Some Tests of Specification for panel data: Monte Carlo Evidence and an Application to Employment Equations. *Rev. Econ. Stud.* **1991**, *58*, 277–297. [CrossRef]
65. Arellano, M.; Bover, O. Another look at the Instrumental Variable Estimation of Error-components Model. *J. Econ.* **1995**, *68*, 29–51. [CrossRef]
66. Roodman, D. How to do xtbond2: An Introduction to Difference and System GMM in Stata. *Stata J.* **2009**, *9*, 86–136. [CrossRef]
67. Rodrik, D. Unconditional Convergence in Manufacturing. *Q. J. Econ.* **2013**, *128*, 165–204. [CrossRef]

Article

Energy Consumption Analysis for Vehicle Production through a Material Flow Approach

Fernando Enzo Kenta Sato [1,2] and Toshihiko Nakata [2,*]

[1] Cyclical Resource Promotion Division, Honda Motor Co., Ltd., Wako 351-0114, Japan; kenta_b_sato@hm.honda.co.jp

[2] Department of Management Science and Technology, Graduate School of Engineering, Tohoku University, Sendai 980-8579, Japan

* Correspondence: nakata@tohoku.ac.jp

Received: 10 April 2020; Accepted: 7 May 2020; Published: 11 May 2020

Abstract: The aim of this study is to comprehensively evaluate the energy consumption in the automotive industry, clarifying the effect of its productive processes. For this propose, the material flow of the vehicles has been elaborated, from mining to vehicle assembly. Initially, processes where each type of material was used, and the relationship between them, were clarified. Subsequently, material flow was elaborated, while considering materials input in each process. Consequently, the consumption of energy resources (i.e., oil, natural gas, coal, and electricity) was calculated. Open data were utilized, and the effects on the Japanese vehicle market were analyzed as a case study. Our results indicate that the energy that is required for vehicle production is 41.8 MJ/kg per vehicle, where mining and material production processes represent 68% of the total consumption. Moreover, 5.23 kg of raw materials and energy resources are required to produce 1 kg of vehicle. Finally, this study proposed values of energy consumption per mass of part produced, which can be used to facilitate future material and energy analysis for the automotive industry. Those values can be adopted and modified as necessary, allowing for possible changes in future premises to be incorporated.

Keywords: vehicle; productive process; energy consumption; material consumption

---

1. Introduction

Climate change is considered to be one of the major social drawbacks of the last decades. To combat it, the Paris Agreement on climate change was established in December 2015, for which 195 nations have unified its environmental goals and agreed to maintain a global temperature increase well below 2 °C [1]. In this sense, different strategies and studies regarding the efficient use of energy are continuously conducted by governmental as well as private entities, demonstrating a global conscience and strong necessity to change the current high energy and resource consumption of society.

The transportation sector accounts for 25% of global energy consumption [2], and it is one of the most challenging sectors for fulfilling the proposed goals. Therefore, several studies centered on the fuel consumption of the vehicle have been conducted over the past few decades. New technologies, such as alternative propulsion methods (hybrid electric vehicles, battery electric vehicles, plug-in hybrid electric vehicles, and fuel cell vehicles) and lightweight materials, have also been developed.

A widely known method to assess the environmental effect of a vehicle is through its life cycle, and previous studies estimate that the production phase constitutes 7–22%, and the use phase 78–93%, of the energy consumption and CO_2 emission of a vehicle's life cycle [3,4], whereas the end-of-life vehicle (ELV) phase is considered to be almost negligible. Thus, improving the energy efficiency of the use phase of the vehicle was prioritized and the production and ELV phases have usually been considered less important. However, comprehensively understanding the environmental impact of the transportation sector is also indispensable for correctly evaluating the impact of both phases [5].

Nemry et al. [3] evaluated possible environmental advantages of the transportation sector in Europe, O'Reilly et al. [6] proposed a lightweigh optimization method, Sato et al. [5,7] evaluated the environmental impact of the ELV phase, Lane [8], and Vinoles-Cebolla et al. [9], Messagie et al. [10], and Yang et al [11] evaluated the environmental impact of electric vehicles, all using life-cycle assessment (LCA). However, even those studies included the effect of the production phase, basing its analysis on external energy consumption or CO_2 emission constant coefficients; in some cases, even the precedence of the data used for the calculations were clarified. Those coefficients are usually presented as an approximation of the energy that is required to produce a vehicle part and defined per unit of mass of the material that composes it. However, they are usually close values and, in this sense, premises considered, the processes included in their calculation did not make the level of accuracy of the proposed values transparent [12–15]. It is worth mentioning that the International Organization for Standardization [16,17] specifies the necessity of clarifing the sytem boundary and also lists data-quality requirements to ensure the transparency of the LCA.

This study aims to comprehensively evaluate the energy consumption in the automotive industry, clarifying the effect of its productive processes. This study focuses on developing a process-by-process breakdown analysis and elaborates the material flow of vehicle production, from raw material mining to vehicle assembly. Moreover, the results obtained for energy and material consumption have been assessed per unit of produced vehicle as well as per mass of product. This approach is based on open data, and the effects on the Japanese vehicle market were analyzed as a case study.

The results presented in this study allow for a comprehensive understanding of the production phase of the vehicle and proposed values of energy consumption that can be used for upcoming vehicle life-cycle studies, contributing to the improvement of future vehicle production and recycling assessments. Moreover, those values can be adopted and modified, depending on necessity, allowing for possible changes in premises to be incorporated.

2. Methodology

Figure 1 shows the boundary of this study, where material and energy consumption from raw-material mining to vehicle assembly were considered. Moreover, this study considered the seven main materials (i.e., steel, iron, plastic, glass, rubber, aluminum, copper) that represent 85–96% of a vehicle's mass in the analysis [4,18–20].

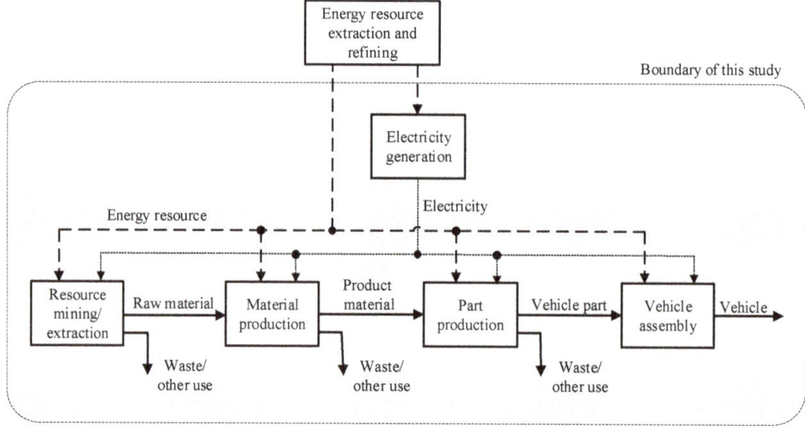

Figure 1. Analysis boundary of the vehicle production supply chain.

Initially, processes where each type of material used is clarified and material flow elaborated, considering materials, were input. Consequently, energy consumption by energy resource (i.e., oil, natural gas, coal) and electricity were calculated in each phase of the flow. Finally, the results were analyzed and compared for discussion to create an in-depth understanding of the industry.

2.1. Material Flow Elaboration

Firstly, the part production processes were analyzed, and Figure 2 was elaborated based on previous studies [5,18,19] while considering the material composition of a generic vehicle (Honda Accord, Internal combustion engine vehicle, 2011 [18]). Here, the mass percent of the material composition of the vehicle and the principal part production process they are subjected are clarified. The mass of the vehicle parts made by determined processes can be calculated through Equation (1):

$$G_{m,i} = G_{veh} * GR_m * GR_{m,i},\tag{1}$$

where $G_{m,i}$ is the mass of vehicle parts made by material m and formed through productive process i, G_{veh} is the mass of the vehicle, adopted as 1481 kg [18], GR_m is the mass ratio of material m of a vehicle, and $GR_{m,i}$ is the mass ratio of material m of a vehicle that is subjected to part production process i.

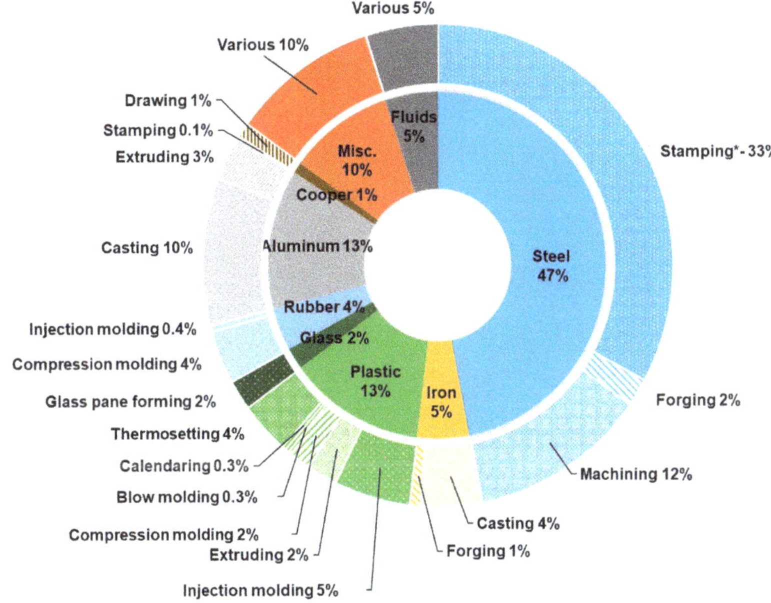

Figure 2. Material composition of a vehicle and its principal part production process.

Moreover, the material that is consumed in each analyzed part of the production process is calculated through Equation (2):

$$GMC_{m,i} = G_{m,i} * MC_{m,i},\tag{2}$$

where $GMC_{m,i}$ is mass of material m consumed in productive process i, $MC_{m,i}$ is the mass of material m consumed in production process i per mass of product (process output), as shown in Table 1.

Table 1. Material and energy consumption of each production process.

| Flow | Process | Material Consumption Per Mass of Process Output (Kg/Kg) | | | Energy Consumption Per Mass Of Process Output (MJ/kg) | | | | | Ref. |
		Material	Amout	Ref.	Oil	Natural Gas	Coal	ElectriCity	Internal Process *	
Steel	Iron ore extraction and processing				0.206	0.186		1.327		(e)
	Limestone mining				0.019		0.004			(b)
	Lime production	Calcium carbonate	2.072	(b)	0.119	0.244	3.489	0.221		(b)
	Coke Production, Sintering, Blast Furnace, Basic Oxygen Furnace and On-site Generation processe	Lime	0.060	(b)	1.192	0.356	16.258	1.256	−1.477	(b)
		Calcium carbonate	0.050	(b)						
		Iron ore	1.150	(b)						
	Hot rolling	Slab	1.031	(a)		0.665		0.743	1.399	(e)
	Skin mill	Hot rolled strip	1.015	(a)				0.044	0.035	(e)
	Cold rolling	Hot rolled strip	1.054	(a)				1.477	0.622	(e)
	Galvanizing	Rolled sheet	1.000	(a)				0.734	1.364	(e)
	Stamping	Rolled sheet	1.000	(a)		4.545		1.208		(e)
	Rod and bar mill	Billet	1.000	(a)		2.275		1.137		(e)
	Forging	Billet	1.000	(f)		40.404		1.357		(c)
	Machining	Bar, rod, others	1.000	(a)				0.628		(c)
Iron	Iron recycling				1.314			0.099		(a)
	Coke production						37.314	0.398	−4.472	(e)
	Forging	Scrap iron/steel	1.000	(b)		34.415		1.248		(a)
	Casting	Scrap iron/steel	1.000	(b)						(a), (b)
		Coke	0.840	(a)						
	Machining	Iron	1.000	(b)				0.570		(a)
Plastic	Plastic fabrication				15.136	36.007		1.275		(e), (f) **
	Injection molding	Pellets	1.139	(a)	1.207	0.858		7.546		(c)
	Extrusion	Pellets	1.002	(a)	0.692	0.039		1.944		(c)
	Compression molding	Pellets	1.000	(a)				1.501		(c)
	Blow molding	Pellets	1.000	(a)				6.152		(c)
	Calendaring	Pellets	1.155	(a)	0.239	0.156		1.822		(c)
	Molding thermoset	Resin	1.000	(a)				1.501		(c)
Glass	Limestone mining				0.019		0.004			(b)
	Dolomite mining				0.158			0.010		(b)
	Trona mining				0.206			1.327		(f) ***
	Sodium carbonate production	Trona	0.907	(b)	0.442		4.220			(b)
	Float glass fabrication	Sand	0.721	(a)		13.143		0.875		(c)
		Calcium carbonate	0.099	(a)						
		Dolomite	0.183	(a)						
		Sodium carbonate	0.232	(a)						
Rubber	Styrene-butadiene rubber fabrication				19.771	19.771		0.395		(b)
	Molding rubber	Styrene-butadiene	1.000	(a)		5.265		2.365		(c)
	Injection molding	Styrene-butadiene	1.031	(a)	8.150			4.950		(c)

Table 1. *Cont.*

Flow	Process	Material Consumption Per Mass of Process Output (Kg/Kg)			Energy Consumption Per Mass Of Process Output (MJ/kg)					Ref.
		Material	Amout	Ref.	Oil	Natural Gas	Coal	ElectriCity	Internal Process *	
Aluminum	Sodium brine production				0.116	0.717		0.232		(b)
	Sodium hydroxide production	Sodium brine	5.830	(b)	0.002	8.141	0.663	6.978		(b)
	Bauxita mining				0.592			0.017		(a)
	Limestone mining				0.019		0.004			(b)
	Lime production	Calcium carbonate	2.072	(b)	0.119	0.244	3.489	0.221		(b)
	Alumina production	Bauxita	2.881	(b)	3.105	13.624	1.412	0.677		(a)
		Sodiun hydroxide	0.306	(a)						
		Lime	0.078	(a)						
	Alumina reduction	Alumina	1.935	(b)				49.354		(a)
	Ingot casting	Aluminium	1.020	(b)	0.146	0.695		0.221		(a)
	Srap preparation	Aluminium scrap	1.010	(b)		0.791		0.369		(b)
	Secondary ingot casting	Aluminium scrap	0.970	(b)		4.347		0.359		(b)
		Aluminium	0.080	(b)						
	Hot rolling	Aluminum ingot	1.035	(b)		3.457		0.371		(a)
	Cold rolling	Aluminum ingot	1.000	(b)		1.993		1.195		(a)
	Stamping	Rolled sheet	1.000	(a)		4.545		1.208		(c)
	Extrusion	Aluminum ingot	1.000	(f)	0.692	0.039		1.944		(c)
	Shape casting	Aluminum ingot	1.000	(f)		27.495		8.046		(c)
	Machining	Aluminium	1.000	(a)				0.628		(c)
Copper	Copper ore mining				0.006			0.007		(e), (f), (d)
	Copper production	Copper ore	169.586	(d)	1.452	9.075	3.448	6.897		(e)
	Wire drawing	Copper	1.000	(b)	0.887		0.021	1.711		(e)

Refereces:
(a) GREET Excel model platform [21]
(b) GREET 2018 Net software [22]
(c) Sullivan, 2010 [19]
(d) Ophardt, 2003 [23]
(e) Keoleian, 2012 [24]
(f) Author estimation

* Blast furnace and coke oven gas; not considered in the energy consumption calculation
** Considered values of Polypropilene
*** Considered same as Iron ore mining

It is worth mentioning that 6.2% of the stamped and 54.7% of the forged steel parts; 95.8% of the casted and 100% of the forged iron parts; and, 91.4% of the casted and 3.4% of the extruded aluminum parts were also subjected to a machining process [19].

Secondly, the material flow of the material production processes was analyzed. Figure 3 was elaborated based on Sullivan et al. [19], GREET Excel Model Platform [21], Greet 2018 Net software [22], Ophardt [23], and Keoleian et al. [24], Brunham et al. [25]. The left side of the figure indicates the upstream of the productive supply chain (mining), where its output (raw material) is subjected to material productive processes before entering part production processes and it is finally assembled as a part of a vehicle in the assembly plant. This flow is quantified when considering both equations proposed above and through the following ones. The materials used to produce parts could be supplied by different material production processes, as indicated by Equation (3). Moreover, different quantities of raw materials are required to produce each material, as indicated by Equation (4).

$$GMC_{m,i} = \sum_j GMP_{m,j}, \tag{3}$$

$$GMC_{n,j} = GMP_{m,j} * MC_{n,j}, \tag{4}$$

where $GMP_{m,j}$ is mass of material m produced in productive process j, $GMC_{n,j}$ is mass of material n consumed in productive process j, and $MC_{n,j}$ is the mass of material n consumed in productive process j per mass material produced (process output), as shown in Table 1.

In the same way, the flow is extended upstream to cover all of the productive processes of the materials. Pressed steel parts are produced by hot rolling, cold rolling, and galvanized steel sheets. The first one represents 21.1%, the second one 19.1%, and the last one 59.8% of the final product mass [21]. Moreover, casted aluminum and iron parts, such as engine blocks, engine/exhaust components, and brake rotors, were used for its production recycled material. Here, it is considered that 85% of casted aluminum parts and the total of casted iron parts contain recycled materials.

Finally, the material flow of the vehicle-assembly phase was elaborated, when considering that the body of each vehicle is produced through the welding and painting of pressed steel parts. Subsequently, the rest of the supplied parts were added to it in the line, before a final verification of the entire vehicle to ensure its quality and functionality.

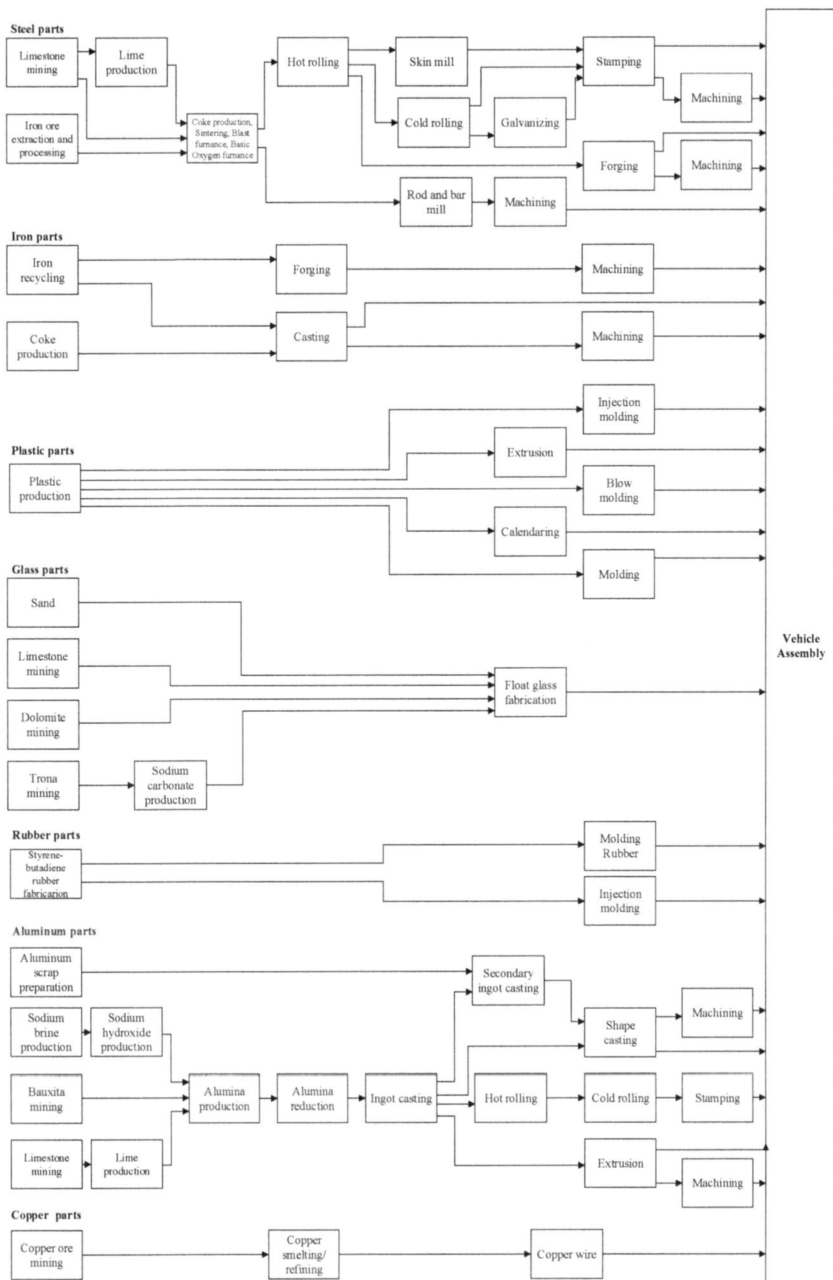

Figure 3. Production flow for a vehicle.

2.2. Energy Consumption Analysis

The energy consumption for vehicle production was calculated while considering the product (output) of each productive process. Equation (5) represents the consumption in the part production processes, and Equation (6) represents types of consumption, from resource mining to material production.

$$ECP_{m,i} = G_{m,i} * \sum_e ECP_{e,i}, \tag{5}$$

$$ECM_{n,j} = GMP_{n,j} * \sum_e ECM_{e,j}, \tag{6}$$

where $ECP_{m,i}$ is the energy that is consumed in the part production process i to produce vehicle parts made by material m, $ECM_{n,j}$ is the energy consumed in the mining or material production process j to produce material n, $ECP_{e,i}$ is the energy resource or electricity e consumed in the part production process i per mass of part (process output), $ECM_{e,j}$ is the energy resource or electricity e consumed in the mining or material production process j per mass of product (process output), as shown in Table 1.

The total energy consumption to produce a determinate part can be calculated as the sum of the energy that is consumed by each productive stage, from material mining to part production, as shown in Equation (7).

$$TECP_{m,i} = ECP_{m,i} + \sum_j ECM_{n,j}, \tag{7}$$

where $TECP_{m,i}$ is the total energy consumed to produce parts made from material m and formed by productive process i.

Finally, the effect of the vehicle assembly plant was added, per unit of vehicle, based on energy consumption data of Sullivan et al. in order to calculate the total energy consumption required to produce a vehicle [19].

3. Results and Discussions

3.1. Results of the Energy and Material Consumption Analysis

Figure 4 shows the material flow for vehicle production elaborated in this study. Here, materials that are necessary for the production of vehicle parts, as well as energy consumed in its production processes, are represented. The mass of oil was considered to be 22.6 g/MJ, natural gas 27.5 g/MJ, and coal 34.4 g/MJ [26,27]. Moreover, the mass of the electricity was estimated as 56.2 g/MJ, when considering the Japanese grid mix, which is generated through oil (19.2%), natural gas (37.5%), coal (32.8%), and others (10.3%) [28]. The efficiency of the generation facilities was considered to be between 42% and 60%, depending on the energy resource utilized in transformation [28]. Plastics and rubbers were made by raw material that were derived from crude oil, and those feedstocks are also represented in the figure as energy resources.

The proposed flow emphasizes the necessity of a considerable amount of resources and material for the production of a vehicle. As raw material, copper ore is the most consumed, due to its low concentration of copper material, followed by iron ore and bauxite. On the other hand, energy resources are mostly consumed in the production of steel and aluminum parts. Figure 5 summarizes those values, where it can be observed that more than 7762 kg of raw material and energy resources is consumed in order to produce a vehicle of 1,481 kg. This means that 5.23 kg of resources are necessary to produce 1 kg of vehicle. Here, copper ore has the highest percentage values, with 2.29 kg of raw material per kg of vehicle (3391 kg per vehicle), followed by energy resources, with 1.46 kg of them being consumed per kg of vehicle (2165 kg per vehicle). The values presented in Figure 5 are also included in Figure 4, where the total raw material and energy resources on the left side of the figure are transformed in stages to a final vehicle on the right.

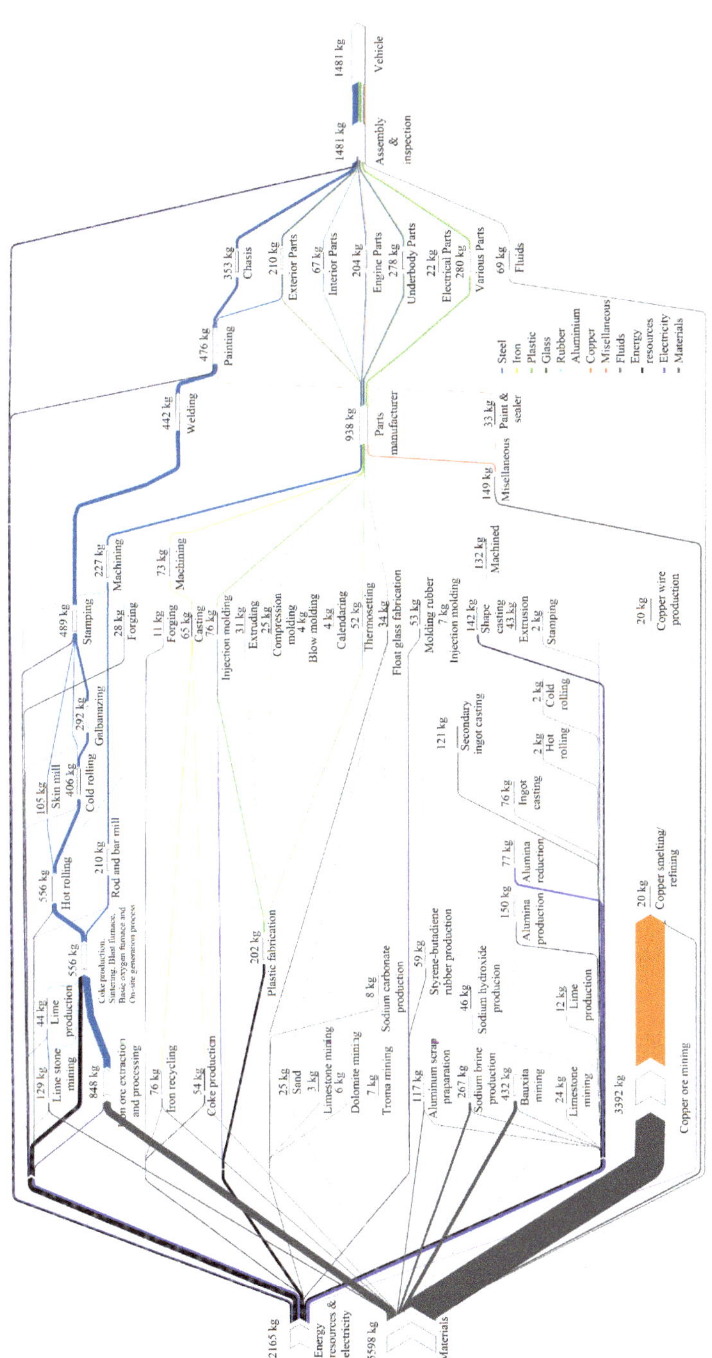

Figure 4. Material flow for vehicle production.

Figure 6 summarizes the results related to energy consumption. The total energy consumed to produce a vehicle was calculated as 62 GJ (41.8 MJ/kg of vehicle). Figure 6a shows that steel parts are the most representative, encompassing 35% of the total. Moreover, even copper parts consume a high quantity of raw material; due to the low concentration of copper on its ore, the energy that is required in its production processes is not as high as could be expected. It can be observed from Figure 6b that natural gas is the highest consumed energy resource in vehicle production, and Figure 6d shows that its consumption is almost equally distributed in aluminum, steel, and plastic parts production, as well as in vehicle assembly. Finally, Figure 6c shows that the energy that is consumed in the production phase of a vehicle is dominated by the mining and material production processes, which represent 68% of total consumption, followed by the part production processes, at 19%, and vehicle assembly, at 13%.

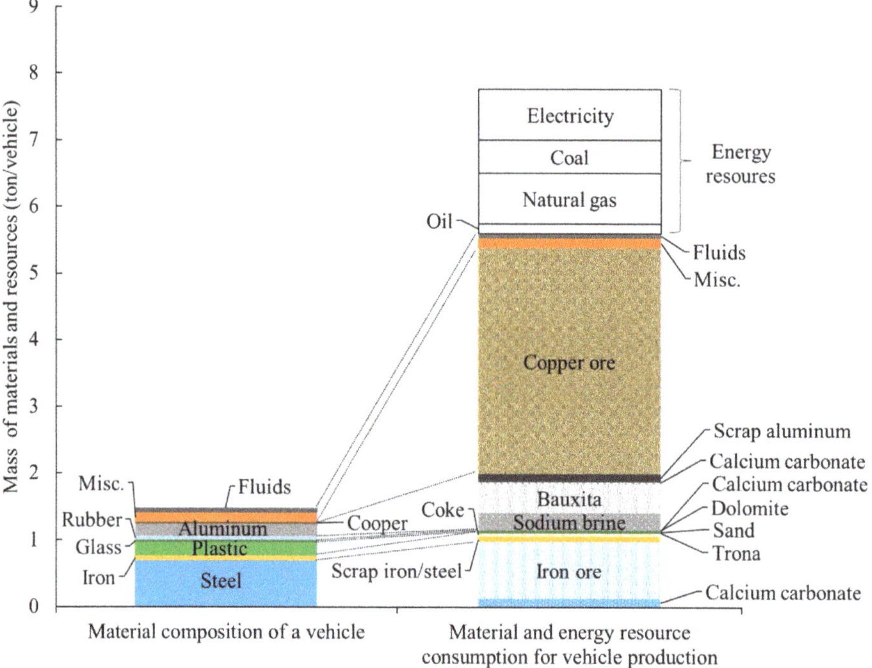

Figure 5. Mass of materials and resources consumed in automobile production.

Figure 7 shows the energy consumption of each productive process of vehicle production. The figure is divided into mining-material production, part production, and vehicle assembly processes. It can be seen that 82% of the total coal is consumed in the steel production processes, 28% electricity in the alumina reduction process, and 26% natural gas in the plastic fabrication processes, showing a demand concentration of determinate resources in specific facilities.

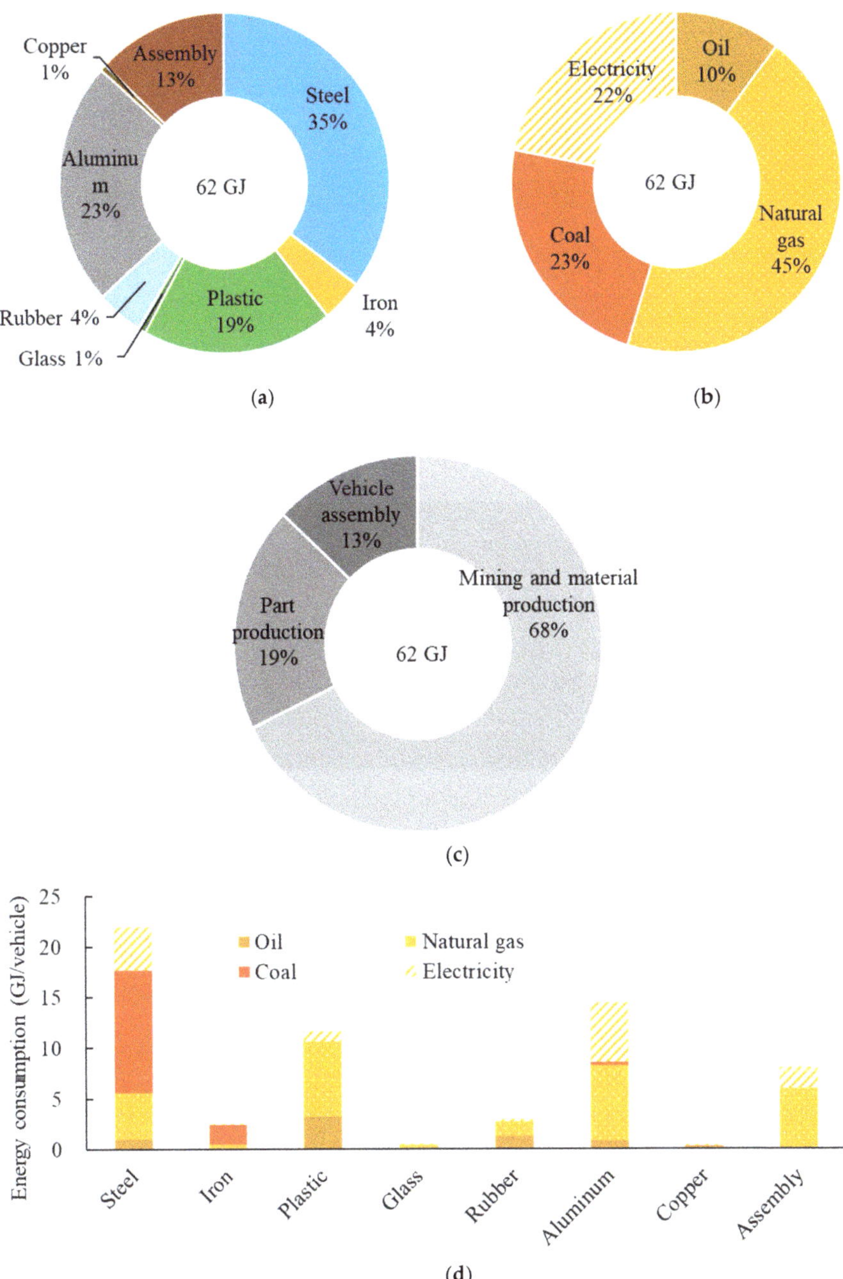

Figure 6. Energy consumption in vehicle production:(**a**) by material; (**b**) by energy resource; (**c**) by productive phase; (**d**) by material and energy resource.

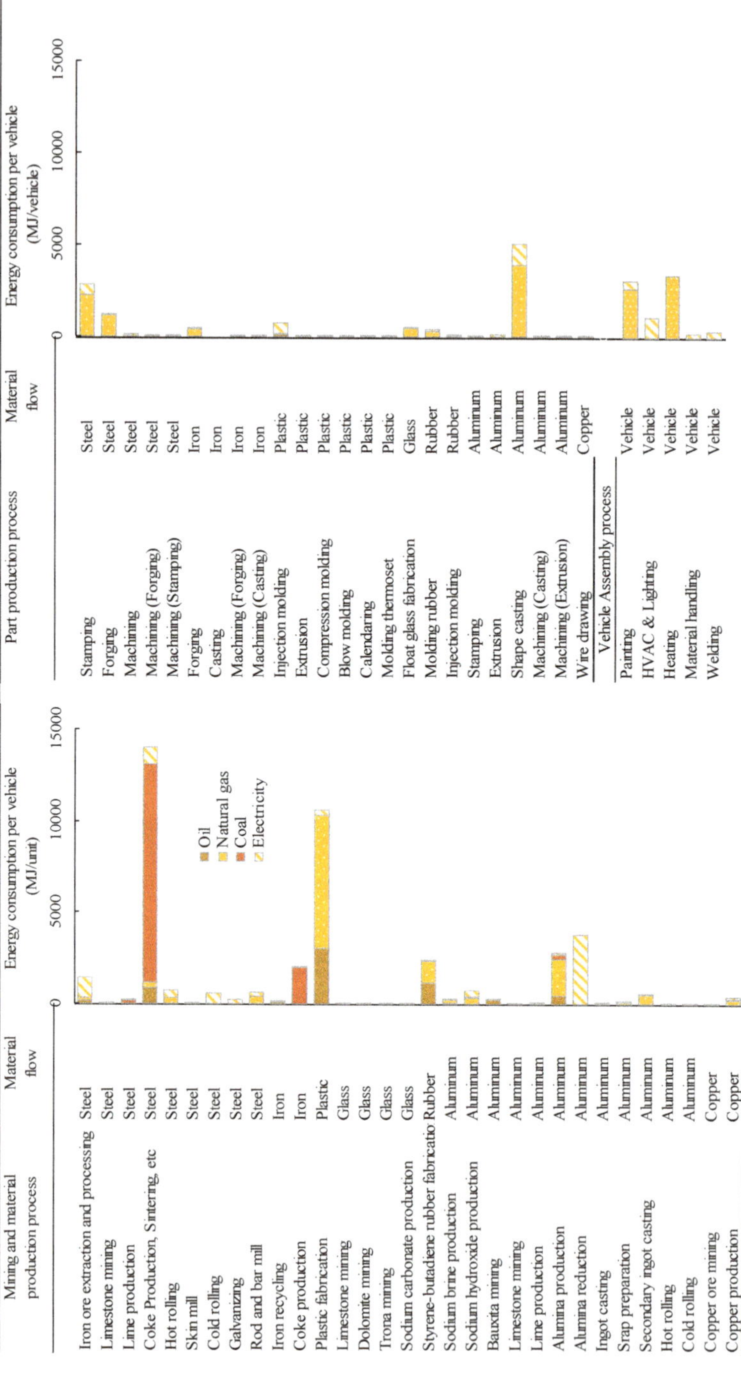

Figure 7. Energy consumption in each productive process.

Finally, the first chart of Figure 8 shows the energy that is required to produce each type of vehicle part per kg of material; those constants have generally been defined in previous studies as embodied energy [5,15]. The proposed energy consumption values could vary widely by part, despite being produced by the same material. More conspicuous are the parts that are made by steel, where the energy that is required to produce forged products doubles that needed to elaborate the stamped ones. Moreover, aluminum parts are the most energy-intensive parts. Figure 8 shows the energy that is required to produce each type of part per unit of vehicle. It can be seen that the stamped steel parts consume the major volume of energy (23%) necessary for the production of vehicles, followed by cast and machined aluminum products (13%).

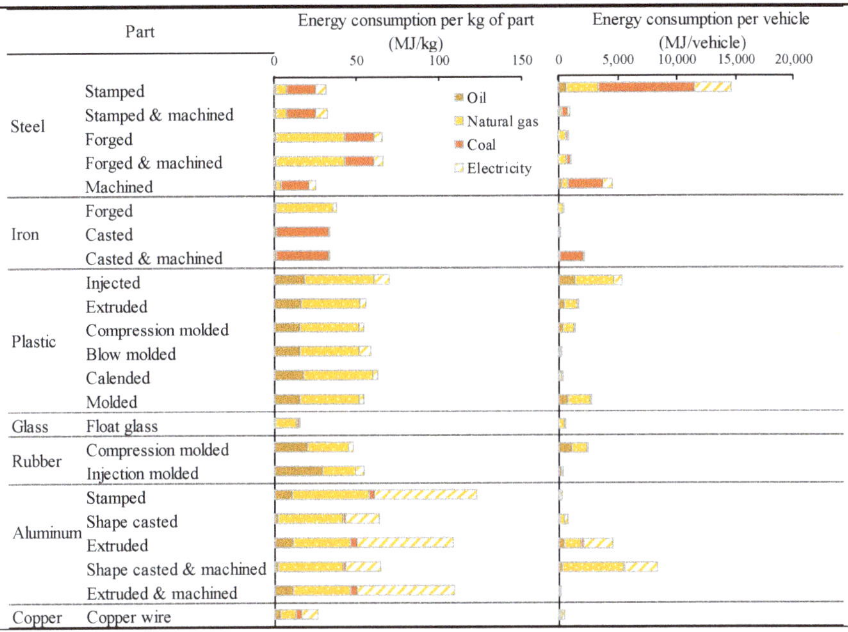

Figure 8. Energy required for the production of each type of vehicle part.

3.2. Energy and Material Consumption for the Entire Japanese Market

Three representative aspects were considered to estimate the total energy consumption for the Japanese automotive industry: the average mass of a passenger car in Japan (1354 kg/vehicle) [29], the number of passenger cars produced annually in the country (9,729,594 vehicles) [30], and the energy that is required for the production of a vehicle (41.8 MJ/kg of vehicle) calculated in this study. It has been calculated, though the product of the above values, that the energy consumption that is related to the automotive industry is 0.55 EJ per year in Japan. Moreover, Figure 9a compares the obtained consumption values and the total energy consumption for different sectors. It can be seen that the energy consumption of the automotive industry represents 15% of the energy consumption of the Japanese industy. This also indicates that strategic decision- or policy-making through a comprehensive analysis of this phase could generate national-level energy benefits, emphasizing the importance of the approach that was proposed in this study. The energy consumption of the automotive industry is included in the "transportation equipment" sub-sector of industrial demand; however, in contrast to the values that are presented in this study, the material production processes are not included. In the referenced report [31], those values are distributed in the respective material production sub-sectors (i.e., material production processes of steel parts are included in the iron and steel sub-sector, material production processes of plastic parts are included in the chemistry sub-sector, etc.).

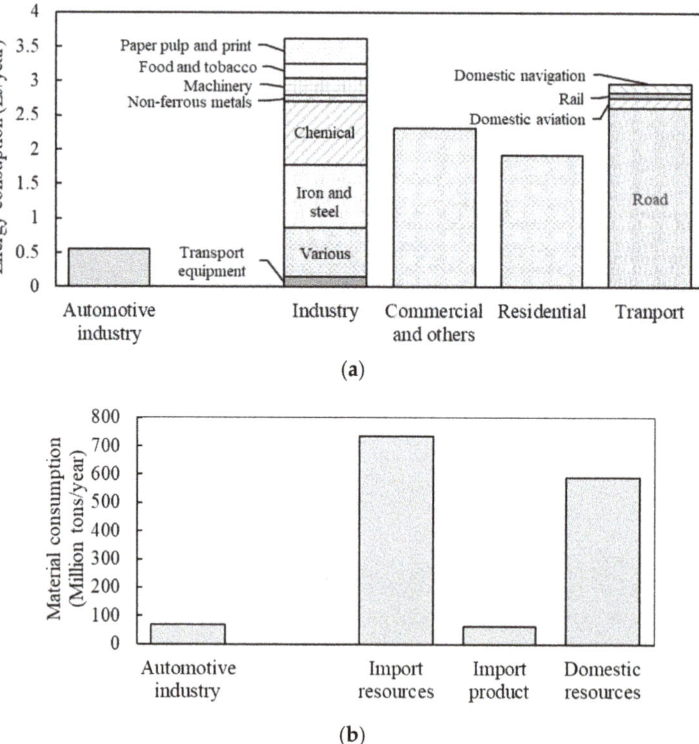

Figure 9. Effect of the automotive industry on Japanese energy and material consumption: (**a**) energy consumption [31]; (**b**) material consumption [32]

On the other hand, the materials and resources consumed in the industry were calculated as 69 million tons per year, representing more than 9.4% of the annual imported resources of Japan, as shown in Figure 9b.

3.3. Primary Assumptions and Limitations

Firstly, the energy required for energy resource extraction and refining, as well as the water consumption in each productive process, have not been included in this study. Water is usually consumed for refrigeration, and the internal reuse of it is a standard operation in the industry. Moreover, thermal energy has been considered to be an internal process of each facility, which is produced by the input energy resources that are listed in the study.

Secondly, this approach bases its calculation on internal combustion engine vehicles (ICEV), which represent more than 63% of vehicle sales in Japan. Moreover, hybrid vehicles represent 31% of the total sales. Future studies will extend this approach to electric vehicles (EV), which are even more energy-intensive products than our base scenario. On the other hand, the material composition of the vehicle varies depending on the model and the year of production. Thus, final energy and material consumption values per vehicle can vary moderately, but they are also actualized when considering the energy that is required per unit of mass, as shown in Figure 8.

Thirdly, even this approach estimated the total energy consumption of the automotive industry when considering the Japanese market as a case study; not all the productive processes are carried in domestic facilities. Nonetheless, the main conclusions of this study will not change.

Finally, our analysis was centered on the seven principal materials. Miscellaneous materials are expected to vary widely, depending on the analyzed vehicle model (i.e., leader in the case of high-spec vehicle seats, electric and audio equipment, wood in high-end vehicles, and others).

3.4. Comparison with Results of Previous Studies

In this section, simple comparisons with previous studies are proposed in order to evaluate the obtained energy consumption values. Our results were compared with the values calculated in previous life-cycle approaches that were conducted by Nemry et al. [3] and Schweimer et al. [33]. The first study is a report for the European Union, which analyzed the potential ways of reducing the life-cycle impact of the transportation sector in Europe. Here, the results of the material and part production processes were included, but the analysis was based on external data. The second study analyzed 1999-year Golf A4 vehicles, centering the analysis on the assembly phase. Here, inventory data of Volkswagen plants were analyzed in detail, including material and energy inputs. However, it did not expand, to the same degree, on the materials and part production processes.

A rough simulation of energy consumption in the use and ELV phase of the studied vehicle was proposed. The energy that is consumed in the use phase can be calculated when considering the fuel economy of the vehicle, as shown in Equation (8).

$$E_U = FE * d * \delta_{gas} * HHV_{gas} \tag{8}$$

where E_U is energy consumed in the use phase, FE is fuel economy, e.g., of Honda Accord 2011, 9.046 l/100 km [34], d is the total traveled distance, 100,000 km, HHV_{gas} is the higher heating value of gasoline, 46.4 MJ/kg [35], and δ_{gas} is the density of gasoline, 0.75 kg/l [35].

The energy that is consumed in the disposal process of the ELV is calculated while using Equation (9).

$$E_{ELV} = ED * G_{veh} \tag{9}$$

where E_{ELV} is energy consumed in the ELV disposal process and ED is disposal energy, 0.602 MJ/kg [36].

The first column of Table 2 shows the life-cycle values that were proposed in this study. The second and third columns compare the obtained results with previous approaches, demonstrating the compatibility between them. It is also worth mentioning that the energy consumption per mass of vehicle in the production phase is slightly lower when compared to previous studies. This can be explained by the fact that the decrease in energy consumption due to the use of recycled materials is included, and that the effects of miscellaneous materials and fluids are not included in our approach.

Table 2. Comparison of vehicle life cycle energy consumption.

	Energy Consumption Values Proposed in Our Approach		Energy Consumption Values from Nemry et al. [3]		Energy Consumption Values from Schweimer et al. [33]	
	MJ/kg of vehicle	Percentage	MJ/kg of vehicle	Percentage	MJ/kg of vehicle	Percentage
Production	41.8	16.4%	53	9%	81	26%
Use	213	83.4%	557	91%	226	73%
ELV	0.6	0.2%	0	0%	-	-
Total	255.4	100%	610	100%	307	100%

3.5. Application of the Results

This study presents a whole picture of the energy and material consumption of the automotive industry, allowing for automakers, part makers as well as researchers, and government bodies to comprehensively understand the production phase of the vehicle. Here, productive processes that have the highest effect in the industry can be identified. Efforts could focus on improving the efficiency of those energy-intensive facilities and processes to elevate the energy efficiency of the industry.

Energy-consumption results tha are obtained from this approach are divided into productive processes, but also per energy resources required for each of them. In this sense, future studies could focus on proposing optimal energy supply systems for the industry. The potential for changing the electricity consumed from the grid to renewable energy could be exploited to improve the environmental aspects of the sector.

This approach also allows researchers and the automotive industry to easily calculate the total energy impact of vehicle production, contributing to upcoming vehicle life-cycle studies and material and energy analysis of the automotive industry. When compared to constant embodied energy values proposed by previous studies, the values presented in this approach not only focus on the automotive industry but also clarify the material flow and processes that are considered in it. This allows for an easy recalculation and adjustment of the values, depending on the changes or differences in production technologies. Moreover, understanding the material flow of the industry enables new approaches for the industry, such as the environmental evaluation of closed-loop recycling, which can identify the process where recyclable material comes back for reprocessing.

Finally, evaluating the automotive industry through a material flow approach also allows one to assess the environmental impact of material required in mining and resource-extraction processes (i.e., the devastation of mining sites, disruption of natural habitats, groundwater contamination, and landscape changes at the extraction site [37]). Moreover, the proposed approach can be applied in risk-evaluation analysis of materials that are supplied to the automotive industry.

4. Conclusions

This study presents a whole picture of the automotive industry in terms of energy and material consumption, allowing for us to comprehensively understand the production phase of the vehicle. For this study, the material flow of the automotive industry has been elaborated. The main conclusions are listed below.

- It has been calculated that for the production of 1 kg of vehicle, at least 5.23 kg of raw materials and energy resources are required. Copper ore has the highest percentage value of 2.29 kg/kg of vehicle, followed by energy resources, with 1.46 kg/kg of vehicle.
- Energy consumption for the production of a vehicle was calculated as 62 GJ (41.8 MJ/kg of vehicle). Mining and material production processes dominate consumption, representing 68% of the total, followed by the part production processes, at 19%, and vehicle assembly, at 13%.
- Natural gas is the most consumed energy resource, representing 44% of the total energy consumption for the automotive industry. This consumption is centered on the plastic fabrication processes, for which 26% of this resource is required. Moreover, 82% of the total coal is consumed in the steel production processes, and 28% of the electricity in the alumina reduction process, showing a demand concentration of determinate resources in specific facilities.
- The energy consumption that is related to the automotive industry is 0.55 EJ per year in Japan, representing 15% of the industrial energy consumption of the country. Moreover, the materials and resources consumed in the industry were calculated as 69 million tons per year, representing more than 9.4% of the annual imported resources for Japan.

Finally, this study proposed values of energy consumption per mass of part that can be used for upcoming material and energy analysis of the automotive industry. Moreover, these values can be adopted and modified, depending on the necessity, allowing for possible changes in premises to be reflected.

Author Contributions: Conceptualization, F.E.K.S.; methodology, F.E.K.S.; formal analysis, F.E.K.S.; investigation, F.E.K.S.; writing—review and editing, F.E.K.S. and T.N.; supervision, T.N. All authors have read and agreed to the published version of the manuscript.

Energies **2020**, *13*, 2396

Funding: This research received no external funding.

Conflicts of Interest: The authors declare no conflict of interest.

References

1. United Nations. United Nations Framework Convention on Climate Change. Historic Paris Agreement on Climate Change: 195 Nations Set Path to Keep Temperature Rise Well Below 2 Degrees Celsius, Announcement/13 December 2015. Available online: https://unfccc.int/news/finale-cop21 (accessed on 30 September 2018).
2. U.S. Energy Information Administration. International energy outlook 2016. DOE/EIA-0484(2016); 2016. Available online: https://www.eia.gov/outlooks/ieo/pdf/0484(2016).pdf (accessed on 30 September 2018).
3. Nemry, F.; Leduc, G.; Mongelli, I.; Uihlein, A. *Environmental Improvement of Passenger Cars (IMPRO-Car)*; EUR—Scientific and Technical Research Reports; OPOCE: Brussels, Belgium, 2008.
4. Kobayashi, O. *Car Life Cycle Inventory Assessment. SAE Technical Paper 971199*; SAE: Warrendale, PA, USA, 1997.
5. Sato, F.E.K.; Furubayashi, T.; Nakata, T. Application of energy and CO_2 reduction assessments for end-of-life vehicles recycling in Japan. *Appl. Energy* **2019**, *237*, 77–794. [CrossRef]
6. O'reilly, C.J.; Göransson, P.; Funazaki, A.; Suzuki, T.; Edlund, S.; Gunnarsson, C.; Ludow, J.; Cerin, P.; Cameron, C.J.; Wennhage, P.; et al. Life cycle energy optimisation: A proposed methodology for integrating environmental considerations early in the vehicle engineering design. *J. Clean. Prod.* **2016**, *135*, 750–759. [CrossRef]
7. Sato, F.E.K.; Furubayashi, T.; Nakata, T. Energy and CO_2 benefit assess- ment of reused vehicle parts through a material flow approach. *Int. J. Automot. Eng.* **2019**, *9*, 91–98. [CrossRef]
8. Lane, B. Life Cycle Assessment of Vehicle Fuels and Technologies. London: Eco- Lane Transport Consultancy and London Borough of Camden. 2006. Available online: http://ww.seeds4green.org/sites/default/files/Camden_LCA_Report_FINAL_10_03_2006.pdf (accessed on 12 February 2020).
9. Vinoles-Cebolla, R.; Bastante-Ceca, M.J.; Capuz-Rizo, S.F. An integrated method to calculate an automobile's emissions throughout its life cycle. *Energy* **2015**, *83*, 125–136. [CrossRef]
10. Messagie, M.; Boureima, F.; Matheys, J.; Sergeant, N.; Timmermans, J.-M.; Macharis, C.; Van Mierlo, J. Environmental performance of a battery electric vehicle: A descriptive Life Cycle Assessment approach. *World Electr. Veh. J.* **2010**, *4*, 782–786. [CrossRef]
11. Yang, Z.; Wang, B.; Kui, J. Life cycle assessment of fuel cell, electric and internal combustion engine vehicles under different fuel scenarios and driving mileages in China. *Energy* **2020**, *198*, 117365. [CrossRef]
12. Nishimura, K.; Hondo, H.; Uchiyama, Y. Comparative analysis of embodied liabilities using an inter-industrial process model: Gasoline- vs. electro-powered vehicles. *Appl. Energy* **2011**, *69*, 307–320. [CrossRef]
13. Rydh, C.J.; Sun, M. Life cycle inventory data for materials grouped according to environmental and materials properties. *J. Clean. Prod.* **2005**, *13*, 1258–1268. [CrossRef]
14. Nishimura, K.; Hondo, H.; Uchiyama, Y. Estimating the embodied carbon emissions from the material content. *Energy Convers. Manag* **1997**, *38*, 589–594. [CrossRef]
15. Das, S.; Curlee, R.T.; Rizy, C.G.; Schexnayder, S.M. Automobile recycling in the United States: Energy impacts and waste generation. *Resour. Conserv. Recycl.* **1995**, *14*, 265–284. [CrossRef]
16. International Organization for Standardization. *Environmental Management—Life Cycle Assessment—Principles and Framework. ISO 14040*; International Organization for Standardization: Geneva, Switzerland, 2006.
17. International Organization for Standardization. *Environmental Management—Life Cycle Assessment—Requirements and Guidelines. ISO 14044*; International Organization for Standardization: Geneva, Switzerland, 2006.
18. Singh, H. *Mass Reduction for Light-Duty Vehicles for Model Years 2017–2025*; U.S. Department of Transportation and the National Highway Traffic Safety Administration; Report No. DOT HS 811 666; 2012. Available online: https://www.nhtsa.gov/corporate-average-fuel-economy/research-supporting-2017-2025-cafe-final-rule (accessed on 6 September 2016).
19. Sullivan, J.L.; Burnham, A.; Wang, M. Energy-Consumption and Carbon-Emission Analysis of Vehicle and Component Manufacturing. Argonne National Laboratory ANL/ESD/10-6; 2010. Available online: https://publications.anl.gov/anlpubs/2010/10/68288.pdf (accessed on 12 February 2020).

20. Ahmed, E. *Advances Composite Materials for Automotive Applications, Structural Integrity and Crashworthiness, Chapter 3*; Wiley: Hoboken, NJ, USA, 2014.
21. GREET Excel Model Platform, 2019. GREET2 Model. Argonne National Laboratory. Available online: https://greet.es.anl.gov/greet.models (accessed on 17 December 2019).
22. GREET 2018 Net software v1.3.0.13395, 2018. Greet Life Cycle Model 2018Argonne. Argonne National Laboratory. Available online: https://greet.es.anl.gov (accessed on 28 October 2018).
23. Ophardt, C.E. Asarco mining Operations in Arizona Mission Mine, Tucson. Virtual Chembook. Elmhurst College. 2003. Available online: http://chemistry.elmhurst.edu/vchembook/330copper.html (accessed on 12 February 2020).
24. Keoleian, G.; Miller, S.; De Kleine, R.; Fang, A.; Mosley, J. Life Cycle Material Update for Greet Model. University of Michigan, Center for Sustainable System Report No. CSS12-12. 2012. Available online: http://css.umich.edu/sites/default/files/css_doc/CSS12-12.pdf (accessed on 12 February 2020).
25. Brunham, A.; Wang, M.; Wu, Y. *Development an Application of Greet 2.7—The Transportation Vehicle-Cycle Model*; Argonee National Laboratory ANL/ESD/06-5; U.S. Department of Energy Office of Scientific and Technical Information: Oak Ridge, TN, USA, 2006.
26. University of California. Fuel Energy Conversion Factors. 2020. Available online: http://w.astro.berkeley.edu/~{}wright/ (accessed on 16 February 2020).
27. The Canadian Natural Gas Vehicle Alliance. Comparing Natural Gas to Diesel. Energy Content. 2014. Available online: http://cngva.org/wp-content/uploads/2017/12/Energy-Content-Factsheet-FINAL-EN.pdf (accessed on 31 March 2020).
28. Takita, Y.; Furubayashi, T.; Nakata, T. Development and analysis of an energy flow considering renewable energy potential, 2015. *Trans. JSME* **2015**, *81*, 15–00164.
29. MOE. Ministry of the Environment, Government of Japan. Change in Size (Weight) of Passenger Cars (for Private/Business Use). 2015. Available online: www.env.go.jp/doc/toukei/data/2015_6.17.xls (accessed on 12 February 2020).
30. Japan Automobile Manufacturers Association. Active Matrix Database System, Passenger Car Production 2018. Available online: http://jamaserv.jama.or.jp/newdb/index.html (accessed on 12 February 2020).
31. International Energy Agency. IEA Sankey Diagram, Japan Balance. 2017. Available online: https://www.iea.org/sankey/#?c=Japan&s=Final%20consumption (accessed on 3 March 2020).
32. Amemiya, T. Current State and Trend of Waste and Recycling in Japan. *Int. J. Earth Environ. Sci.* **2018**, *3*, 155. [CrossRef] [PubMed]
33. Schweimer, G.W.; Levin, M. *Life Cycle Inventory for the Golf A4. Research, Environment and Transport*; Volkswagen, A.G., Ed.; Wolfsburg and Center of Environmental Systems Research; University of Kasse: Kassel, Germany, 2000; Available online: http://www.wz.uw.edu.pl/pracownicyFiles/id10927-volkswagen-life-cycle-inventory.pdf (accessed on 22 April 2020).
34. U.S. Environmental Protection Agency. The Official U.S. Government Source for Fuel Economy Information, 2011 Honda Accord 2.4L, 4cyl, Manual 5-spd. 2017. Available online: http://www.fueleconomy.gov/feg/Find.do?action=sbs&id=30684 (accessed on 3 August 2017).
35. Demirel, Y. *Energy: Production, Conversion, Storage, Conservation, and Coupling*; Springer Science & Business Media: Berlin, Germany, 2012; Volume 69.
36. Schuckert, M.; Saur, K.; Florin, H.; Eyerer, P.; Beddies, H. Life cycle analysis: Getting the total picture on vehicle engineering alternatives. *Automot. Eng.* **1996**, *104*, 49–52.
37. European Environment Agency. Environmental Signals 2000, 16. Developments in Indicators: Total Material Requirement (TMR). 2016. Available online: https://www.eea.europa.eu/publications/signals-2000/page017.html (accessed on 17 February 2020).

Article

Energy Use Efficiency Past-to-Future Evaluation: An International Comparison

Chia-Nan Wang [1], Thi-Duong Nguyen [1,*] and Min-Chun Yu [2]

[1] Departments of Industrial Engineering and Management, National Kaohsiung University of Sciences and Technology, Kaohsiung 80778, Taiwan; cn.wang@hi-p.com

[2] Departments of Business Administration, National Kaohsiung University of Sciences and Technology, Kaohsiung 80778, Taiwan; yminchun@cc.kuas.edu.tw

* Correspondence: duongyennguyen@gmail.com; Tel.: +886908566040

Received: 19 September 2019; Accepted: 6 October 2019; Published: 8 October 2019

Abstract: Despite the many benefits that energy consumption brings to the economy, consuming energy also leads nations to expend more resources on environmental pollution. Therefore, energy efficiency has been proposed as a solution to improve national economic competitiveness and sustainability. However, the growth in energy demand is accelerating while policy efforts to boost energy efficiency are slowing. To solve this problem, the efficiency gains in countries where energy consumption efficiency is of the greatest concern such as China, India, the United States, and Europe, especially, emerging economies, is central. Additionally, governments must take greater policy actions. Therefore, this paper studied 25 countries from Asia, the Americas, and Europe to develop a method combining the grey method (GM) and data envelopment analysis (DEA) slack-based measure model (SMB) to measure and forecast the energy efficiency, so that detailed energy efficiency evaluation can be made from the past to the future; moreover, this method can be extended to more countries around the world. The results of this study reveal that European countries have a higher energy efficiency than countries in Americas (except the United States) and Asian countries. Our findings also show that an excess of total energy consumption is the main reason causing the energy inefficiency in most countries. This study contributes to policymaking and strategy makers by sharing the understanding of the status of energy efficiency and providing insights for the future.

Keywords: energy efficiency; data envelopment analysis; super-SBM; grey model; energy consumption

---

1. Introduction

In recent decades, energy is considered as the basic input of numerous productions; therefore, energy is one of the key indicators of economic growth. According to Barney et al. [1], energy consumption is the central operation of modern economies and drives economic productivity as well as industrial development with at least half of industrial growth based on energy consumption [1]. However, consuming energy emits greenhouse gases, which are directly related to global warming and climate change as well as environmental pollution [2]. According to International Energy Agency (IEA) [3], the global energy-related carbon dioxide (CO_2) emissions in 2018 increased by 1.7% and reached its historic highest growth rate since 2013 with a total amount of CO_2 emissions of 33.1 gigatons (Gt), which is equal to 70% higher than the average increase since 2010.

Despite the many benefits that energy consumption brings to the economy, consuming energy also leads nations to expend more resources on environmental pollution [2]. Therefore, energy efficiency, which has featured in national and international policy for more than 40 years, has been proposed as a solution, namely as a highly effective pathway, to improve the economic competitiveness and sustainability of every economy, lower emissions, reduce energy dependency, and increase the security of supply as well as job creation [4]. The idea that energy efficiency should be an important part of government energy policy developed in response to the first oil price crisis in 1973, when reducing energy demand was seen as a route to greater energy security in many developed countries. Thus, energy efficiency is already understood as a means by which to reach a variety of ends and its role in policy making is increasing [5].

According to the IEA report [3], despite the progress on energy efficiency, the growth in energy demand is accelerating. To solve this problem, the efficiency gains in countries where energy consumption efficiency is of the greatest concern such as China, India, the United States, and Europe, especially emerging economies, is central. Additionally, to obtain the targets of environmental protection and economic growth, many countries have been implementing a suite of policies to improve energy efficiency [3]. However, the current policy efforts to boost energy efficiency are slowing down in a time when energy efficiency could deliver significant economic, social, and environmental benefits, but only if governments take greater policy action.

To create efficiency gains, the right policies and greater policy actions are necessary. The throughout energy efficiency evaluation and forecast is helpful in enhancing the understanding of the current status and outlook for the energy efficiency of different economies, which can help in making the right policies to boost energy efficiency. Therefore, focusing on the importance of energy efficiency evaluation in policy making, this study used the top 25 energy consuming countries to develop a method to measure and forecast the energy efficiency, from which a detailed energy efficiency evaluation can be made from the past to the future. Furthermore, this method can be extended to more countries around the world.

Energy efficiency in European countries has always captured the great attention of researchers. Therefore, many previous studies measuring energy efficiency and energy efficiency policies in European countries can be found in the literature [6–16]. Calvet et al. [6] evaluated the environmental performance of the European Union (EU) over the period 1993–2010. In that study, a two-stage DEA analysis was applied to obtain the research objectives. The results of that paper indicated that the eco-efficiency indicator has improved over the last two decades; however, in the case of traditional indicators such as CO_2 emissions, the abatement opportunities are still remarkable.

Energy efficiency in Asia, where there are many emerging economies, is the hot issue for policymakers. However, not many studies have evaluated the energy efficiency of Asian countries except in those related to China [17–21]. China is the largest country in terms of energy consumption and related-energy CO_2 emissions, which is why energy efficiency is one of China's greatest concerns. Yang et al. [17] measured the energy efficiency of 30 Chinese provinces in 2013 and 2014 by applying the DEA Super-SBM model and found that China's overall energy efficiency was low and had decreased when taking the undesirable outputs into consideration.

The attention to energy efficiency has not been given in Europe and China, but also in other regions and cross-countries, as shown in the study by Zhou et al. [22], who measured the energy efficiency of the Asia-Pacific Economic Cooperation region (APEC) and Guo et al. [23], who evaluated the energy efficiency of Organisation for Economic Co-operation and Development (OECD) countries. Studies measuring the energy efficiency of different countries from different regions can also be found in the literature, for example, Zhang et al. [24] investigated the energy efficiency of 23 different developing countries; Pang et al. [25] evaluated the environmental efficiency of different countries; Wang et al [26] used the DEA super-SBM model and Malmquist productivity index (MPI) to measure the energy efficiency and efficiency improvement of 17 countries; and Wang et al. [27] measured the energy efficiency of the top 25 countries by CO_2 emissions in 2017.

Despite many studies in this field, no study forecasting the energy efficiency cross-country was found in the literature. Therefore, our study is expected to be the first empirical study to use a hybrid model to measure and forecast energy efficiency, which sheds new light in the literature for a new research aspect of energy efficiency. The rest of this paper is organized as follows. Section 2 describes the research methodology measuring and forecasting energy efficiency. Section 3 presents the detailed empirical results. Section 4 presents the discussion and the conclusions drawn from the research.

2. Materials and Methods

In this section, the hybrid method is introduced to evaluate the energy efficiency of different countries from the past to the future. First, the DEA slack-based measure model (SBM) is applied to obtain the efficiency of the selected countries. Second, the grey model GM (1,1) is used for forecasting the values of the inputs and outputs over future period. Finally, the DEA SBM is employed again to evaluate efficiency in future years, then a comparison can be made between the results in the past and future.

Due to the presence of undesirable outputs in this study, the SBM model, which can deal with bad outputs, is the best choice, while the grey model GM (1,1) is a suitable forecast model for this study because it does not require a large amount of input data.

2.1. Grey Model GM (1,1)

In recent years, the grey prediction model has been applied in many research fields thanks to its popularity and computational efficiency. GM (1,1) represents the time-series prediction model in the first order of placing a variable [28], and is the most popular forecast model used by scientists because a part of GM (1,1) can deliver a relatively high predictive rate while not requiring an entire set of historical data except for a small amount of input data (at least four). This is the reason why GM (1,1) is suitable for almost all fields and different areas.

In this study, as the period of the past data collection was only five years, the selection of this model to predict future results is perfectly appropriate.

The model structure of GM (1,1) is described as follows:

Denote the original form of GM (1,1) as in Equation (1)

$$x^{(0)} = \left(x^{(0)}(1), \ \ldots, \ x^{(0)}(n)\right), \ n \geq 4 \tag{1}$$

The one-time accumulated generating operation (1-AGO) of the original sequence $x^{(0)}$ is defined as:

$$x^{(1)} = \left(x^{(1)}(1), \ \ldots, \ x^{(1)}(n)\right), \ n \geq 4 \tag{2}$$

Consider Equation (3) as the original form of the GM (1,1) model, where the symbol GM (1,1) stands for the first order grey model in variables.

$$x^{(0)}(k) + ax^{(1)}(k) = b \tag{3}$$

Consider Equation (4) as the basis form of this model.

$$x^{(0)}(k) + az^{(1)}(k) = b, \ k = 1, \ 2, \ \ldots, n \tag{4}$$

where $z^{(1)}(k) = 0.5x^{(1)}(k) + 0.5x^{(1)}(k-1)$, $k = 1, \ 2, \ \ldots, \ n$. a, b are the coefficients; in grey system theory terms, a is said to be a developing coefficient and b the grey input; and $x^{(0)}(k)$ is the grey derivative that maximizes the information density for a given series to be modeled.

According to the least square method, we have $\hat{a} = \begin{bmatrix} a \\ b \end{bmatrix} = (B^T B)^{-1} B^T Y_N$.

Therefore,

$$B = \begin{bmatrix} -z^1(2) & 1 \\ -z^1(3) & 1 \\ \vdots & \vdots \\ -z^1(n) & 1 \end{bmatrix} \quad Y_N = \begin{bmatrix} x^{(0)}(2) \\ x^{(0)}(3) \\ \vdots \\ x^{(0)}(n) \end{bmatrix} \tag{5}$$

Here, **B** is called a data matrix.

By considering the following equation $dx^{(1)}/dt + ax = b$ as a shadow for $x^{(0)}(k) + az^{(1)}(k) = b$, then the response equations for GM (1,1) are as follows:

$$\hat{x}^{(1)}(k+1) = \left(\left(x^{(0)}(1) - \frac{b}{a}\right)e^{-ak} + \frac{b}{a}\right), \quad k = 1, 2, 3, \ldots, n. \tag{6}$$

$$\hat{x}^{(0)}(k+1) = \hat{x}^{(1)}(k+1) - \hat{x}^{(1)}(k) \tag{7}$$

2.2. DEA Slack-Based Measure Model

With the presence of undesirable outputs, there is one popular DEA model that can deal with this problem called Slack-based measure model (SBM), which was proposed by Tone in 2003 [29] and extended the SBM model proposed in 2001 by Tone [30]. In this study, we followed the equations proposed by Tone (2003) and are detailed as follows.

Suppose that n represents the number of decision-making units (DMUs) and each DMU has inputs, desirable outputs, and undesirable outputs.

Let us decompose the output matrix Y into $\left(Y^g, Y^b\right)$, where Y^g, Y^b denote good (desirable) and bad (undesirable) output matrices, respectively. For a DMU (x_o, y_o), the decomposition is denoted as $\left(x_o, y_o^g, y_0^b\right)$.

The production possibility set is defined by:

$$P = \left\{\left(x, y^g, y^b\right) \middle| x \geq X\lambda, y^g \leq Y^g\lambda, y^b \geq Y^b\lambda, L \leq e\lambda \leq U, \lambda \geq 0\right\} \tag{8}$$

where λ is the intensity vector while L and U are the lower and upper bounds of the intensity vector, respectively. Then, a DMU $\left(x_o, y_o^g, y_o^b\right)$ is efficient in the presence of bad outputs, if there is no vector $\left(x, y^g, y^b\right) \in P$ such that $x_o \geq x$, $y_o^g \leq y^g$, $y_o^b \geq y^b$ with at least one strict inequality.

In accordance with this definition, we modified the SBM in Tone (2001) as follows.

$$[SBM]\rho^* = min \frac{1 - \frac{1}{m}\sum_{i=m}^{m} S_i^-}{1 + \frac{1}{S_1 S_2}\left(\sum_{r=1}^{S1} \frac{S_r^g}{y_{ro}^g} + \sum_{r=1}^{S2} \frac{y_r^b}{y_{ro}^b}\right)} \tag{9}$$

Subject to

$$x_o = X\lambda + S^-$$
$$y_o^g = Y^g\lambda - s^g$$
$$y_o^b = Y^b\lambda + S^b$$
$$L \leq e\lambda \leq U$$
$$S^- \geq 0, \ S^g \geq 0, \ S^b \geq 0, \ \lambda \geq 0$$

where vector $S^- \in R^m$: excesses in inputs and $S^b \in R^{S2}$: excesses in bad outputs and $S^g \in R^{S1}$ is the shortage in the good outputs. With the presence of bad outputs, the DMU $\left(x_o, y_o^g, y_o^b\right)$ is efficient if and only if $\rho^* = 1$, i.e., $S^{-*} = 0$, $S^{g*} = 0$, and $S^{b*} = 0$.

3. Empirical Results

3.1. Data Collection

There are different approaches to energy efficiency in the national and international literature as well as in various scientific disciplines. Traditionally, energy efficiency is defined as the use of energy in an optimum to achieve the same service that could have been achieved using a common less efficient manner [3], however, many opinions have claimed that energy alone cannot produce any economic output; therefore, energy efficiency should be interpreted as the use of non-energy inputs and energy to produce economic desirable outputs while reducing greenhouse gases emissions. Energy efficiency can be achieved by meeting the requirements of human, institutional, legal, technical, and financial capacities, or in general, the right energy efficiency policy combines all of the above requirements.

In this study, we followed the concept that using energy along economic inputs can produce desirable economic outputs (i.e., gross domestic production-GDP), however, at the same time, they emit undesirable outputs (i.e., greenhouse gas emissions such as CO_2). Regarding the economy, labor force and capital stock are two popular indicators where the data are available; therefore, these two economic indicators were selected as the economic inputs while the total energy consumption represented the energy inputs. GDP was selected as the desirable output while CO_2 emissions were considered as the undesirable output. Additionally, since no direct source provides data of capital stock, in this study, we used gross capital formation to present capital.

Due to the incomplete data for some countries for 10 years, 25 countries that provided sufficient data were chosen as the sample in this study. Data used in this study were collected from two main sources: the Enerdata Yearbook [31] and data from the World Bank [32]. Since data for 2018 are not available, data from the 10 years between 2008–2017 were used to measure the energy efficiency in the first stage. In the second stage, we used data from 1990 to 2017 to forecast the efficiency for period 2018–2023. The flow of this study is described in Figure 1.

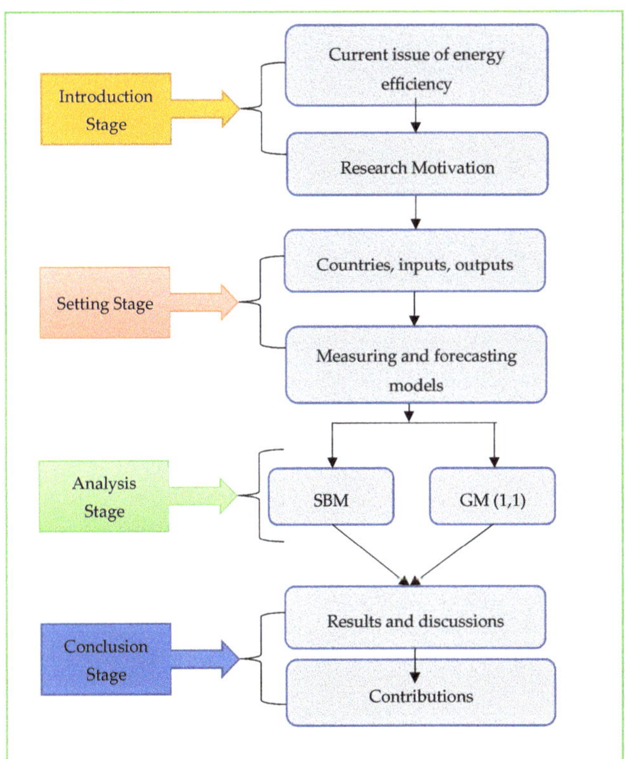

Figure 1. Research flow.

3.2. Energy Efficiency from 2008 to 2017

In this stage, DEA-Solver-Pro software was used to measure the energy efficiency through the DEA SBM model. The results obtained from 2008–2017 are shown in Table 1 and the graphical illustration of the average score of each country can be seen in Figure 2.

Table 1 indicates the low efficiency score of the 25 countries with the average score ranging from the lowest at 0.64 in 2011 to the highest of 0.70 in 2010. As can be observed, the average score for each year from 2008 to 2017 remained stable with insignificant change.

When considering the whole observation period from 2008 to 2017, eight out of the total of 25 selected countries were efficient in terms of energy with a corresponding score of 1 such as France, Italy, Japan, Norway, Portugal, Sweden, the United Kingdom, and the United States. The average score of Germany was 0.96 (relatively efficient), while the average scores of other countries were lower than 1 and ranged from the lowest of 0.17 (India) to 0.84 (the Netherlands). India had the lowest efficiency for all of the 10 observed years with a score under 0.2, followed by China and Indonesia with average scores around 0.2.

By examining the separate years, it can be seen that some countries were efficient for several years from 2008 to 2017 such as Brazil, Belgium, Malaysia, Romania, and the Netherlands. On average, the efficiency scores of these countries for the whole period of 2008–2017 were lower than 1; however; these countries did have an efficiency score of 1 for at least one or more than one year. Other countries suffered a poor efficiency score for all of the 10 observed years.

Table 1. Energy efficiency score during 2008–2017.

Countries	2008	2009	2010	2011	2012	2013	2014	2015	2016	2017	Average
Belgium	0.73	0.64	0.65	0.67	0.73	1.00	0.73	0.66	0.63	0.64	0.71
Brazil	0.66	0.64	0.64	0.57	0.56	0.54	0.51	0.53	1.00	1.00	0.67
Canada	0.62	0.61	0.62	0.60	0.60	0.60	0.61	0.63	0.63	0.61	0.61
China	0.20	0.18	0.19	0.18	0.19	0.20	0.21	0.22	0.22	0.23	0.20
Czech Republic	0.60	0.62	0.60	0.57	0.55	0.54	0.54	0.55	0.55	0.57	0.57
France	1.00	1.00	1.00	1.00	1.00	1.00	1.00	1.00	1.00	1.00	1.00
Germany	0.90	0.89	0.90	0.91	1.00	1.00	1.00	1.00	1.00	1.00	0.96
India	0.17	0.15	0.16	0.15	0.16	0.17	0.17	0.18	0.19	0.19	0.17
Indonesia	0.31	0.21	0.22	0.23	0.23	0.24	0.23	0.24	0.24	0.24	0.24
Italy	1.00	1.00	1.00	1.00	1.00	1.00	1.00	1.00	1.00	1.00	1.00
Japan	1.00	1.00	1.00	1.00	1.00	1.00	1.00	1.00	1.00	1.00	1.00
Korea, Rep.	0.35	0.35	0.36	0.34	0.35	0.36	0.36	0.36	0.35	0.33	0.35
Malaysia	1.00	1.00	1.00	0.38	0.34	0.33	0.33	0.34	0.34	0.35	0.54
Mexico	0.39	0.37	0.39	0.37	0.38	0.39	0.39	0.40	0.40	0.40	0.39
Netherlands	0.80	0.77	0.77	0.80	1.00	1.00	1.00	0.73	0.78	0.75	0.84
Norway	1.00	1.00	1.00	1.00	1.00	1.00	1.00	1.00	1.00	1.00	1.00
Poland	0.40	0.42	0.42	0.38	0.41	0.43	0.41	0.41	0.43	0.43	0.41
Portugal	1.00	1.00	1.00	1.00	1.00	1.00	1.00	1.00	1.00	1.00	1.00
Romania	0.52	1.00	1.00	0.51	0.49	0.51	0.52	0.54	0.55	0.57	0.62
Spain	0.71	0.73	0.75	0.73	0.75	0.78	0.75	0.72	0.72	0.69	0.73
Sweden	1.00	1.00	1.00	1.00	1.00	1.00	1.00	1.00	1.00	1.00	1.00
Thailand	0.30	0.34	0.34	0.27	0.25	0.25	0.27	0.28	0.31	0.30	0.29
Turkey	0.43	0.43	0.45	0.41	0.44	0.46	0.44	0.44	0.42	0.41	0.43
United Kingdom	1.00	1.00	1.00	1.00	1.00	1.00	1.00	1.00	1.00	1.00	1.00
United States	1.00	1.00	1.00	1.00	1.00	1.00	1.00	1.00	1.00	1.00	1.00
Average	0.68	0.69	0.70	0.64	0.66	0.67	0.66	0.65	0.67	0.67	0.67

The average score was the lowest in 2014 and reached its peak in 2008. The low efficiency scores were mainly caused by the excess of inputs. According to the analysis, for inefficient countries, there was no shortage in good outputs (GDP), but there was an excess mostly in labor forces and energy consumption as well as in CO_2 emissions.

Taking China and India as examples, in 2017, China had a labor force of 790 million with a US $4795 billion gross capital formation along with 3105 million tons of energy to create a GDP of US $10,161 billion along with 9297 metric tons of CO_2 emissions. Consequently, the efficiency score of China was 0.23. The calculated excess for labor force, gross capital formation, and energy consumption were 87%, 53%, and 66%, respectively while the excess in CO_2 emissions was approximately 73%. To be efficient, China needs to reduce its excess in inputs and undesirable outputs. Regarding India, it had the lowest efficiency score of 0.17. In 2017, the GPD of India was US $2660 billion while its CO_2 emissions was 2234 metric tons and the total energy consumption in that year was around 933 million tons, causing an excess of 82.53% in energy used and 84.63% in CO_2 emissions.

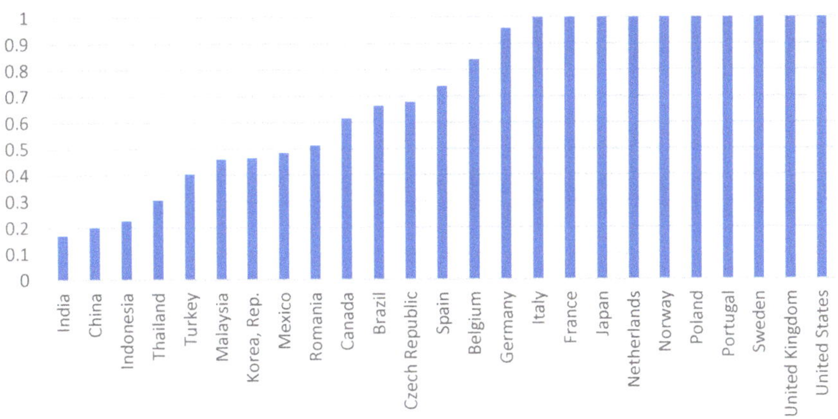

Figure 2. Average energy-efficiency by country from 2008 to 2017.

As observed, the efficiency scores of European countries are higher than those from the Americas and Asia. From 2008 to 2017, the average score of European countries was 0.81, while the average score of the Americas and Asia was 0.69 and 0.40, respectively as shown in Figure 3. The average efficiency scores of Asian countries remained stable with a very low score from 2008 to 2017. As observed, the average score was around 0.4. The low score was driven by China, India, and Indonesia, whose scores was around 0.2 and 0.3. Among the Asian countries, Japan was the only country that was efficient, thanks to the reduction in energy consumption during the observation period.

As illustrated in Figure 3, it is clear that there was a big gap between the average scores of Asia vs. the Americas and Europe. While European countries and the United States tried to reduce their amount of energy consumption, Asian countries such as China, India, and Indonesia have consumed more energy to achieve their economic development targets.

It is also worth noting that the average score of European countries slightly fluctuated with an up and down trend. However, the average score was quite stable and around 08. Of the 14 selected European countries, six countries were efficient with a corresponding score of 1, while the score of two countries such as Germany was 0.96. Among the European countries, Turkey had the lowest score (0.4), followed by the Czech Republic (0.57) and Romania (0.63). The average score of European countries remained stable during 2008–2017, which can be explained by the actions taken by the European Union, which has stressed the economic case for increasing resource efficiency including energy efficiency.

The average score of the countries from the Americas was stable during 2008–2015, then notably increased from 2016 to 2017. Among the selected countries from the Americas, two northern countries had higher scores than the two southern ones from 2008 to 2015. From 2008 to 2017, the United States was always efficient with a corresponding score of 1, while Canada and Mexico remained stable with average scores around 0.6 and 0.4, respectively. Regarding the data in Table 2, the average score of Brazil fluctuated and ranged from 0.5 to 0.66 from 2008 to 2015, then rapidly increased to 1 over 2016–2017, which helped the average score of the Americas become higher in those two years and narrowed the distance with the average score of European countries.

As a whole observation, the average energy-efficiency score of 25 countries was around 0.67 during 2008–2017, illustrated in Figure 4.

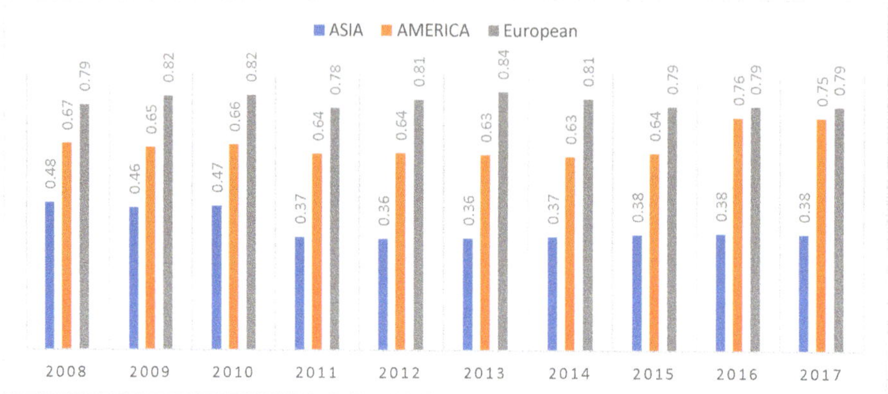

Figure 3. Energy efficiency of Europe, the Americas, and Asia from 2008 to 2017.

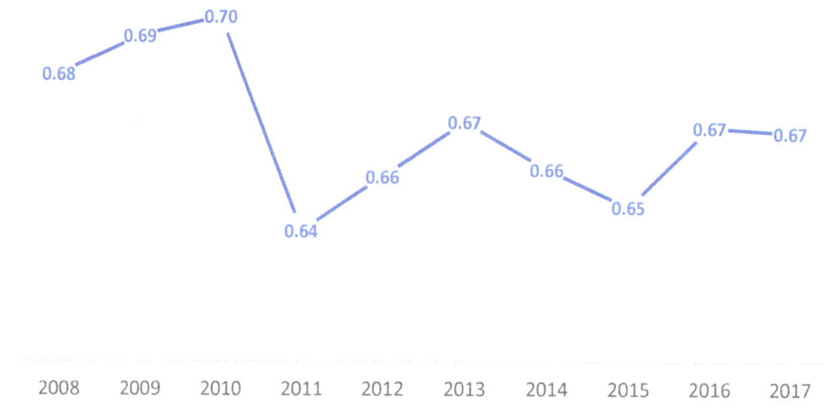

Figure 4. Average energy-efficiency score of the 25 countries from 2008 to 2017.

3.3. Forecasting Inputs and Output for 2018 to 2023

In this stage, the grey model GM (1,1) was employed to forecast the input and output data for the future period of 2018–2023 based on the data of the past period of 1990–2017. Then, the forecasted data were used to obtain the efficiency scores for the future period. However, before using the DEA to measure energy efficiency over the period 2018–2023, an accuracy test must be conducted to ensure that the forecasted data are reliable. Accuracy is controversial and of concern whenever a forecasting is produced since an error always exists. Therefore, this study measured the accuracy by using the mean absolute percent error (MAPE), which is applied commonly in many prediction studies.

MAPE is the mean average absolute percent error that measures the accuracy in a fitted time series value in statistics, specifically trending [33].

$$MAPE = \left(\frac{1}{n} \sum_{k=1}^{n} \left| \frac{x^{(0)}(k) - \hat{x}^{(0)}(k)}{x^{(0)}(k)} \right| \right) \times 100\%$$ (10)

where n is the forecasting number of steps.

The parameters of the MAPE state the forecasting ability as follows:

- MAPE < 10% represents Excellent.
- 10% < MAPE < 20% is Good.
- 20% < MAPE < 50% is reasonable.

Table 2. Average mean absolute percent error (MAPE) of all DMUs.

Countries	Labor Force	Gross Capital Formation	Energy Consumption	GDP	CO$_2$ Emissions	Average
Belgium	5.2%	0.3%	3.1%	0.7%	2.5%	2.4%
Brazil	3.1%	0.4%	8.2%	3.4%	5.5%	4.1%
Canada	3.7%	0.1%	4.7%	0.8%	7.8%	3.4%
China	3.4%	0.1%	4.0%	1.4%	3.6%	2.5%
Czech Republic	1.6%	0.3%	5.5%	2.0%	2.4%	2.4%
France	1.0%	0.2%	2.8%	0.7%	2.4%	1.4%
Germany	1.7%	0.2%	3.7%	1.1%	1.8%	1.7%
India	1.0%	0.8%	3.9%	0.8%	1.5%	1.6%
Indonesia	1.5%	0.5%	6.4%	0.5%	5.1%	2.8%
Italy	2.2%	0.5%	5.9%	1.5%	2.6%	2.5%
Japan	1.5%	0.6%	3.3%	0.9%	3.0%	1.9%
Korea, Rep.	2.8%	0.4%	4.0%	0.6%	2.8%	2.1%
Malaysia	1.5%	1.0%	3.7%	0.9%	2.3%	1.9%
Mexico	1.8%	0.4%	2.6%	1.1%	1.8%	1.5%
Netherlands	3.8%	0.4%	6.4%	1.3%	2.0%	2.8%
Norway	2.1%	0.5%	3.3%	0.7%	2.3%	1.8%
Poland	2.7%	0.4%	3.8%	0.7%	1.9%	1.9%
Portugal	3.2%	0.5%	8.8%	2.0%	4.8%	3.9%
Romania	1.6%	0.5%	6.2%	3.4%	4.8%	3.3%
Spain	2.3%	0.7%	8.2%	2.5%	3.8%	3.5%
Sweden	2.6%	0.2%	4.5%	1.5%	3.1%	2.4%
Thailand	2.4%	0.8%	6.7%	1.1%	1.7%	2.5%
Turkey	2.5%	0.3%	5.1%	2.0%	3.3%	2.6%
United Kingdom	1.5%	0.1%	4.3%	1.2%	2.8%	2.0%
United States	1.1%	0.4%	3.6%	0.9%	1.3%	1.5%
Average	2.3%	0.4%	4.9%	1.3%	3.1%	2.4%

The MAPE results are displayed in Table 2, which shows that the MAPE results of all inputs and outputs ranged from the lowest of 0.1% to the highest at 8.2% and the average MAPE of all inputs and outputs was 2.4%. As the MAPE values obtained were all smaller than 10%, it confirmed that the GM (1,1) has good prediction accuracy in this research and that the forecasted data can be used in the further step of obtaining efficiency scores.

The input "Total energy consumption" of Indonesia is used as an example to illustrate the generation of the forecast data. The sequence of raw data during 2008–2017 is as follows:

$$x^{(0)} = \left(x^{(0)}(1), \ldots, x^{(0)}(10)\right)$$
$$= (58.3,\ 55.9,\ 60.1,\ 56.2,\ 53.8,\ 55.9,\ 52.9,\ 53.2,\ 56.7,\ 55.6) \tag{11}$$

Simulate this sequence by respectively using the following three GM (1,1) models and comparing the simulation accuracy:

From $x^{(0)}(k) + ax^{(1)}(k) = b$; compute the accumulation generation of $x^{(0)}$ as follows:

$$x^{(1)} = \left(x^{(1)}(1),\ \ldots,\ x^{(1)}(n)\right)$$
$$= (58.3,\ 114.2,\ 174.3,\ 230.5,\ 284.3,\ 340.2,\ 391.1,\ 446.3,\ 503.0,\ 558.6) \tag{12}$$

In the next stage, the different equations of GM (1,1) are created with the mean equation:

$$z^{(1)}(2) = 0.5(58.3 + 114.2) = 86.22 \tag{13}$$

$$z^{(1)}(10) = 0.5(503 + 558.6) = 530.8 \tag{14}$$

To continue, the values for coefficients a and b are found

$$B = \begin{bmatrix} -116.3 & 1 \\ -127.3 & 1 \\ -228.4 & 1 \\ -284.5 & 1 \\ -340.5 & 1 \\ -396.6 & 1 \\ -452.6 & 1 \\ -508.7 & 1 \\ -564.8 & 1 \end{bmatrix} \qquad Y_N = \begin{bmatrix} 55.9 \\ 60.1 \\ 56.2 \\ 53.8 \\ 55.9 \\ 52.9 \\ 53.2 \\ 56.7 \\ 55.6 \end{bmatrix} \tag{15}$$

By using the least square estimation, we can obtain the sequence of parameters $[a, b]^T$ as follows:

$$\hat{a} = \begin{bmatrix} a \\ b \end{bmatrix} = (B^T B)^{-1} B^T Y_N = \begin{bmatrix} 0.00548567 \\ 57.4599547 \end{bmatrix} \tag{16}$$

Compute the simulated value of $x^{(0)}$, the original series according to the accumulated generating operation by using

$$\hat{x}^{(0)}(k+1) = \hat{x}^{(1)}(k+1) - \hat{x}^{(1)}(k)$$
$$= 54.24\ (forecasted\ for\ 2018) \tag{17}$$

The same was used to forecast the inputs and outputs of other countries over the period 2018–2023.

3.4. Energy Efficiency from 2008 to 2017

It was observed that during 2018–2023, the average efficiency score will be stable with a score of 0.68. This stable trend can be applied for all observed countries as the score of the later years will remain relatively the same with the score for the previous years. Table 3 shows the forecasted efficiency over 2018–2023 along with a graphical illustration in Figure 5, which compares the scores in the past years and those for the future years.

As a whole, the average efficiency scores of the 25 countries over the period 2018–2023 will be low at 0.68, with the number of efficient countries increasing to nine as one inefficient country in the past will become efficient 2018–2023 (Brazil). From the results, we also noted that the most inefficient countries will witness a lower energy-efficiency score in the future period of 2018–2023 when compared to those in the past period of 2008–2017. However, the change will be small and insignificant, as can be seen in Figure 5, except in the case of Brazil. The higher average efficiency score of Brazil can be explained by the faster growth of GPD than the growth of inputs and undesirable outputs.

Besides the nine efficient countries (Brazil, France, Italy, Japan, Norway, Portugal, Sweden, the United Kingdom, and the United States), the efficiency score of Germany will remain the same as its score over 2008–2017 (0.96), and nearly reach the efficiency frontier with a score ranging from 0.94 to 0.97 caused by around a 1.5% excess of labor force and approximately 2.3% redundancy in energy consumption leading to around a 3.5% higher amount of CO_2 emissions yearly from 2018 to 2023. To be efficient in terms of energy, Germany should consider cutting down its energy and labor force.

Regarding inefficient countries, the average score over 2018–2023 will range from 0.19 to 0.79. The lowest efficiency scores will be found in India (0.19), Indonesia (0.22), and China (0.26). It is noted that the efficiency of China from 2018 to 2023 was found to be higher than this during 2008–2017 with 0.26 for the former and 0.20 for the later, indicating its energy improvement, while the average score of Indonesia will decrease from 2018 to 2023, leading Indonesia to replace China as the second worst country in terms of energy efficiency.

Table 3. Efficiency scores from 2018 to 2023.

Countries	2018	2019	2020	2021	2022	2023	Average
Belgium	0.77	0.77	0.76	0.75	0.75	0.74	0.76
Brazil	1.00	1.00	1.00	1.00	1.00	1.00	1.00
Canada	0.62	0.62	0.62	0.62	0.62	0.62	0.62
China	0.25	0.25	0.26	0.27	0.27	0.28	0.26
Czech Republic	0.62	0.63	0.63	0.64	0.64	0.65	0.64
France	1.00	1.00	1.00	1.00	1.00	1.00	1.00
Germany	0.97	0.97	0.96	0.96	0.95	0.94	0.96
India	0.18	0.19	0.19	0.20	0.20	0.21	0.19
Indonesia	0.23	0.22	0.22	0.22	0.21	0.21	0.22
Italy	1.00	1.00	1.00	1.00	1.00	1.00	1.00
Japan	1.00	1.00	1.00	1.00	1.00	1.00	1.00
Korea, Rep.	0.36	0.36	0.36	0.36	0.37	0.37	0.36
Malaysia	0.33	0.32	0.32	0.31	0.30	0.30	0.31
Mexico	0.38	0.38	0.38	0.37	0.37	0.37	0.38
Netherlands	0.79	0.79	0.79	0.79	0.79	0.79	0.79
Norway	1.00	1.00	1.00	1.00	1.00	1.00	1.00
Poland	0.43	0.43	0.44	0.44	0.44	0.44	0.44
Portugal	1.00	1.00	1.00	1.00	1.00	1.00	1.00
Romania	0.61	0.62	0.63	0.64	0.65	0.66	0.64
Spain	0.68	0.68	0.67	0.67	0.66	0.66	0.67
Sweden	1.00	1.00	1.00	1.00	1.00	1.00	1.00
Thailand	0.30	0.30	0.30	0.30	0.30	0.30	0.30
Turkey	0.41	0.41	0.40	0.40	0.40	0.40	0.40
United Kingdom	1.00	1.00	1.00	1.00	1.00	1.00	1.00
United States	1.00	1.00	1.00	1.00	1.00	1.00	1.00
Average	0.68	0.68	0.68	0.68	0.68	0.68	0.68

Additionally, by comparing the average efficiency of each country over two periods, it reveals that 10 countries (Belgium, Brazil, Canada, China, Czech Republic, India, Korea Republic, Poland, Romania, and Thailand) will experience an increase in efficiency while six countries will suffer the opposite trend (Indonesia, Malaysia, Mexico, the Netherlands, Spain, and Turkey).

By analyzing countries with an increase in their efficiency score, we found that the increase was caused by the faster growth of GDP despite the growing inputs. For example, in the case of China, in 2017, China had a labor force of 790 million with US \$4795 billion gross capital formation along with 3105 million tons of energy to create US \$10,161 billion in GDP, along with 9297 metric tons of CO_2 emissions, where the calculated excess for labor forces, gross capital formation, and energy consumption were 87%, 53%, and 66%, respectively while the excess in CO_2 emissions was approximately 73%. However, in 2018, China had 821 million in labor (approximately 3.9% higher than 2017) along with US \$5425 billion of gross capital formation and 3641 million tons of energy equally higher at 13.1% and 17.3%, respectively, to generate a US \$12,128 billion in GDP and emit 11,367 metric tons of CO_2 emission. Therefore, the excess in inputs will be 85.8%, 51.4%, and 62.4% for labor force, gross capital formation, and energy consumption, respectively, while the redundancy in CO_2 emissions will be 71.3%.

Regarding the six countries with a decreasing efficiency score, we found that the faster growth in energy consumption was the main reason for most countries, with the exception of Turkey. In 2018, the total amount of energy consumed by Turkey decreased when compared to 2017; however, the significant increase in gross capital formation was the reason behind the lower score.

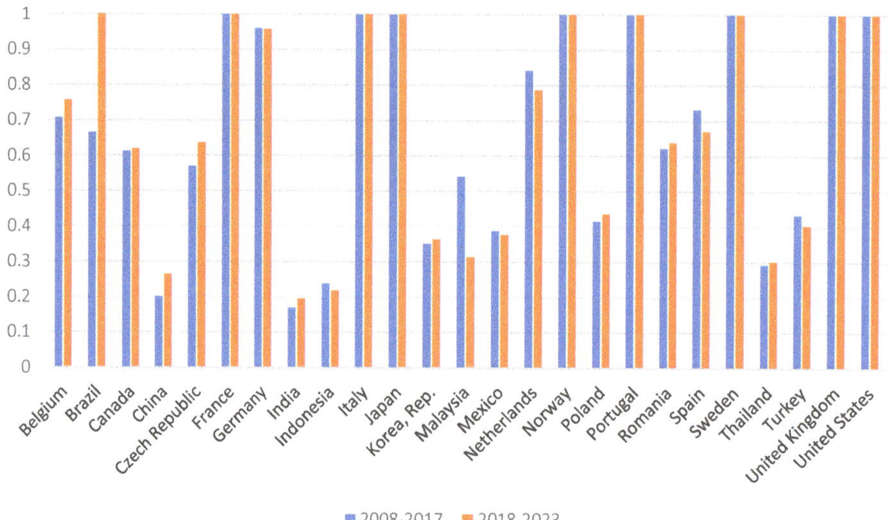

Figure 5. Energy-efficiency score from 2008 to 2017 vs. 2018 to 2023.

As above-mentioned, there is a large gap between the average score of European and countries in the Americans versus those of Asian countries from 2008 to 2017. This will not disappear during 2018–2023, as the average score of European, American, and Asian countries will be 0.81, 0.75, and 0.38, respectively. The gap between European countries and that of the Americas will be narrower while the gap between European and Asian countries will widen.

As shown in Figure 6, the average score of Asian countries will continue to be stable over the period 2018–2023 with the very low score of 0.38 driven by the poor score of all selected Asian countries with the exception of Japan. The score of these countries ranged from the lowest at 0.19 (India) to the highest at 0.36 (Republic of Korea). The average score of countries in the Americas will be higher thanks to the increase in Brazil and Canada, while the score for European countries will remain stable and unchanged.

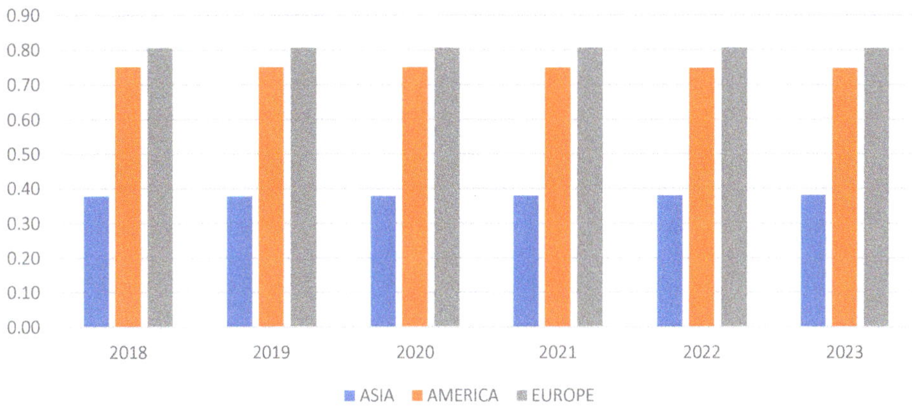

Figure 6. Efficiency score from 2018 to 2023.

4. Discussions and Conclusions

This paper focused on measuring and forecasting energy efficiency. First, the energy efficiency for 2008 to 2017 was analyzed, and the results at this stage revealed that the 25 selected countries showed an inefficiency in terms of energy. Of these 25 countries, eight countries were efficient during 2008–2017 and will continue to be efficient from 2018 to 2023 (France, Italy, Japan, Norway, Portugal, Sweden, the United Kingdom, and the United States), indicating the good balance between economic growth and environmental protection. Germany was the only inefficient country, with a very high score of 0.96 over the period 2008–2023, caused by an approximately 5% excess in energy consumption, leading to around 6.5% higher CO_2 emissions. Additionally, Brazil was inefficient from 2008 to 2017, but will become efficient over the period 2018–2023.

The findings of this study also suggest higher energy-efficiency scores for European countries than those of countries from the Americas and Asia. The higher efficiency score of Europe is the result of constantly reducing the amount of energy consumption in most countries in Europe. On the other hand, the low score of Asian countries is the consequence of a higher demand of energy used in industries. Using more energy can accelerate the growth of economic development. However, more greenhouse gases have a negative impact on the environment. The United States, the second nation in terms of energy consumption, was always efficient due to its the high GDP and its reduction of energy consumption in recent years, while Canada had a score of around 0.6 to 0.7, caused by the increasing energy demand due to climate change. Like Canada, the energy consumption of Brazil will continue to grow in future; however, thanks to the faster growth of GPD, while inefficient during 2008–2017, it will become efficient. However, consuming more energy to promote economic growth is not a sustainable solution.

The analyzed results found that the excess of total energy consumption was the main reason causing the energy inefficiency of most countries. Therefore, these countries should reconsider their energy infrastructure as well as reduce the amount of energy used in order to reach the efficiency-frontier. It was also observed that among the 25 countries, India suffered the lowest energy efficiency score, followed by China, Indonesia, and Thailand. The share of renewables in the total energy consumption of India and China ranges from 14% to 18%, much higher than those of the United Kingdom, the United States, and Japan; however, the scores of these countries were still much lower than countries that have a moderate share of renewable energy. In fact, the increase in the use of renewable energy instead of unrenewable energy can help in reducing the greenhouse gases emitted into the environment, which works for every country, even China and India, as the huge and increasing amount of energy used in industrial zones in these countries is the main reason causing the inefficiency. Furthermore, the price of labor in these two countries is cheaper when compared to Europe and some countries in the Americas, causing a higher number in labor force, but lower productivity, which is the other reason for a low efficiency score. As observed, the total amount of energy used by China in 2017 was approximately 3105 metric tons and that for India was around 933 metric tons while the consumption of most European countries was less than 300 metric tons. Additionally, the total energy consumption of China and India increased year by year from 2014 to 2017 and will continue to grow from 2018 to 2023, while European countries showed a decrease in the amount of energy used year by year, not only during 2008–2017 but also during 2018–2023.

The results of this study also reveal that emerging countries such as China, India, Indonesia, Malaysia, Mexico, and Brazil had a low efficiency score from 2008 to 2009, which made these countries capture the great attention of policymakers. As indicated by the IEA [3], the efficiency gains in these emerging countries is the center of energy efficiency. However, by forecasting the performance of these countries over the period 2018–2023, the results found that with the exception of Brazil, other countries will not have significant efficiency gains without greater policy actions, as evidenced by the very low efficiency scores.

It is clear that implementing the right energy policies could help improve energy efficiency, which benefits in lowering the energy bell, improving air quality, reducing greenhouse gases, energy security, etc. By measuring and forecasting the energy efficiency of different countries, this study helps in not only sharing the understanding of the current status of how efficient different countries are in terms of energy, but also provides a clear picture for the future. Therefore, this study makes a core contribution to policymaking and strategy makers by providing useful and important information. Energy efficiency is pointed one of the most important criteria for sustainable development, therefore understanding and having an outlook for the future in this area are very helpful in considering the various policy strategies.

Author Contributions: C.-N.W. contributed to the research framework, checked, and revised draft paper; T.-D.N. collected the data and wrote the draft manuscript, checked, and revised the paper; M.-C.Y. supervised. All authors read and approved the final manuscript.

Funding: This research was partly supported by National Kaohsiung University of Science and Technology, and project number 108-2622-E-992-017-CC3 from the Ministry of Sciences and Technology in Taiwan.

Acknowledgments: The authors appreciate the support from National Kaohsiung University of Science andTechnology and Ministry of Sciences and Technology in Taiwan.

Conflicts of Interest: The authors declare no conflict of interest.

References

1. Barney, F.; Franzi, P. *The Future of Energy from Future Dilemmas: Options to 2050 for Australia's Population, Technology, Resources and Environment*; CSIRO Sustainable Ecosystems: Canberra, Australia, 2002; pp. 157–189.
2. EPA. Sources of Greenhouse Gas Emissions. Available online: https://www.epa.gov/ghgemissions/sources-greenhouse-gas-emissions (accessed on 10 May 2019).
3. IEA. Energy Efficiency 2018—Analysis and Outlook to 2040. Available online: https://www.iea.org/efficiency2018/ (accessed on 10 May 2019).
4. Owusu, P.A.; Asumadu-Sarkodie, S. A review of renewable energy sources, sustainability issues and climate change mitigation. *Cogent Eng.* **2016**, *3*, 1167990. [CrossRef]
5. Fawcett, T.; Killip, G. Re-thinking energy efficiency in European policy: Practitioners' use of multiple benefits' arguments. *J. Clean. Prod.* **2018**. [CrossRef]
6. Gomez-Calvet, R.G.; Conesa, D.; Gomez-Calvet, A.R.; Tortosa-Ausina, E. On the Dynamics of Environmental Performance in the European Union. Working Paper. 2014. Available online: https://ideas.repec.org/p/jau/wpaper/2014-20.html (accessed on 8 October 2019).
7. Cornelis, E. History and prospect of voluntary agreements on industrial energy efficiency in Europe. *Energy Policy* **2019**, *132*, 5675–5682. [CrossRef]
8. Hu, J.L.; Honma, S. A Comparative Study of Energy Efficiency of OECD Countries: An Application of the Stochastic Frontier Analysis. *Energy Procedia* **2014**, *61*, 2280–2283. [CrossRef]

9. Boroza, D. Technical and total factor energy efficiency of European regions: A two-stage approach. *Energy* **2018**, *152*, 5215–5232.
10. Rosenow, J.; Bayer, E. Cost and benefits of Energy Efficiency Obligations: A review of European programmes. *Energy Policy* **2017**, *107*, 53–62. [CrossRef]
11. Thonipara, A.; Runst, P.; Ochsner, C.; Bizer, K. Energy efficiency of residential buildings in Europe Union—An exploratory analysis of cross-country consumption patterns. *Energy Policy* **2019**, *129*, 1156–1167. [CrossRef]
12. Marwwues, A.D.; Fuinhas, J.A.; Tomas, C. Energy efficiency and sustainable growth in industrial sectors in European Union countries: A nonlinear ARDL approach. *J. Clean. Prod.* **2019**, *239*, 118045.
13. Makridou, G.; Andriosopoulos, K.; Doumpos, M.; Zopounidis, C. Measuring the efficiency of energy-intensive industries across European countries. *Energy Policy* **2016**, *88*, 5735–5783. [CrossRef]
14. Gomez-Calvet, R.G.; Conesa, D.; Gomez-Calvet, A.R.; Tortosa-Ausina, E. Energy efficiency in the European Union: What can be learned from the joint application of directional distance functions and slacks-based measures? *Appl. Energy* **2014**, *132*, 137–154. [CrossRef]
15. Cucchilla, F.; D'Adamo, I.; Gastaldi, M.; Miliacca, M. efficiency and allocation of emission allowances and energy consumption over more sustainable European economies. *J. Clean. Prod.* **2018**, *182*, 805–817. [CrossRef]
16. Vieites, E.; Vassileva, I.; Arias, J.E. European Initiatives Towards Improving the Energy Efficiency in Existing and Historic Buildings. *Energy Procedia* **2015**, *75*, 1679–1685. [CrossRef]
17. Yang, T.; Chen, W.; Zhou, K.; Ren, M. Regional energy efficiency evaluation in China: A super efficiency slack-based measure model with undesirable outputs. *J. Clean. Prod.* **2018**, *198*, 859–866. [CrossRef]
18. Yu, J.; Zhou, K.; Yang, S. Regional heterogeneity of China's energy efficiency in "new normal": A meta-frontier Super-SBM analysis. *Energy Policy* **2019**, *134*, 110941. [CrossRef]
19. Lin, B.; Zhang, G. Energy efficiency of Chinese service sector and its regional differences. *J. Clean. Prod.* **2017**, *168*, 614–625. [CrossRef]
20. Feng, C.; Wang, M. Analysis of energy efficiency and energy savings potential in China's provincial industrial sectors. *J. Clean. Prod.* **2017**, *164*, 15311–15541. [CrossRef]
21. Yang, L.; Wang, K.L.; Geng, J.C. China's regional ecological energy efficiency and energy saving and pollution abatement potentials: An empirical analysis using epsilon-based measure model. *J. Clean. Prod.* **2018**, *194*, 3003–3008. [CrossRef]
22. Zhou, D.Q.; Meng, F.Y.; Bai, Y.; Cai, S.Q. Energy efficiency and congestion assessment with energy mix effect: The case of APEC countries. *J. Clean. Prod.* **2017**, *142*, 819–828. [CrossRef]
23. Guo, X.; Lu, C.C.; Lee, J.H.; Chiu, Y.H. Applying the dynamic DEA model to evaluate the energy efficiency of OECD countries and China. *Energy* **2017**, *134*, 392–399. [CrossRef]
24. Zhang, X.P.; Cheng, X.M.; Yuan, J.H.; Gao, X.J. Total-factor energy efficiency in developing countries. *Energy Policy* **2011**, *39*, 6446–6450. [CrossRef]
25. Pang, Z.F.; Deng, Z.Q.; Hu, J.L. Clean energy use and total-factor efficiencies: An international comparison. *Renew. Sustain. Energy Rev.* **2015**, *52*, 1158–1171. [CrossRef]
26. Wang, C.N.; Ho, T.H.X.; Hsueh, M.H. An Integrated Approach for Estimating the Energy Efficiency of Seventeen Countries. *Energies* **2017**, *10*, 1597. [CrossRef]
27. Wang, L.W.; Le, K.D.; Nguyen, T.D. Assessment of the Energy Efficiency Improvement of Twenty-Five Countries: A DEA Approach. *Energies* **2019**, *12*, 1535. [CrossRef]
28. Deng, J. Introduction to grey system theory. *J. Grey Syst.* **1989**, *1*, 12–14.
29. Tone, K. A slacks-based measure of efficiency in data envelopment analysis. *Eur. J. Oper. Res.* **2001**, *130*, 498–509. [CrossRef]
30. Tone, K. A slacks-based measure of super-efficiency in data envelopment analysis. *Eur. J. Oper. Res.* **2002**, *143*, 32–41. [CrossRef]
31. Enerdata Yearbook. Available online: https://yearbook.enerdata.net (accessed on 5 April 2019).
32. The World Bank. Available online: https://data.worldbank.org (accessed on 5 April 2019).
33. Deng, J.L. Control problems of grey systems. *Syst. Control Lett.* **1982**, *1*, 288–294.

Article

Effectiveness of a Power Factor Correction Policy in Improving the Energy Efficiency of Large-Scale Electricity Users in Ghana

Samuel Lotsu [1], Yuichiro Yoshida [2,*], Katsufumi Fukuda [3] and Bing He [2]

[1] Akosombo Hydro Station, Hydro Generation Department, Volta River Authority, 28 February Road, Accra GA-145-7445, Ghana

[2] Graduate School for International Development and Cooperation, Hiroshima University, 1-5-1 Kagamiyama, Higashi-Hiroshima, Hiroshima-ken 739-8529, Japan

[3] Center for Far Eastern Studies, University of Toyama, 3190 Gofuku, Toyama 930-8555, Japan

* Correspondence: yuichiro@hiroshima-u.ac.jp

Received: 22 May 2019; Accepted: 1 July 2019; Published: 4 July 2019

Abstract: Confronting an energy crisis, the government of Ghana enacted a power factor correction policy in 1995. The policy imposes a penalty on large-scale electricity users, namely, special load tariff (SLT) customers of the Electricity Company of Ghana (ECG), whose power factor is below 90%. This paper investigates the impact of this policy on these firms' power factor improvement by using panel data from 183 SLT customers from 1994 to 1997 and from 2012. To avoid potential endogeneity, this paper adopts a regression discontinuity design (RDD) with the power factor of the firms in the previous year as a running variable, with its cutoff set at the penalty threshold. The result shows that these large-scale electricity users who face the penalty because their power factor falls just short of the threshold are more likely to improve their power factor in the subsequent year, implying that the power factor correction policy implemented by Ghana's government is effective.

Keywords: energy efficiency; power factor; regression discontinuity design

---

1. Introduction

The power factor is a relevant measure of the efficiency of electrical energy use. A higher power factor implies efficient energy use and simultaneously ensures the safe, smooth, and efficient operation of electrical utilities. The power factor is the ratio between active power and its vector sum with reactive power, where active power executes actual work such as producing heat, illumination, and moving vehicles and machines. Reactive power is used to maintain the electromagnetic field that ensures the accomplishment of the work of active power. Summing the active and reactive power vectorially produces an amount of apparent power, which defines the total required energy capacity. A high power factor therefore implies an efficient use of the electricity generation capacity, while a low power factor reflects an abysmal use of energy. Facilities with a low power factor use more electricity than actually needed to conduct useful work [1]. A poor power factor should be corrected to reduce waste and the production costs of energy and thereby to help save the environment. There are ways to improve a low power factor to a desirable level of efficiency. The most mature, economical, and simple method is through an investment by energy consumers in capacitor banks [2]. These banks compensate for the inductive load demand of reactive power and thereby minimize the stress or burden on the electricity supply system [3].

Another alternative would be the use of synchronous alternators (or synchronous compensators). By exciting or de-exciting the magnetic fields, these inject reactive power into the network so that the voltage profiles of the system can be improved and the losses can be reduced. However, their high installation and maintenance costs make them unsuitable for such applications in developing nations. Firms can further improve their power factor by carefully considering the design of processes and load cycles that have a repetitive nature during the design phase. Such a technique typically uses a computer algorithm to determine the ideal compromise in the relevant design parameters for improved energy efficiency and power factors. This approach, however, achieves a higher power factor at the sacrifice of reduced output levels in terms of the firms' production.

A higher power factor results in lower energy-related costs to users. According to the literature, improving power factors in industrial set-ups results in 10% to 30% cost savings. A study [4] estimated the cost savings of an independent power producer that carried out a power factor correction measure of its facility. However, the necessary investment in capacitor banks places a financial burden on users, making them reluctant to improve their power factor. Regardless, given the growing concern about climate change and increasing demand for electricity consumption, power factor correction has the potential to affect long-term economic and environmental gains to society [5]. Many countries, including Ghana, therefore urge electricity users, especially larger ones, to improve their power factor through this policy. Since 1995, the government in Ghana has imposed a policy that penalizes large-scale electricity users, labeled as special load tariff (SLT) customers, of the Electricity Company of Ghana (ECG) whose power factor falls short of 90%. This paper investigates the effectiveness of this power factor correction policy implemented by the Ghana government.

Historically, Ghana has been heavily dependent on hydro power supply, only in 2016 was it exceeded by thermal generation. Prior to that point, hydro power had accounted for the majority of the electricity supply of Ghana over history. As late as 2010, for example, the share of hydro power was approximately 70%, while that of thermal was just approximately 30%. Power from renewable energy (e.g., solar) entered the supply in 2013, but it accounts only for a negligible portion of the entire power supply [6].

In 1983, Ghana was hit by a severe drought that continued into 1984. This situation led the Akosombo Dam to experience a shortfall of water inflows, reaching below 15% of the long-term projected total [6]. This shortfall impacted the electricity supply, as the country's only electricity source was the Akosombo and Kpong hydroelectric power stations. Volta Lake did not fully recover from the 1983 drought before it was hit by another drought in 1993–1994, which again led to a reduction in electricity to consumers. This crisis exposed the country's vulnerability and security issues in terms of hydroelectric power [7].

Previous research has revealed that power factor correction measures by industrial plants, mining establishments, firms, and large commercial buildings can release 20 megawatts of tied-up electricity in a year [7]. At that time, Ghana's electricity capacity was approximately 900 megawatts, with a sizable share exported to the Republics of Togo and Benin. In addition, SLT customers represented only a part of the entire set of electricity users in Ghana. This release of 20 megawatts capacity from SLT customers was therefore considered to be relevant by the government of Ghana in that context. To improve the situation by raising the power factor among its users, in 1994, the government of Ghana announced that it would restructure tariffs to penalize those that do not adopt power factor correction measures, and it enacted this policy in January 1995 [7]. According to the policy, a user whose power factor is below a certain threshold is charged a penalty. The penalty policy is applied to large-scale electricity users, namely, the ECG's SLT customers. Tariff categories in Ghana are classified into three main groups, namely, (i) residential, (ii) nonresidential, and (iii) SLT customers [8]. Residential consumers are domestic users and nonresidential customers use electricity for commercial purposes with a capacity less than 100 kV [6]. SLT customers are defined as those that use energy for industrial purposes with loads greater than or equal to 100 kVA. According to the policy, SLT customers of ECG whose actual power factor falls short of 90% are to be penalized in proportion to the gap and to their size, measured in terms of the maximum electricity demand. For example, if the actual power factor falls short of the threshold by 5%, the penalty is 5% of the electricity bill, determined based on the maximum electricity demand of the users. Users whose power factor is above or equal to 90% are not subject to this penalty.

A relatively large body of literature has investigated the impact of energy efficiency enhancement policies on economic outcomes. For example, Costantini et al. [5] claim that sectoral energy efficiency gains display a negative effect on employment growth. They showed that this negative effect is stronger, in particular, in energy-intensive industries using data from 15 European countries. Lee and Min [9] found a negative correlation between green R&D and the financial performance of Japanese manufacturing firms. A study by Lee and Min [9] is, in the literature, one of a few that analyzes at the firm level to obtain the relationship between policy and firm economic performance.

On the one hand, these studies seem to indicate a negative relationship between energy efficiency improvement and business and economic performance. On the other hand, some of the more theory-based analyses derive opposite results. For example, by using the ASTRA model, or a dynamic, integrated macroeconomic, transport and environmental impact model, Ringel [10] concludes that enhanced green energy policies in Germany trigger tangible economic benefits in terms of GDP growth and new jobs, even in the short term. Hartwig et al. [11] use an input–output analysis-based model to show the positive growth effects and employment of energy efficiency policy in Germany. Henriques and Catarino [12] also use an input–output-based model, called the impact of sector technologies (ImSET) model, to conclude that green investment has the potential to increase employment and wage income. Allan et al. [13] use a computable general equilibrium (CGE) model to measure the impact of energy efficiency improvement in the UK.

In turn, there is a relatively smaller body of literature investigating the impact of policy on improvements in energy efficiency. Tanaka [14] provided an extensive review of energy efficiency policies implemented in the member countries of the International Energy Agency. Cox et al. [15] provided a review of non-energy- related policies. Among these studies, Xiong et al. [16] claim that a policy to restructure the industrial organization would have a large positive impact on provincial industrial energy efficiency in China. They used a slacks-based measure (SBM) that is a sophisticated variation of data-envelopment analysis (DEA), or a linear programing approach, where they allowed the existence of undesirable outputs to address the environmental burden due to inefficiency. They used a Tobit regression to reveal the positive association between industrial organization and energy efficiency as the second-stage regression following the efficiency measurement by SBM. Villca-Pozo & Gonzales-Bustos [17] found, at a provincial level, that tax policies to modernize the energy efficiency of housing in Spain have a nonsignificant impact on energy efficiency. At a more micro level, Anderson & Newell [18] found that manufacturing plants that receive government-sponsored energy audits have improved energy efficiency.

While these papers use a reduced form regression, such as the Tobit, to obtain a relationship between policy and energy efficiency, there are more model-based studies that construct a theoretical model to describe the mechanism between policy and energy use. Ringel [10] investigated the linkage between green energy policies in Germany and the country's primary energy consumption and greenhouse gases using the ASTRA model. Li et al. [19] used the VALDEX index that is a measure of energy efficiency based on value added, to find the impact of policies on eliminating low-efficiency production capacity and improving the energy efficiency of energy-intensive industries in China. Using a bottom-up approach, Fleiter et al. [20] investigated the impact of grants for small and medium enterprises in Germany to carry out energy audits of their facilities on their energy efficiency improvements.

Most of these studies, however, either showed a correlation between policy and energy efficiency or derived the results based on models that are dependent on their assumptions. For example, the Tobit regression conducted by Xiong et al. [16] shows a correlation and does not reveal the causation between policy and energy efficiency. In fact, relatively few studies have used appropriate empirical strategies to determine policy impact on energy efficiency. Yu & Zhang [21] investigated the impact of a "smart city policy" implemented in China on energy efficiency using a difference-in-differences (DID) approach at a city level. However, DID only addresses the time-invariant heterogeneity and is still based on the selections-on-observables assumption. In addition, many of these studies investigated the impacts of energy efficiency policies at an aggregated level. To our knowledge, no study has investigated the impact of a power factor correction policy on its improvement by applying a reliable identification strategy to more micro-level data.

This paper, therefore, aims to add to the literature by identifying the causal impact of the power factor correction policy on energy efficiency improvement using firm-level data. Specifically, this study applies a regression discontinuity design (RDD), with a cutoff at the 90% penalty threshold stipulated by the policy, to five years of panel data consisting of 183 SLT customers of ECG for the years following the policy announcement in 1994 to 1997 as well as in 2012. The paper shows the effectiveness of the power factor correction policy in Ghana: firms strictly improved their power factor if it fell short of the required threshold. This paper is organized as follows. Section 2 discusses the data and identification strategies. Section 3 presents the estimation results. Finally, Section 4 concludes the paper.

2. Empirical Strategy

2.1. Data

A series of power crises in Ghana, mainly due to low water inflows, were recorded during multiple periods. The first power crisis dates back to the early 1980s, and this was followed by another in 1998. During these periods, many large-scale electricity users did not operate at their full capacity. Hence, we constructed our time-series data set starting from the year just before the policy enforcement up to the year just before the second power crisis took place in 1998.

Due to power crises, some large-scale electricity users ceased operations either temporarily or permanently. We excluded those SLT users from our data and limited it to the companies that sustained their business during our sample period to construct strongly balanced panel data. Thus, we constructed panel data for 183 SLT companies for five years, namely, from 1994 to 1997 and for 2012, which comprised 915 observations. Our data included power factor values of these SLT companies obtained from the ECG. Data collection took place in 2017.

Table 1 shows the summary statistics for these 183 companies. Our outcome was a dummy variable that indicated whether the firm improved its power factor compared to the previous year; it took a value of one if the firm's power factor improvement from the previous year was strictly positive, and zero otherwise. SLT customers are categorized in terms of the voltage with which their electricity is supplied. High-voltage SLT customers are those firms that are supplied electricity at 33,000 volts; medium-voltage customers are supplied at 11,000 volts; and low-voltage SLT customers are supplied at 415 volts. The observed panel data from the ECG provide information for SLT customers in Greater Accra, which consists of Tema, East and West Accra, as well as the Western Region of Ghana. Tema is the industrial hub of Ghana, and most SLT customers are located in these study areas.

Figure 1 provides a box plot of the power factor distribution over the sample period and clearly shows the overall improvement of the power factor among the SLT customers over the years. The policy was first announced in 1994 and enacted in 1995. Table 1 shows that as many as 88.5% of firms improved their power factor between 1994 and 1995, immediately after the policy announcement. The mean and minimum values of the power factor among these firms continuously improved over the period. By 2012, the average was above the penalty threshold of 90% and the minimum value had reached as high as 74%.

Table 1. Summary statistics.

Variable	N. Obs.	Mean	Std. Dev.	Min	Max
Power factor	915	0.821	0.127	0.35	0.98
Year 1994 only	183	0.765	0.127	0.35	0.96
Year 1995 only	183	0.799	0.139	0.36	0.97
Year 1996 only	183	0.810	0.132	0.37	0.98
Year 1997 only	183	0.830	0.124	0.37	0.98
Year 2012 only	183	0.903	0.042	0.74	0.97
High-Voltage SLT Customers only	395	0.818	0.142	0.35	0.98
Medium-Voltage SLT Customers only	340	0.812	0.118	0.49	0.98
Low-Voltage SLT Customers only	180	0.847	0.103	0.59	0.98
SLT Customers in Greater Accra only	520	0.842	0.106	0.44	0.98
SLT Customers in Tema only	275	0.861	0.099	0.44	0.98
SLT Customers in Western Region only	395	0.794	0.145	0.35	0.97
Improvement of power factor (*a dummy variable*)	732	0.701	0.458		
from year 1994 to year 1995	183	0.885	0.320		
from year 1995 to year 1996	183	0.623	0.486		
from year 1996 to year 1997	183	0.634	0.483		
from year 1997 to year 2012	183	0.661	0.475		

Notes: The data includes a balanced five-year panel of 183 firms. The power factor data are provided by Electricity Company of Ghana and measured in the incremental unit of 0.01. Year 1994 is before the policy enforcement and years 1995 to 1997 and 2012 are after policy enforcement. Improvement of power factor is a dummy variable that takes a value of one for the firm whose power factor has strictly improved since the previous year, and zero otherwise.

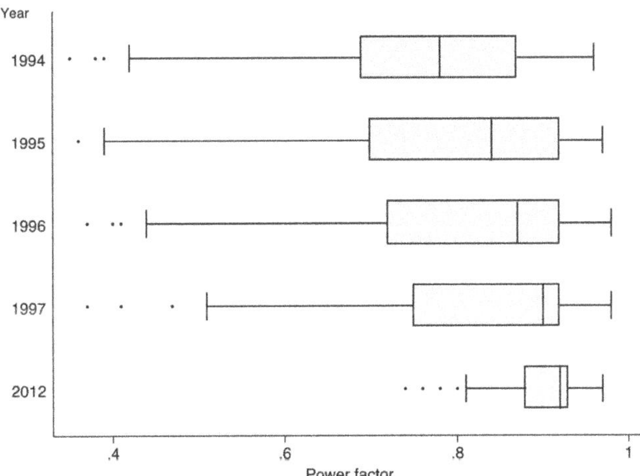

Figure 1. Box plot of power factor distribution over the sample period. A line in the middle of each box indicates the median of the sample in each year. The left and right edges of each box show the first and the third quartiles, respectively. Whiskers show the upper and lower adjacent values. The upper (or lower) adjacent value is the largest (or the smallest) observation that is within 1.5 times the interquartile range from each edge of the box. Dots are the units outside of the adjacent values.

2.2. Identification Strategies

The summary statistics above seem to indicate the effectiveness of the power factor correction policy introduced by the government of Ghana. However, the unobserved heterogeneity between the efficient and inefficient units may well confound our outcome in terms of whether they improved their power factor when facing the penalty set forth by the policy. That is, we cannot simply compare the outcomes of inefficient firms to those of more efficient firms above the threshold when investigating the impact of the power factor correction policy on the improvement exhibited by these firms.

We bypassed this issue of potential endogeneity by conducting an RDD with a cutoff at the penalty threshold set by the policy. RDD is a quasi-experimental method used to identify the average treatment effect of those units around the threshold called a cutoff (please see Appendix A for more details). In the context of RDD, treatment assignment is performed according to whether the observation units are below or above the cutoff. RDD enables a local average treatment effect (LATE) to be identified for those units around the cutoff, even under a situation like ours where conducting a pure randomized experiment or randomized controlled trial (RCT) is not possible (see, for example, Moscoe et al. [22] for details.) When the treatment assignment is unambiguously determined by whether the unit is above or below the cutoff, it is called the sharp RDD. In our case, units will face a penalty if the power factor is below the threshold but not if the power factor is above or equal, without exceptions. Therefore, we must utilize the sharp RDD strategy. In turn, if the treatment assignment depends, but not solely, on whether the unit lies below or above the cutoff, one must use the fuzzy RDD. That is, if by the research design, noncompliers exist, fuzzy RDD is the appropriate identification strategy.

According to the policy, firms whose power factor is below 90% are charged a penalty. Note that firms cannot independently observe their power factor precisely, which suggests that it would be difficult for firms to precisely manipulate their power factor values. Instead, the power factor is measured periodically by the electricity company. In the case of Ghana, the ECG reports their power factor to the firms along with the amount of the penalty, if any. At the end of each month, ECG delivers the electricity bills along with the amount of penalty and the power factor to the customers. The firm then decides whether to invest in power factor improvement. Our data only contain the annual average of the power factor reported to the customers in our sample. To allow for this time lag in decision making by the firms, we used the power factor in the previous year as a running variable that indicated whether the firm faced the penalty in the previous year. Although the penalty scheme designed by the policy in Ghana exhibits a kink at the threshold because our outcome is a binary variable, we still adopt an ordinary discontinuity design, rather than a kink RDD. Because our outcome variable is an indicator capturing whether the firm has improved its own power factor from the previous year, the same company should be tracked to enable our identification. As mentioned earlier, our outcome was a binary variable that took a value of one if a firm strictly improved its power factor independently from the previous year, and zero otherwise. This empirical strategy with lagged and differenced variables made our working sample a four-year panel from 1995 to 1997 plus 2012 for 183 firms, consisting of 732 observations.

ECG measures the power factor of SLT customers in the incremental unit of 1%. Thus, any firm whose power factor is equal to or less than 89% faces the penalty, while those at 90% or above do not. We, therefore, set the cutoff of the running variable at 89.5% in our RDD to determine whether this policy implemented by Ghana's government indeed induced SLT customers of the ECG to improve their power factor.

3. Results

Table 2 presents the RDD estimation results. Column (1) shows the impact of the penalty policy, measured as a gap in the power factor improvement probability between those two types of firms, namely, those whose power factor in the previous year was just above and those whose power factor was below the threshold, estimated using the full sample. Column (2) shows the same impact estimated using only the subsample from 1995 to 1997. We refer to them as Models 1 and 2, respectively.

The coefficients were significantly negative in both models, which indicated that, relative to those immediately below the cutoff, those firms whose power factor was immediately above the cutoff in the previous year exhibited a strictly smaller probability of improving their power factor in the current year. In other words, the firms that were penalized were more likely to improve their power factor in the following year.

Table 2. Sharp regression discontinuity design (RDD) estimation results.

Dependent Variable	(1) Full Sample	(2) Subsample of Years from 1995 to 1997
	Power Factor Improvement	Power Factor Improvement
Treatment	−0.455 *** (0.104)	−0.598 *** (0.127)
N. obs.	732	549
N. obs. above the cutoff	278	183
N. obs. below the cutoff	454	366
Effective N. obs. above the cutoff	116	76
Effective N. obs. below the cutoff	52	44
Bandwidth	0.020	0.020

Notes: Standard errors are in parentheses. Statistical significance at 1% is indicated as ***. The dependent variable is a dummy variable that takes the value of one if a firm strictly improved its power factor since the previous year, and zero otherwise. The running variable is the power factor in the previous year with a cutoff at 0.895. The model in column (1) uses the entire sample of four years, namely, three years from 1995 to 1997 as well as 2012. The model in column (2) uses the subsample of three years from 1995 to 1997 only. Bandwidths in both models are selected by a mean-squared error optimal bandwidth selector. Kernel type of a triangular function is used in both models.

This result is also clearly shown in Figure 2, where the probability of power factor improvement is shown on the vertical axis over the level of the power factor in the previous year on the horizontal axis. Panel (a) shows the figure for Model 1, while panel (b) does the same for Model 2. In both figures, there is a clear discontinuity at the penalty threshold, which is shown as a vertical line. The results show that firms slightly below (i.e., on the left of the cutoff) were much more likely to improve their power factor than firms that were slightly above the threshold line (or on the right of the vertical line). Indeed, the probability of power factor improvement was approximately 90% for firms immediately below the threshold, while only approximately half of the firms immediately above the threshold exhibited an improvement (the gray dot shows the local average within each bin).

A closer look at the result reveals that the effect is stronger in Model 2 than in Model 1. The estimate of the gap in the probability of power factor improvement between firms immediately below and those immediately above the cutoff is as large as 59.8% in Model 2. Figure 2 also shows that the gap is larger in Model 2. The plot of the probability of power factor improvement in Model 2 depicted in panel (b) of Figure 2 shows that firms that were slightly above the threshold were less likely to improve their power factor than the firms in Model 1 depicted in panel (a). In Model 2, we have only three consecutive years; therefore, the resulting greater impact in Model 2 suggests a strong short-term effect of the policy. These results imply that the policy was effective in giving a strong incentive to firms, particularly those that fell slightly below the penalty threshold, to improve their power factor, even in the short term.

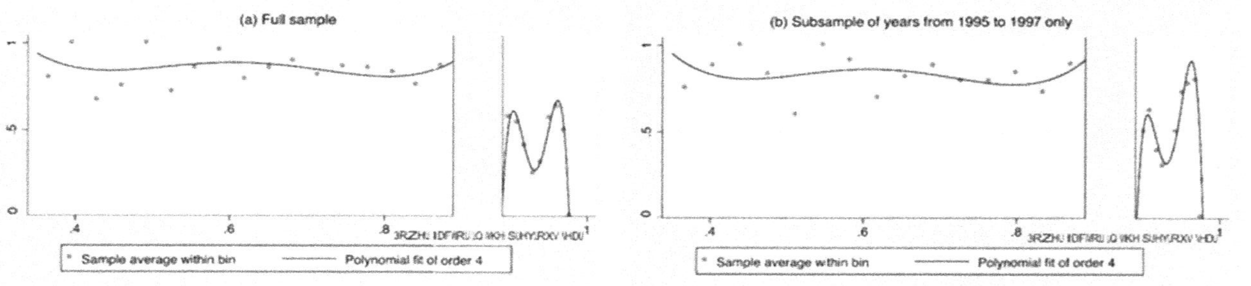

Figure 2. Probability of power factor improvement and its discontinuity at the penalty threshold. The probability of power factor improvement is on the vertical axis, and the level of the power factor in the previous year is on the horizontal axis. The cutoff is shown as a vertical line at 0.895 of the horizontal axis. The number of observations used in panel (**a**) is 732, of which 454 are on the left and 278 are on the right of the cutoff. The number of observations used in panel (**b**) is 549, of which 366 are on the left and 183 are on the right of the cutoff.

4. Concluding Remarks

In confronting its energy crisis, the government of Ghana enacted a power factor correction policy in 1995. The policy imposes a penalty on large-scale electricity users, labeled as SLT customers of the ECG, when their power factor falls below the threshold of 90%. This paper investigated the policy-induced improvement in these firms' power factor by applying RDD to panel data for 183 SLT customers. Our sample period ranged from 1994, when the policy was first announced, to 1997 and also included data for 2012. Specifically, we defined our running variable as the value of the power factor in the previous year, with the cutoff being the penalty threshold. Our outcome was a binary variable indicating whether the firm strictly improved its power factor since the previous year. The results show that the SLT customers whose power factor fell slightly below the threshold in the previous year were indeed more likely to improve their power factor, and the effect was stronger when we limited our sample period to a shorter term. This finding suggests that the power factor correction policy implemented by Ghana's government had an immediate impact on energy efficiency improvement by the country's large-scale electricity users.

Over the last few decades, the total power generation capacity has been constantly increasing in Ghana, and the majority of such change is attributable to the growth of thermal power generation. As a result, the installed capacity of hydro power is now 1584 megawatts and that of thermal is 3456 megawatts, while renewable energy is only approximately 42 megawatts [23]. With the thermal share of the generation mix now being at 68% and hydro set to reduce further, electricity prices will largely depend on international fuel prices; Ghana will thus be more vulnerable to global energy shocks. Energy efficiency has become even more relevant, and power factor improvement is one of the most important, simplest, and inexpensive strategies for achieving efficiency. For Ghana, like other new emerging economies with low energy efficiency, an efficient and stable electricity supply is a prerequisite for achieving its goal of consolidating its middle-income status via industrialization.

Author Contributions: Conceptualization, S.L. and Y.Y.; Methodology, Y.Y.; Software, Y.Y.; Validation, Y.Y. and K.F.; Formal Analysis, Y.Y.; Investigation, Y.Y.; Resources, S.L.; Data Curation, S.L. and B.H.; Writing-Original Draft Preparation, S.L.; Writing-Review & Editing, Y.Y. and K.F..; Visualization, Y.Y.; Supervision, Y.Y.; Project Administration, Y.Y.; Funding Acquisition, Y.Y.

Funding: This work is partly supported by the Ministry of Education of Japan Grant-in-Aid for Scientific Research #16K03628.

Acknowledgments: We thank Takahiro Ito for his constructive comments.

Conflicts of Interest: The authors declare no conflicts of interest.

Appendix A

A.1. Regression Discontinuity Design

Let *Ti* be the treatment variable for firm *i* such that

$$T_i = \begin{cases} 0, & if\ z_i \geq \bar{z} \\ 1, & if\ z_i < \bar{z} \end{cases}$$

where *zi* is a running variable for firm *i*, and $\bar{z}$ is the cutoff of the running variable. In our context, *Ti* is an indicator that firm *i* faces the penalty, and *zi* is the power factor of firm *i* in the previous year.

We then describe the outcome *Yi* as

$$Y_i = \beta_0 + \alpha T_i + \sum_{k=1}^{K} \beta_k (z_i - \bar{z})^k + \sum_{k=1}^{K} \gamma_k (z_i - \bar{z})^k T_i + u_i$$

where α gives the causal effect of the treatment.

In estimating the causal effect α, we weight observations according to the distance to the cutoff on the domain of the running variable. The weight is computed automatically by optimizing the mean-squared error of the estimation.

A.2. Estimation Code

We conducted the actual estimation using STATA 15. The actual code is as follows:

$$\text{rdrobust dpfpositive pf\_1, c(0.895).}$$

Here, rdrobust is the command to conduct an RDD estimation. The variable dapfpositive is an indicator that takes a value of one if the change in power factor is strictly positive, while pf_1 is the power factor in the previous year. The last term, c(0.895), indicates that the cutoff $\bar{z}$ in our case is 0.895, implying that the firms with a power factor of 89% or below face the penalty, while the firms with 90% power factor and above do not.

References

1. Oo, M.T.; Cho, E.E. Improvement of power factor for industrial plant with automatic capacitor bank. In Proceedings of the World Academy of Science Engineering and Technology, Boston, MA, USA, August 2008; Volume 32.
2. Umesh, S.; Venkatesha, L.; Usha, A. Active power factor correction technique for single phase full bridge rectifier. In Proceedings of the 2014 International Conference on Advances in Energy Conversion Technologies, Manipal, India, 23–25 January 2014; pp. 130–135.
3. Kueck, J.; Kirby, B.; Rizy, T.; Li, F.; Fall, N. Reactive Power from Distributed Energy. *Electr. J.* **2006**, *19*, 27–38. [CrossRef]
4. Adefarati, T.; Oluwole, A.S.; Olusuyi, O.; Sanusi, M.A. Economic and Industrial Application of Power Factor Improvement. *Int. J. Eng. Res. Technol.* **2013**, *2*, 1945–1955.
5. Costantini, V.; Crespi, F.; Paglialunga, E. The employment impact of private and public actions for energy efficiency: Evidence from European industries. *Energy Policy* **2018**, *119*, 250–267. [CrossRef]
6. Kumi, E.N. *The Electricity Situation in Ghana: Challenges and Opportunities*; Center for Global Development: Washington, DC, USA, 2017.
7. Edjekumhene, I.; Amadu, B.M.; Brew-Hammond, A. *Power Sector, Reform in Ghana: The Untold Story*; Kumasi Institute of Technology and Environment (KITE): Accra, Ghana, 2001.
8. Ajodhia, V.; Mulder, W.; Slot, T. *Tariff Structures for Sustainable Electrification in Africa*; European Copper Institute (ECI): Arnhem, The Netherlands, 2012.
9. Lee, K.H.; Min, B. Green R&D for eco-innovation and its impact on carbon emissions and firm performance. *J. Clean. Prod.* **2015**, *108*, 534–542.
10. Ringel, M.; Schlomann, B.; Krail, M.; Rohde, C. Towards a green economy in Germany? The role of energy efficiency policies. *Appl. Energy* **2016**, *179*, 1293–1303. [CrossRef]
11. Hartwig, J.; Kockat, J.; Schade, W.; Braungardt, S. The macroeconomic effects of ambitious energy efficiency policy in Germany–Combining bottom-up energy modeling with a non-equilibrium macroeconomic model. *Energy* **2017**, *124*, 510–520. [CrossRef]
12. Henriques, J.; Catarino, J. Motivating towards energy efficiency in small and medium enterprises. *J. Clean. Prod.* **2016**, *139*, 42–50. [CrossRef]
13. Allan, G.; Hanley, N.; McGregor, P.; Swales, K.; Turner, K. The impact of increased efficiency in the industrial use of energy: A computable general equilibrium analysis for the United Kingdom. *Energy Econ.* **2007**, *29*, 779–798. [CrossRef]
14. Tanaka, K. Review of policies and measures for energy efficiency in industry sector. *Energy Policy* **2011**, *39*, 6532–6550. [CrossRef]
15. Cox, E.; Royston, S.; Selby, J. *The Impacts of Non-Energy Policies on the Energy System: A Scoping Paper*; UKERC: London, UK, 2016.
16. Xiong, S.; Ma, X.; Ji, J. The impact of industrial structure efficiency on provincial industrial energy efficiency in China. *J. Clean. Prod.* **2019**, *215*, 952–962. [CrossRef]

17. Villca-Pozo, M.; Gonzales-Bustos, J.P. Tax incentives to modernize the energy efficiency of the housing in Spain. *Energy Policy* **2019**, *128*, 530–538. [CrossRef]
18. Anderson, S.T.; Newell, R.G. Information programs for technology adoption: The case of energy-efficiency audits. *Res. Energy Econ.* **2004**, *26*, 27–50. [CrossRef]
19. Li, L.; Wang, J.; Tan, Z.; Ge, X.; Zhang, J.; Yun, X. Policies for eliminating low-efficiency production capacities and improving energy efficiency of energy-intensive industries in China. *Renew. Sustain. Energy Rev.* **2014**, *39*, 312–326. [CrossRef]
20. Fleiter, T.; Gruber, E.; Eichhammer, W.; Worrell, E. The German energy audit program for firms—A cost-effective way to improve energy efficiency? *Energy Effic.* **2012**, *5*, 447–469. [CrossRef]
21. Yu, Y.; Zhang, N. Does smart city policy improve energy efficiency? Evidence from a quasi-natural experiment in China. *J. Clean. Prod.* **2019**, *229*, 501–512. [CrossRef]
22. Moscoe, E.; Bor, J.; Bärnighausen, T. Regression discontinuity designs are underuti-lized in medicine, epidemiology, and public health: A review of current and best practice. *J. Clin. Epidemiol.* **2015**, *68*, 132–143. [CrossRef] [PubMed]
23. Energy Commission. *Electricity Supply Plan for the Ghana Power System*; Energy Commission: Accra, Ghana, 2019.

Article

The Technological Progress of the Fuel Consumption Rate for Passenger Vehicles in China: 2009–2016

Jihu Zheng, Rujie Yu *, Yong Liu, Yuhong Zou and Dongchang Zhao

Automotive Data Center, China Automotive Technology and Research Center, Tianjin 300300, China; zhengjihu@catarc.ac.cn (J.Z.); liuyong@catarc.ac.cn (Y.L.); zouyuhong@catarc.ac.cn (Y.Z.); zhaodongchang@catarc.ac.cn (D.Z.)
* Correspondence: yurujie@catarc.ac.cn

Received: 27 May 2019; Accepted: 19 June 2019; Published: 21 June 2019

Abstract: China has set stringent fuel consumption rate (FCR) targets to address the serious environmental and energy security problems caused by vehicles. Estimating the technological progress and tradeoffs between FCR and vehicle attributes is important for assessing the viability of meeting future targets. In this paper, we explored the relationship between vehicle FCR and other attributes using a regression model with data from 2009–2016. We also quantified the difference in the tradeoff between local and joint venture brands. The result showed that from 2009 to 2016, if power and curb mass were held constant, 2.3% and 2.9% annual technological progress should have been achieved for local and joint venture brands, respectively. The effectiveness of fuel-efficient technologies for joint venture brands is generally better than that of local brands. Impacts of other attributes on FCR were also assessed. The joint venture brands made more technological progress with FCR improvement than that of local brands. Even if 100% of technological progress (assume the technological progress in the future were the same as that of 2009–2016) investment were used to improve actual FCR after 2016, it would be difficult to meet 2020 target. Accelerating the adoption of fuel-efficient technologies, and controlling weight and performance, are both needed to achieve the 2020 and 2025 targets.

Keywords: fuel consumption; trade-off; technological progress; passenger vehicle

---

1. Introduction

The total number of vehicles in China has increased dramatically over the past decades and reached 231.2 million in 2018 (excluding 9.06 million three-wheeled vehicles and low-speed trucks) [1]. The rapid growth in vehicle numbers has caused serious environmental and energy security problems. In 2015, nearly 700 million tons of carbon dioxide (CO_2) was produced from China's road traffic, and these emissions are increasing [2]. China's dependency on foreign countries for oil reached 68.4% in 2017 [3]. Automotive gasoline and diesel accounted for 80% of all gasoline and diesel consumption in 2017. To reduce the fuel consumption from passenger cars, China has issued a series of fuel consumption rate (FCR) standards and regulations for passenger vehicles. China released its first FCR limit standard for passenger vehicles in 2004 [4]. The corporate average FCR standard for passenger vehicles was released in 2011 [5] and updated in 2014 [6]. China also set the FCR targets for new passenger vehicles to 5.0 L/100 km in 2020 [7] and 4.0 L/100 km in 2025.

Globally, the transport sector contributes about one-fourth of total fossil fuel greenhouse gas (GHG) emissions, about three-quarters of that amount come from road transport [8]. To curb the GHG emissions from road transport, ten countries and regions including China, U.S., EU, and Japan, etc., have established mandatory or voluntary standards for light-duty vehicles [9,10].

There are two pathways to achieve the increasingly stringent passenger vehicle FCR target for carmakers. One is to produce more new energy vehicles (NEVs), as the electricity consumption of NEVs is calculated as zero, and they have multipliers in the Phase IV standard. In the Phase IV standard [6], the equivalent energy consumption of battery electric vehicles (BEV), the electric-drive part of plug-in hybrid vehicles (PHEV) and fuel cell vehicles (FCV) are calculated as zero. The multiplier of NEVs is set at five from 2016 to 2017, decreasing to three from 2018 to 2019, and two in 2020. The other pathway for compliance is to improve the FCR of conventional vehicles by deploying fuel-efficient technologies or adjusting portfolios by producing more smaller and lighter vehicles. As there are still many barriers to the promotion of NEVs, such as high retail prices, short electric ranges and a shortage of charging infrastructure [11], improving the FCR of conventional vehicles has become one of the necessary paths for carmakers to comply with their FCR targets.

With the tightening passenger vehicle FCR targets and regulations, China achieved 1.7% FCR improvement annually for the period 2009–2017. The market penetration of fuel-efficient technologies is increasing rapidly. The adoption rate of gasoline direct injection (GDI) and turbocharging for new passenger vehicles in China reached 39.39% and 45.11% in 2017 [3], from 0.5% and 3.4% in 2009 [12], respectively.

Previous research found that the official tested FCR is highly correlated with curb weight, power, acceleration time, and other characteristics. By using the features of the U.S. car from 1975 to 2009, the research found that a 1% increase in weight results in a 0.69% increase in FCR, and a 1% reduction in 0–97 km/h acceleration time results in a 0.44% increase in FCR when holding all else equal [13]. Similar results were also found by using the technical specifications and fuel consumption information of automobiles for sale in Europe from 1975 to 2015 [14]. After reviewing relevant studies, the fuel-mass coefficients (the ratio of FCR change (%) and weight change (%)) was found to be in the range of 0.315–0.71 [15]. By evaluating a wide range of vehicle case studies of gasoline turbocharged cars, which represent the 2015 European market, the fuel reduction value coefficient (the ratio of FCR achieved through mass reduction to vehicle mass reduction) was found within the range of 0.159–0.237 L/100km*100kg for the mass reduction only and 0.252–0.477L/100km*100kg for the secondary effect [16]. The combination of life cycle assessment with the traditional design procedure was also proposed to assess the environmental performances of automotive component light weighting [17]. Fuel economy standards and regulations aim to improve the FCR of new passenger vehicles. In addition to FCR, consumers also pay attention to vehicle performance, etc., which are highly correlated with FCR. Thus, to analyze the feasibility of achieving the targets of 5.0 L/100 km in 2020 and 4.0 L/100 km in 2025 for new passenger vehicles in China, it is necessary to quantify the tradeoff between FCR and other attribute parameters of passenger vehicles in China.

Many studies have been conducted on the trend of passenger vehicle FCRs and the tradeoff between official tested FCR and other vehicle attributes. A new index called the Performance-Size-Fuel economy Index (PSFI) was proposed by An, which is defined as the product of the vehicle performance, size, and fuel economy [18]. The PSFI was used to assess the technical efficiency improvement rates of cars and trucks from 1977 to 2005 in the U.S. The PSFI showed good correlations and appeared quite linear for both cars and trucks by using the 1977 to 2005 data from the Fuel Economy Trends Report of the Environmental Protection Agency (EPA). The PSFI provides a way to measure the technological progress of vehicles, but it simply sets the coefficient values of the three variables to one, which requires further study and explanation. Compared with recent research results [13,14], the impact of curb weight on FCR could be exaggerated in the definition of PSFI. To better estimate the technological progress and the tradeoff between FCR and other attributes, Knittel built a regression model on fuel economy, weight, engine power, and torque, and also introduced the production possibilities frontier (PPF) to capture the technological progress. The result showed that if the power, torque and curb weight of the light-duty vehicle in the U.S. stayed at the same level as 1980, the fuel economy from 1980 to 2006 could have improved by nearly 50% for both passenger cars and light trucks [19]. Klier and Linn expanded Knittel's analysis by matching engine data to vehicle production data. The changes in the rate were examined, and the results showed that recent changes in the U.S. and European fuel economy standards had increased the speed of technology adoption [20]. MacKenzie and Heywood extended

Knittel's econometric approach by adopting both vehicle system attributes and consumer amenities as independent variables. They found that between 1975 and 2009, per-mile fuel consumption could have been reduced by approximately 70%, or an average of 3.4% per year, if not for reductions in acceleration time and the introduction of new attributes and functionality to vehicles [13]. To the best of our knowledge, this topic for China has not been carefully studied before due to a lack of sufficient data.

In this paper, we mainly aimed to address the following points:

1. Summarize the trend of FCR and vehicle characteristics for Chinese passenger vehicles for 2009–2016 based on a comprehensive database built by China Automotive Technology and Research Center (CATARC).
2. Construct a quantitative model between FCR and vehicle attributes to estimate the impact of vehicle characteristics on FCR and the technological progress for local and joint venture manufacturers (Chinese domestic carmakers are generally divided into the joint venture and local corporations. Joint venture corporations are generally jointly invested by foreign carmakers and Chinese carmakers, such as SAIC-Volkswagen, which was established by Volkswagen and Shanghai Automotive Industry Corporation (SAIC). In the joint venture corporations, the technology and brand are usually provided by the foreign side. The Chinese side usually provides land use rights and funds. The vehicle models produced by the joint venture corporation is called joint venture brand cars. Local corporations are usually 100% owned by Chinese companies, with completely independent product modification rights and brand operation rights. The vehicle models produced by local corporations are called local brands, such as BYD and Geely. Since the local brands and joint venture brands have significant differences in the quota of different vehicle types and technical characteristics, they are studied separately in this paper), respectively.
3. Analyze the differences between the joint venture and the local brands in the tradeoff between FCR and vehicle attributes.

The rest of the paper is organized as follows: The data used in this paper is discussed in Section 2. The methodology is presented in Section 3. The results are detailed in Section 4. The conclusion is presented in Section 5.

2. Data

The data used in this paper were from the Automotive Data Center Database of CATARC, including FCR in L/100 km, vehicle level characteristics (such as curb weight, vehicle acceleration time, etc.), engine attributes (such as engine power, torque, etc.), fuel-efficient technology configuration (such as engine aspiration type, fuel injection mode, etc.), and production by model and year for all new passenger vehicles (according to GB/T 3730.3–2001, the passenger vehicle in China is defined as a vehicle designed and constructed for the carriage of passengers and having a maximum design mass not exceeding 3.5 tons) produced by domestic manufacturers between 2009 and 2016. Gasoline vehicles dominated the conventional vehicle market of China and other fuel types of vehicles, such as diesel vehicles, bi-fuel vehicles, etc., accounted for less than 2% of the total market. In this paper, we focused only on gasoline passenger vehicles. The total number of gasoline passenger vehicle models was 14,183, which were all used to run the regression model introduced in the methodology section.

Figure 1 illustrates the production-weighted FCR of domestic passenger vehicles from 2009 to 2016. As shown in Figure 1, the average official tested FCR of gasoline passenger vehicles in China improved from 7.71 L/100 km in 2009 to 6.82 L/100 km in 2016, with an annual improvement rate of 1.7% (the trend is the result of both the improvement of vehicle fuel efficiency and change of market share for different vehicle types and classes). The unadjusted average fuel economy for cars in the U.S. market increased from 32.1 MPG (equivalent to 7.32 L/100 km after unit conversion) in 2009 to 36.9 MPG (6.37 L/100 km) in 2016 [21], with an annual improvement rate of 2.0%. In the EU market, the average CO_2 emissions from new gasoline passenger cars decreased from 146.6 g/km (equivalent to 6.27 L/100 km) in 2009 to 121.7 g/km (equivalent to 5.20 L/100 km) in 2016 [22], with an annual improvement rate of 2.62%. As the passenger vehicle FCR in both the EU and China was tested under the New European Driving Cycle (NEDC) between 2009 and 2016, the FCRs of passenger vehicles in the EU and China were comparable. We found that the FCR improvement in the EU was faster than China, and there was a big gap between the FCR of passenger vehicles in China and the EU. The FCR gap between China and the EU might be due to two main reasons. Firstly, the curb weight in the EU was lower than that of China. Secondly, the EU had a higher adoption rate of fuel-efficient technologies for passenger vehicles than China.

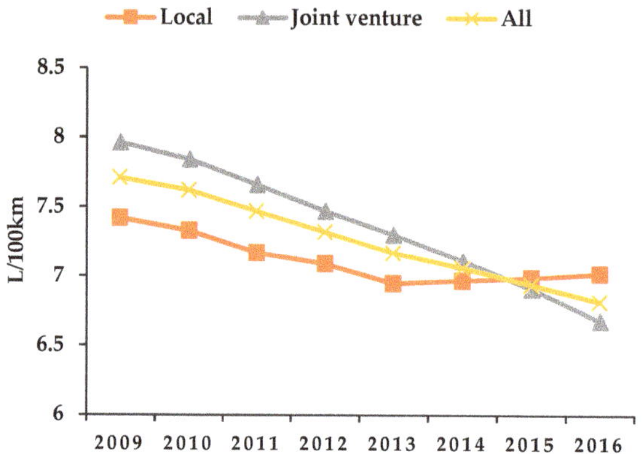

Figure 1. Production-weighted fuel consumption rate (FCR) of China's domestic passenger vehicles in 2009–2016.

Curb weight is the basis for the current FCR standard for passenger vehicles in China. As illustrated in Figure 2, the production-weighted passenger vehicle curb weight increased by 13.1%, from 1222 kg in 2009 to 1382 kg in 2016. The increasing curb weight may be mainly due to the increasing percentage of sales and stock for sport utility vehicles (SUVs) and multi-purpose vehicles (MPVs). The SUVs and MPVs stock dramatically increased from 10.3% and 4.3% in 2012 to 20.0% and 6.9% in 2016, respectively [23].

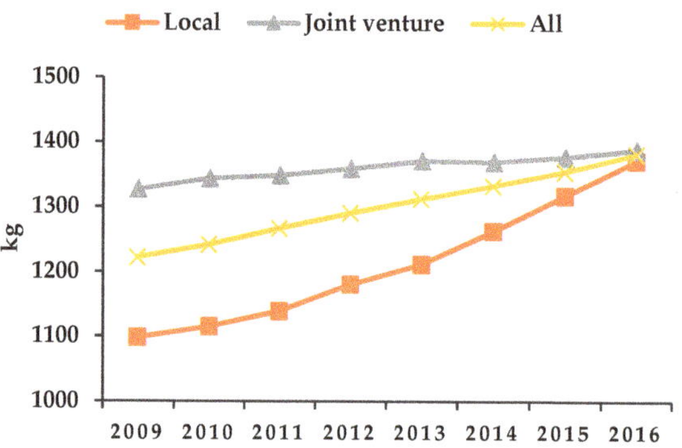

Figure 2. Production-weighted curb weight of China's domestic passenger vehicles in 2009–2016.

As illustrated in Figure 3, the production-weighted power increased from 78.7 kW in 2009 to 101.7 kW in 2016, with an annual growth rate of 3.7%. The negative effect of increasing power on the FCR of passenger vehicles has already been realized in the U.S. and EU markets. To better estimate the effect of vehicle attributes on FCR, it is necessary to explore the relationship between FCR and other attributes of passenger vehicles in the Chinese market by using an econometric model.

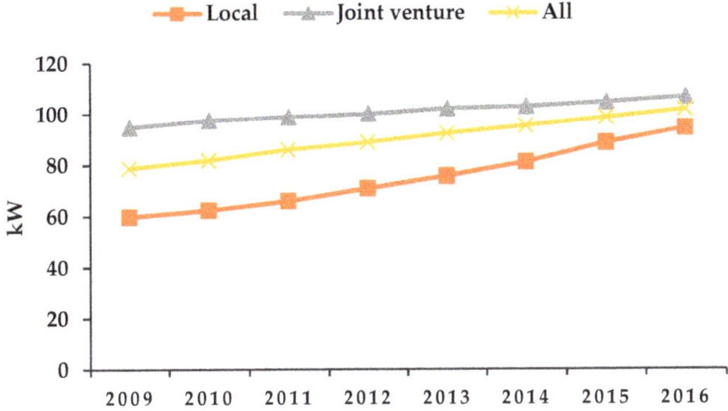

Figure 3. Production-weighted power of China's domestic passenger vehicles in 2009–2016.

3. Methodology

There are generally two ways to achieve the desired goal in this paper. The engineering simulation method could simulate the relationship between FCR and the influencing factors. Based on engineering simulation analysis, the FCR can also be predicted based on the change of influencing factors. However, the simulation method requires high quantity and quality of input data, and the simulation results for certain models may not represent the nationwide fleet. A linear regression method based mainly on the methodology of Knittel [19] and MacKenzie [13] was adopted in this study to analyze the relationship between official tested FCR and observable variables such as power, curb weight, fuel-efficient technology and other unobservable variables, such as vehicle brand, build year, etc., as shown in Equation (1):

$$\ln FC_{it} = \beta_0 + \beta_1 \ln CW_{it} + \beta_2 \ln acc_{it} + \mathbf{X}'_{it}\mathbf{B} + \tau_t + \mu_i + \varepsilon_{it} \tag{1}$$

where FC_{it} represents the FCR of passenger vehicle model i in year t in the unit of L/100 km. β_0 is a constant. CW_{it} is the curb weight in kg. acc_{it} is the 0 to 100 km/h acceleration time in seconds. $\mathbf{X}_{it}$ is a vector of dummy variables including whether the vehicle has a manual transmission, whether the vehicle is an SUV, or whether the vehicle has a turbocharge. τ_t is the year fixed effects to estimate the technological progress by year t. μ_i represents the difference by vehicle brands. We chose the vehicle brand rather than the manufacturer as a variable because the same manufacturer often produces more than one brand, and the vehicle characteristics of different brands vary widely. For instance, Changan Ford Mazda Corporation has four brands: the brand Ford of Sino-U.S., the brand Mazda of Sino-Japan, the brand Volvo of Sino-EU, and the local brand Changan. ε_{it} is the random error term.

As τ_t is a set of variables to estimate the annual technological growth, if $\mathbf{X}_{it}$ does not include any fuel-efficient technology, then τ_t will capture all the technological growth in year t. If vehicle attributes do not change, e^{τ_t} represents the potential FCR improvement in year t compared with the FCR in the base year due to the fuel-efficient technology deployment, as shown in Equation (2):

$$\frac{FC_t}{FC_{base}} = e^{\tau_t} \tag{2}$$

where, FC_t represents FCR in year t. FC_{base} represents FCR in the base year. For small values of τ_t, $e^{\tau_t} \approx 1 + \tau_t$.

4. Results Analysis

4.1. Model Estimation Results

Table 1 shows the estimation results of the regression models for passenger vehicle FCRs as a function of curb weight, power, acceleration time, and other attributes based on Equation (1). The products of joint venture brands and local brands showed great differences in both FCR and other attributes from 2009 to 2016. In this paper, we estimated the joint venture brands and local brands separately and compared them with each other. In each case, we estimated four models with different sets of control variables to explore the technological progress and effects of vehicle attributes on the FCR from 2009 to 2016. The estimated coefficients represent the elasticity coefficients of corresponding variables to FCR. The variables of all models include vehicle types, year fixed effect, vehicle brand, and curb weight. We captured the annual technological growth by using the year fixed effect while holding other variables constant. The estimate results are shown in Table 1. We also explored the effects of curb weight, acceleration time, power, drive type, turbocharging, GDI, and advanced transmissions on the official tested FCR.

Model 1 controlled the curb weight, power, vehicle type, year fixed effect, and brands. The fuel-efficient technologies, such as turbocharging, GDI, and advanced transmissions (Continuously Variable Transmission (CVT), Dual-clutch Transmission (DCT), etc.), were not controlled in Model 1, so that we could capture the technological progress for 2009–2016 through the year fixed effects while holding curb weight, power, vehicle type, and brand constant. The estimation results of Model 1 show that a 1% increase in curb weight results in a 0.86% and 0.85% increase in FCR for local brands and joint venture brands, respectively. A 1% increase in power leads to a 0.061% decrease in FCR for local brands in Model 1. The coefficient of power is negative, which is opposite to similar studies [13,14] but consistent with the results of Figures 4 and 5. Figures 4 and 5 show the linear regression result of power and FCR when holding curb weight constant (actually in a relatively small interval). The effect of power on FCR in Model 1 is because models with larger power are always more expensive and equipped with more fuel-efficient technologies, resulting in lower FCR. The coefficients of curb weight and power will be more reasonable if the model controls more variables of fuel-efficient technologies. More details of the results will be discussed in Model 3 and Model 4. The year fixed effect of local and joint venture brands between 2009 to 2016 are 16.1% and 20.3%, respectively, which means that if the controlling variables were kept constant in the base year of 2009, the local and joint venture brands could achieve the FCR improvement rate of 16.1% and 20.3% in 2016, respectively. We can also conclude that joint venture vehicles showed faster technological progress than that of local brands.

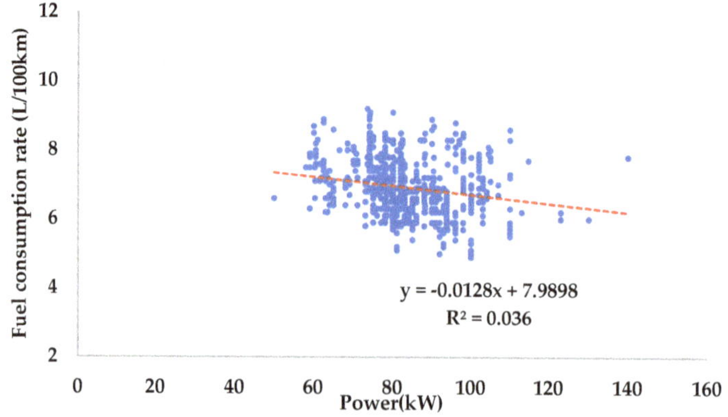

Figure 4. Regression results of power and fuel consumption rate of models in the 1200–1300 kg curb weight class.

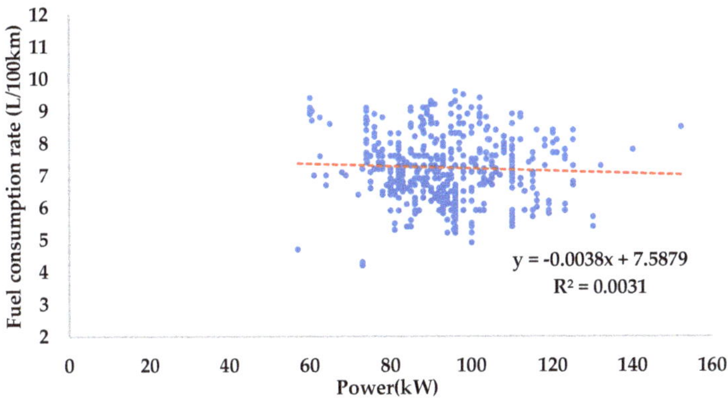

Figure 5. Regression results of power and fuel consumption rate of models in the 1300–1400 kg curb weight class.

Model 2 is similar to Model 1 but replaces power with vehicle acceleration time. The coefficients of FCR to curb weight decrease, which is inconsistent with the findings of MacKenzie and Heywood. It is expected that the increase in weight at a constant power will result in both higher FCR and slower acceleration, but our results show this is not the case. The explanation is that power and acceleration time are both related to fuel-efficient technologies. In Model 2, the decreasing acceleration time may be accompanied by the deployment of fuel-efficient technologies. A comparison of the effects of power and acceleration time on the sensitivity of FCR to curb weight after controlling more variables will be discussed in Models 4 and 5.

Model 3 further calls for fuel-efficient technologies, such as turbocharging, GDI, advanced transmissions, drive type, and specific power deciles based on Model 1. Like the Model 5 proposed by MacKenzie and Heywood, it calls for specific power deciles as dummy variables to reflect the technical level of the engine and to make this model more robust and explanatory. The results show that compared with Model 1, the sensitivity of FCR to curb weight increases to 0.164 for local brands and 0.263 for joint venture brands after controlling more attributes. The sensitivities of the dummy variables automatic transmission (AT) and CVT to the FCR for local brands are 7.8% and 5.0%, which means that, compared with manual transmission (MT), AT and CVT could increase the passenger vehicle FCR by 7.8% and 5.0%, respectively, when other variables are constant. For the joint venture carmakers, the coefficients of AT and CVT are 4.5% and −4.4%. From the above, we can conclude that, compared with joint venture brands, the effectiveness of AT and CVT for local brands needs to be improved. The sensitivities of GDI to FCR are −3.1% for local brands and −6.5% for joint venture brands, which means the GDI of joint venture brands has a better effect than that of local brands.

Compared with Model 3, Model 4 replaces power with vehicle acceleration time. Instead of decreasing the sensitivity of FCR to curb weight in Model 1 and Model 2, this change increases it. After calling for more fuel-efficient technologies, especially those related with power and acceleration time, the change of sensitivity of FCR to curb weight is consistent with the findings of MacKenzie and Heywood and can be explained by the increase in weight at a constant acceleration time, which requires a higher FCR for both weight change and power change.

Table 1. Results of estimating regression models of passenger vehicle FCR as a function of curb weight, power, acceleration time, and other attributes.

	Local Brands				Joint Venture Brands			
	Model 1	Model 2	Model 3	Model 4	Model 1	Model 2	Model 3	Model 4
(Intercept)	−3.706 ***	−2.962 ***	−2.521 ***	−2.324 ***	−4.101 ***	−4.017 ***	−3.277 ***	−3.294 ***
	−0.194	−0.231	−0.19	−0.23	−0.276	−0.255	−0.168	−0.203
ln (curb weight)	0.859 ***	0.764 ***	0.561 ***	0.728 ***	0.852 ***	0.839 ***	0.546 ***	0.815 ***
	−0.036	−0.029	−0.040	−0.025	−0.045	−0.034	−0.036	−0.027
ln (power)	−0.061 *		0.164 ***		−0.007		0.263 ***	
	−0.026		−0.032		−0.026		−0.041	
ln(accel.)		−0.122 ***		−0.251 ***		−0.006		−0.258 ***
		−0.026		−0.038		−0.018		−0.04
4WD/AWD			0.009+	0.008			0.015 *	0.011
			−0.005	−0.006			−0.007	−0.007
Turbo			0.132 ***	0.131 ***			−0.014	−0.019
			−0.021	−0.02			−0.021	−0.022
GDI			−0.031 *	−0.033 *			−0.065 ***	−0.066 ***
			−0.014	−0.015			−0.015	−0.016
AT			0.078 ***	0.030 **			0.045 ***	−0.013
			−0.006	−0.011			−0.005	−0.009
CVT			0.050 ***	0			−0.044 ***	−0.117 ***
			−0.013	−0.018			−0.009	−0.011
DCT			−0.022 *	−0.118 ***			−0.007	−0.107 ***
			−0.009	−0.018			−0.012	−0.022
Vehicle types	√	√	√	√	√	√	√	√
Year fixed effect	√	√	√	√	√	√	√	√
Brand	√	√	√	√	√	√	√	√
Specific power decile [a]			√	√			√	√
Number of observations	8516	8516	8516	8516	5667	5667	5667	5667
R^2	0.821	0.823	0.878	0.876	0.756	0.756	0.874	0.871
Adj. R^2	0.819	0.821	0.877	0.874	0.755	0.755	0.873	0.869

Statistical significance of t-tests on coefficient estimates: *** $p < 0.001$, ** $p < 0.01$, * $p < 0.05$, +$p < 0.1$.[a] The first decile includes those passenger vehicles with engine specific power values in the lowest one-tenth of all passenger vehicles in their build year, etc.

4.2. Technological Growth

Technological progress is captured by the year fixed effect τ_t. To make technological progress more intuitive, we define it as follows:

$$T_{progress} = 1 - e^{\tau_t}. \tag{3}$$

The results of the technological progress estimation are shown in Table 2. The results of these four models show that the technological progress of the joint venture is greater than that of local brands. In Model 1, the joint venture shows an 18.4% improvement from 2009 to 2016, which is faster than local brands, with an improvement of 14.9% from 2009 to 2016. The results are consistent with the penetration of fuel-efficient technologies by brands from 2009 to 2016 in China [12].

Table 2. Technological progress estimates for passenger vehicles in China from 2009 to 2016 (percent).

	Local Brands				Joint Venture Brands			
	Model 1	**Model 2**	**Model 3**	**Model 4**	**Model 1**	**Model 2**	**Model 3**	**Model 4**
2010	−1.6	−1.7	−0.9	−1.0	−3.2	−3.2	−2.1	−2.2
2011	−4.0	−4.2	−2.4	−2.7	−5.5	−5.5	−3.4	−3.5
2012	−6.9	−7.4	−4.5	−4.7	−8.1	−8.1	−5.4	−5.6
2013	−8.7	−9.2	−5.9	−6.1	−10.0	−10.1	−6.9	−7.0
2014	−10.8	−11.5	−7.6	−7.8	−12.7	−12.8	−9.0	−9.1
2015	−12.6	−13.5	−8.9	−9.1	−15.6	−15.7	−10.8	−10.9
2016	−14.9	−15.9	−10.5	−10.7	−18.4	−18.5	−12.7	−12.8

When holding other variables constant at the base year, the mathematical expression of FCR potential ($FC_{potential}$) is:

$$FC_{potential} = FC_{base} \cdot e^{\tau_t} \tag{4}$$

In this paper, we set 2009 as the base year, and the FC_{base} is the actual FCR in 2009. $FC_{potential}$ is regarded as the FCR reduction potential for the target year. τ_t is the year fixed effect.

Figure 6 shows the actual FCR and expected FCR of different models shown in Table 2 by Equation (4). The solid and dotted lines with different colors in Figure 6 indicate the results of local brands and joint venture brands, respectively. The blue lines indicate the actual FCR of new passenger vehicles. The green lines show the FCR estimates if curb weight, power, and the share of vehicle types had remained at the 2009 level (Model 1 in Table 2). The black lines represent the FCR estimates if the curb weight, vehicle acceleration performance, and the share of different vehicle types remained at the 2009 levels (Model 2 in Table 2). The red lines represent the FCR estimates if the weight, power, drive type, powertrain features (turbocharging, GDI, transmission, etc.) and the share of vehicle types remained at the 2009 levels (Model 3 in Table 2).

From Figure 6, we find that, for local and joint venture brands, 1) the actual FCR was reduced by 5.4% and 16.1%, respectively. 2) Holding weight, power, and the share of vehicle types constant at the 2009 levels, the FCR could have been reduced by 14.9% and 18.4%, respectively. 3) If weight, acceleration time, and the share of vehicle types remain unchanged, the FCR could have been reduced by 15.9% and 18.5%, respectively. 4) If weight, power, 4WD, powertrain features, and the share of vehicle types remained constant at 2009 levels, the FCR could have been reduced by 10.5% and 12.7%, respectively.

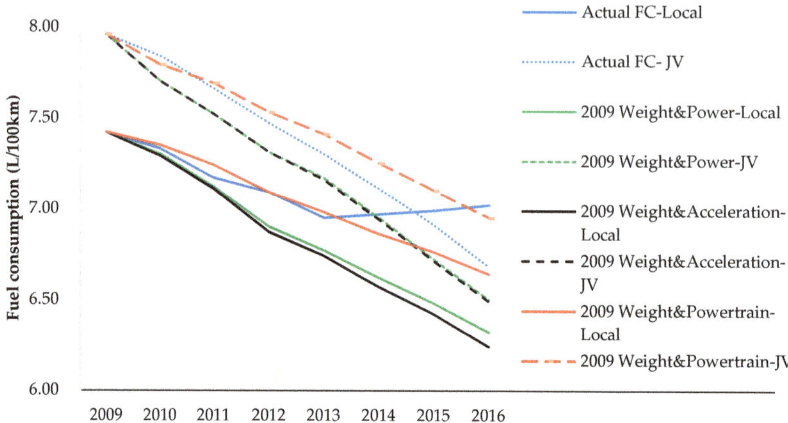

Figure 6. Actual FCR of average new passenger vehicles, and potential FCR if various attributes had remained at 2009 levels for local and joint venture brands.

Table 3 summarizes some results of the analogous studies that focused on the technological progress of U.S. and EU passenger vehicle markets and shows a comparison with the technological progress of passenger vehicles in China. China implemented its first passenger vehicle FCR standard in 2004, and then updated the standard in 2008, 2011 and 2014, respectively. China also set its passenger vehicle FCR target at 5.0 L/100 km by 2020. The compliance pressure from the FCR standard forces carmakers to accelerate the adoption of fuel-efficient technologies in the Chinese market. The technological progress is more rapid than those of the EU and U.S., as shown in Table 3. If we hold weight and power constant, the U.S. and Europe show a 1.7% (from 1975 to 2015) and 1.2% (from 1975 to 2009) annual technological progress, which is lower than the 2.3% and 2.9% for local and joint venture brands from 2009 to 2015. Even if we control more variables, such as powertrain features, the annual technological progress could still reach 1.7% and 1.9% for local and joint venture brands, respectively. It is important to note that technological progress is not distributed evenly over time. The results of MacKenzie show that the U.S. would have reached an annual technological progress of 5% between 1975 and 1990, and 2.1% between 1990 and 2009, if not for changes in acceleration, features, and functionality. The results of Hu and Chen show that the EU had a solid improvement in fuel-efficient technology with an annual technological progress rate of 2.8% from 2005 to 2015. However, there was no noticeable improvement in potential FCR reduction from 1975 to 2005 in the EU.

Table 3. Comparison of annual technological progress between the results of this work and other studies.

Source	Attributes Controlled	Year	Country	Annual Technological Progress
HU	Power, weight	1975–2015	EU	1.2%
HU	Weight, accel.	1975–2015	EU	1.4%
HU	Weight, engine attributes and amenities	1975–2015	EU	0.7%
HU	Weight, engine attributes and amenities	1975–2005	EU	No noticeable improvement
HU	Weight, engine attributes and amenities	2005–2015	EU	2.8%
MacKenzie	Power, weight	1975–2009	The U.S.	1.7%
MacKenzie	Weight, accel.	1975–2009	The U.S.	2.2%
MacKenzie	Accel. Features, functionality	1975–2009	The U.S.	2.9%
MacKenzie	Accel. Features, functionality	1975–1990	The U.S.	5%
MacKenzie	Accel. Features, functionality	1990–2009	The U.S.	2.1%
This work	Power, weight	2009–2016	China	2.3% (local brands) 2.9% (joint venture brands)
This work	Weight, accel.	2009–2016	China	2.4% (local brands) 2.9% (joint venture brands)
This work	Power, weight, powertrain features	2009–2016	China	1.7% (local brands) 1.9% (joint venture brands)

4.3. Comparison of Tradeoffs Between FCR and Other Attributes

Banddivadekar [24] introduced an index to quantify the trade-offs among vehicle fuel consumption, performance, and size, which is called Emphasis on Reducing Fuel Consumption (ERFC). We modified the ERFC equation by converting fuel economy (commonly expressed as miles per gallon) to FCR (expressed as liters per 100 km) in Equation (5):

$$ERFC = \frac{FC_{base} - FC_{cal}}{FC_{base} - FC_{potential}} \tag{5}$$

where FC_{base} is the actual FCR in the base year, FC_{cal} is the actual fuel consumption in the target year, and $FC_{potential}$ is the evaluated fuel consumption, as shown in Table 2, holding the other variables constant. *ERFC* is the index quantifying how much technological progress is used for improving fuel efficiency during the base year and the target year.

Figure 7 shows the comparisons of ERFC between the U.S., EU, and China (based on the results of Model 1) during the various periods. The ERFC of the U.S. is from the results of MacKenzie and Heywood, and the ERFC of Europe was calculated from the estimation results of Hu and Chen. The ERFC values of the U.S. are highly correlated with the Corporate Average Fuel Economy (CAFE) standard and fuel price. It varied between 130%, when the U.S. was facing an oil crisis at the beginning of CAFE standard implementation during 1975–1980, and −25%, during the unchanged stringency of the CAFE standard during 1995–2000. Europe also had a high ERFC value (ranging from 74% to 145% during 2000–2015) after the European Commission adopted a Community Strategy based on three pillars for reducing CO_2 emissions from cars in 1995. China released its first FCR limits standard in 2004 and updated the standard in 2008. Passenger vehicles that do not meet the FCR limit standard were not allowed to be produced and sold. Therefore, the ERFC values of local and joint venture brands in 2009–2012 showed a relatively high value of more than 60%. To achieve the target of 6.9 L/100 km in 2015, China introduced Corporate Average Fuel Consumption (CAFC) management by releasing the Phase III FCR standard for passenger vehicles in 2011, but there were no effective non-compliance penalties until China introduced CAFC and New Energy Vehicle Credits Regulation in 2017 [25]. Local brands showed much lower ERFC than the joint venture brands, especially during 2012–2016 (9% versus 81%).

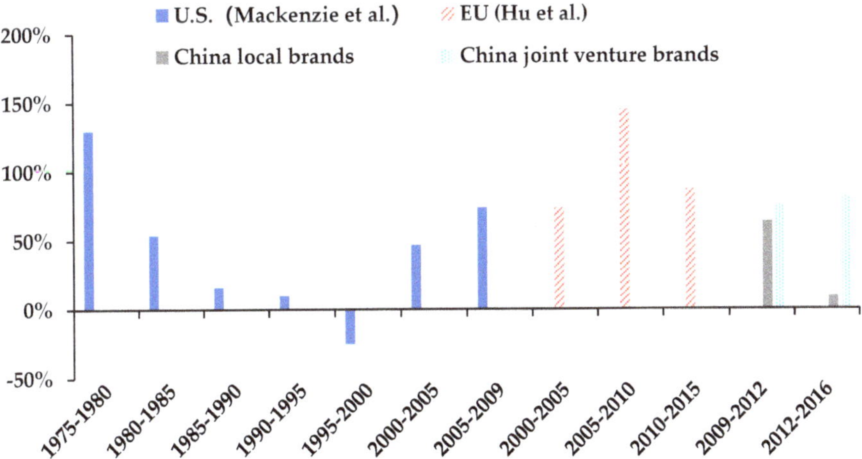

Figure 7. Comparisons of Emphasis on Reducing Fuel Consumption (ERFC) between the U.S., EU, and China.

5. Conclusions

Our analysis showed that China's passenger vehicles underwent significant technological progress due to the rapidly increasing fuel-efficient technology adoption rate. However, to achieve the FCR target of 5.0 L/100 km in 2020 and 4.0 L/100 km in 2025, more than a 5% annual FCR improvement rate is needed from 2016. Our findings show that if we hold power and curb mass constant, a 2.3% to 2.9% annual technological progress was achieved between 2009 and 2016 for local and joint venture brands, respectively, which means an extra 2% technological progress is needed to achieve the future targets. As the NEVs are included in carmakers' CAFC calculations for compliance with the standard, and counted as multiple vehicles, the target for gasoline vehicles in 2020 could be higher than 5.0 L/100 km. We predicted that 1.7 million NEVs will be produced in 2020 (the target of 2 million NEVs including passenger and commercial vehicles) with a ratio of EV to PHEV of 3:1. We found that the FCR target of the conventional vehicle in 2020 is around 5.7 L/100 km, which means an annual FCR improvement of 4.4% is needed based on the year 2016. If the ERFC value of 100% is retained, an extra annual technological progress rate of about 2% is still needed to achieve the 2020 FCR target. Our results indicate that China still faces significant challenges in achieving the FCR targets of 2020 and the future.

Our findings show that the technological progress of local brands is slower than that of joint venture brands. The regression results show the effects of fuel-efficient technologies, such as advanced transmission, GDI and turbocharging, for local brands are smaller than those of joint venture brands, which gives a reasonable explanation for why the technological progress of local brands is slower than that of the joint venture.

Our results show that the ERFC value of China's local brands is decreasing from 63.5% (which means that 63.5% of total FCR reduction potential is used for improving fuel efficiency) between 2009 and 2012 to 9.1% between 2012 and 2016. This is mainly due to three reasons: 1) Local brands are more responsive to the high market demand for larger size and better performance vehicles such as SUVs and MPVs than joint venture brands. 2) The absence of non-compliance penalties of CAFC standards during 2012–2016. 3) Strong NEV incentive measures, such as multipliers, electric consumption considered as zero in the FCR standard, and fiscal incentives, resulted in rapid development of the NEV market, which reduced the willingness of local brands to develop fuel-efficient vehicles.

China introduced the CAFC and NEV Credits Regulation in 2017. All carmakers need to comply with specific CAFC requirements, while companies with large-scale production of conventional passenger vehicles need to comply with both CAFC and NEV targets. Under dual credit regulation, NEV credits can balance the negative CAFC credits. The carmakers with negative CAFC or NEV credits that are unbalanced will be not allowed to produce and sell new models that cannot comply with specific 2020 targets in the Phase IV standard (which is a stepped curve based on curb weight). As the deployment of NEV and fuel-efficient technology are two main compliance paths, in the future, we need not only pay attention to the tradeoff between FCR and vehicle attributes, but also the tradeoff between NEV and conventional vehicle technological progress under the newly introduced dual-credit scheme.

Author Contributions: Conceptualization, J.Z., R.Y. and Y.L.; methodology, J.Z., R.Y. and Y.L.; resources, D.Z.; writing—original draft preparation, J.Z., R.Y., Y.L. and Y.Z.; writing—review and editing, J.Z. and R.Y.

Funding: This research was funded by Energy Foundation China, grant number G–1607–24947 and G–1607–24944.

Acknowledgments: The authors would like to thank the anonymous reviewers for their reviews and comments.

Conflicts of Interest: The authors declare no conflict of interest.

Nomenclature

Abbreviations

FCR	Fuel Consumption Rate
NEV	New Energy Vehicle
BEV	Battery Electric Vehicle
PHEV	Plug-in Hybrid Vehicle
FCV	Fuel Cell Vehicle
GDI	Gasoline Direct Injection
PSFI	Performance-Size-Fuel economy Index
PPF	Production Possibilities Frontier
EPA	Environmental Protection Agency
CATARC	China Automotive Technology and Research Center
MPG	Miles Per Gallon
NEDC	New European Driving Cycle
SUVs	Sport Utility Vehicles
MPVs	Multi-Purpose Vehicles
CVT	Continuously Variable Transmission
DCT	Dual-clutch Transmission
AT	Automatic Transmission
MT	Manual Transmission
ERFC	Emphasis on Reducing Fuel Consumption
CAFE	Corporate Average Fuel Economy
CAFC	Corporate Average Fuel Consumption

Symbols

FC_{it}	Fuel consumption rate
CW_{it}	Curb weight
acc_{it}	0 to 100 km/h acceleration time
$\mathbf{X_{it}}$	A vector of dummy variables
$\mathbf{B}$	A vector of coefficients
τ_t	The year fixed effects to estimate the technological progress
μ_i	The difference by vehicle brands
ε_{it}	The random error term
$T_{progress}$	Technological progress

Subscripts

i	Vehicle model
t	Year
base	Base year
cal	The actual fuel consumption rate
potential	Fuel consumption evaluated

References

1. National Bureau of Statistics (NBS), China's 2018 Year National Economic and Social Development of Statistical Bulletin. Available online: http://www.stats.gov.cn/tjsj/zxfb/201902/t20190228_1651265.html (accessed on 10 June 2019).
2. IEA. *CO$_2$ Emissions from Fuel Combustion*; IEA: Paris, France, 2017.
3. CATARC. *Annual Report on Energy-Saving and New Energy Vehicle in China 2018*; Posts and Telecom Press: Beijing, China, 2018.
4. GAQSIQ. *Limits of Fuel Consumption for Passenger Cars*; GAQSIQ: Beijing, China, 2004; Volume GB 19578–2004.
5. GAQSIQ. *Fuel Consumption Evaluation Methods and Targets for Passenger Cars*; GAQSIQ: Beijing, China, 2011; Volume GB 27999–2011.
6. GAQSIQ. *Fuel Consumption Evaluation Methods and Targets for Passenger Cars*; GAQSIQ: Beijing, China, 2014; Volume GB 27999–2014.

7. The Central People's Government of China, Energy-Saving and New Energy Vehicle Industry Development Plan (2012–2020). Available online: http://www.gov.cn/zwgk/2012-07/09/content_2179032.htm (accessed on 10 June 2019).
8. IEA. *CO₂ Emissions from Fuel Combustion 2016 Highlights*; IEA: Paris, France, 2016.
9. U.S. Energy Information Administration (EIA), Vehicle Standards around the World Aim to Improve FUEL Economy and Reduce Emissions. Available online: https://www.eia.gov/todayinenergy/detail.php?id=23572 (accessed on 10 June 2019).
10. Bandivadekar, Z.Y.A. *Light-Duty Vehicle Greenhouse Gas and Fuel Economy Standards*; ICCT: Washington, DC, USA, 2017.
11. Li, W.; Long, R.; Chen, H.; Geng, J. A review of factors influencing consumer intentions to adopt battery electric vehicles. *Renew. Sustain. Energy Rev.* **2017**, *78*, 318–328. [CrossRef]
12. Zou, Y.; Ren, H.; Yu, R. Energy-saving technology application for conventional energy passenger cars in China. *J. Automot. Saf. Energy* **2017**, *8*, 102–108.
13. MacKenzie, D.; Heywood, J.B. Quantifying efficiency technology improvements in US cars from 1975–2009. *Appl. Energy* **2015**, *157*, 918–928. [CrossRef]
14. Hu, K.; Chen, Y. Technological growth of fuel efficiency in european automobile market 1975–2015. *Energy Policy* **2016**, *98*, 142–148. [CrossRef]
15. Kim, H.C.; Wallington, T.J. Life-Cycle Energy and Greenhouse Gas Emission Benefits of Lightweighting in Automobiles: Review and Harmonization. *Environ. Sci. Technol.* **2013**, *47*, 6089–6097. [CrossRef] [PubMed]
16. Pero, F.D.; Delogu, M.; Pierini, M. The effect of lightweighting in automotive LCA perspective: Estimation of mass-induced fuel consumption reduction for gasoline turbocharged vehicles. *J. Clean. Prod.* **2017**, *154*, 566–577. [CrossRef]
17. Delogu, M.; Maltese, S.; Del Pero, F.; Zanchi, L.; Pierini, M.; Bonoli, A. Challenges for modelling and integrating environmental performances in concept design: The case of an automotive component lightweighting. *Int. J. Sustain. Eng.* **2018**, *11*, 135–148. [CrossRef]
18. An, F.; DeCicco, J. Trends in Technical Efficiency Trade-Offs for the U.S. Light Vehicle Fleet. *J. Engines* **2007**, *116*, 859–873.
19. Knittel, C.R. Automobiles on Steroids: Product Attribute Trade-Offs and Technological Progress in the Automobile Sector. *Am. Econ. Rev.* **2011**, *101*, 3368–3399. [CrossRef]
20. Klier, T.H.; Linn, J. The Effect of Vehicle Fuel Economy Standards on Technology Adoption. *J. Public Econ.* **2016**, *133*, 41–63. [CrossRef]
21. EPA. *Light-Duty Automotive Technology, Carbon Dioxide Emissions, and Fuel Economy Trends: 1975 Through 2017*; EPA: Washington, DC, USA, 2018.
22. EEA. *Monitoring CO₂ Emissions from Passenger Cars and Vans in 2016*; EEA: Copenhagen, Denmark, 2017.
23. Zheng, J.; Zhou, Y.; Yu, R.; Zhao, D.; Lu, Z.; Zhang, P. Survival rate of China passenger vehicles: A data-driven approach. *Energy Policy* **2019**, *129*, 587–597. [CrossRef]
24. Bandivadekar, A.; Bodek, K.; Cheah, L.; Evans, C.; Groode, T.; Heywood, J.; Kasseris, E.; Kromer, M.; Weiss, M. *On the Road in 2035: Reducing Transportation's Petroleum Consumption and GHG Emissions*; Massachusetts Institute of Technology: Cambridge, MA, USA, 2008.
25. MIIT Regulation for Corporate Average Fuel Consumption and New Energy Vehicle Credits of Passenger Vehicles. Available online: http://www.miit.gov.cn/n1146290/n4388791/c5826378/content.html (accessed on 10 June 2019).

Article

Direct Rebound Effect for Electricity Consumption of Urban Residents in China Based on the Spatial Spillover Effect

Ying Han, Jianhua Shi *, Yuanfan Yang and Yaxin Wang

School of Business Administration, Northeastern University, Shenyang 110169, China;
yhan@mail.neu.edu.cn (Y.H.); 1601546@stu.neu.edu.cn (Y.Y.); 1601547@stu.neu.edu.cn (Y.W.)
* Correspondence: 1510506@stu.neu.edu.cn

Received: 29 April 2019; Accepted: 27 May 2019; Published: 30 May 2019

Abstract: Based on methods of price decomposition and spatial econometrics, this paper improves the model for calculating the direct energy rebound effect employing the panel data of China's urban residents' electricity consumption for an empirical analysis. Results show that the global spatial correlation of urban residents' electricity consumption has a significant positive value. The direct rebound effect and its spillover effects are 37% and 13%, respectively. Due to the spatial spillover effects, the realization of energy-saving targets in the local region depends on the implementation effect of energy efficiency policies in the surrounding areas. However, the spatial spillover effect is low, and the direct rebound effect induced by the local region is still the dominant factor affecting the implementation of energy efficiency. The direct rebound effect for urban residents' electricity consumption eliminating the spatial spillover effect does not show a significant downward trend. The main reason is that the rapid urbanization process at the current stage has caused a rigid residents' electricity demand and large-scale marginal consumer groups, which offsets the inhibition effect of income growth on the direct rebound effect.

Keywords: energy efficiency; direct energy rebound effect; spatial spillover effect; price decomposition

---

1. Introduction

China has been the world's largest energy consumer since 2009, accounting for 23% of global energy consumption and 27% of global energy consumption growth in 2016 [1]. Meanwhile, China is currently the world's largest emitter of carbon dioxide and sulfur dioxide. The annual economic losses caused by air and water pollution account for 8–12% of GDP [2]. However, China's electricity production is still dominated by coal, and the carbon dioxide emitted by electricity and heat production is more than 50% of the total fuel emissions, which is contrary to the current economic transformation goals. Among all terminal electricity consumption, residential electricity consumption accounts for a relatively large proportion, and due to the energy substitution policy effect, electric energy substitution will further expand residents' electricity consumption. Therefore, it is necessary to control residents' electricity consumption. Although improving the electricity efficiency through technical means can inhibit the increase of household electricity consumption to a certain extent, improving energy efficiency will induce a direct rebound effect, and its negative effect on energy conservation and emission reduction cannot be ignored [3,4]. In view of the spatial agglomeration of electricity consumption in China, there are two problems that need to be studied in depth: Is there a spatial spillover effect of the direct rebound effect of residential electricity consumption? How to distinguish the direct rebound effect and its spatial spillover effect if there is a spatial spillover effect, so as to accurately and comprehensively examine the magnitude and trend of the direct rebound effect of residential electricity.

1.1. Types of the Energy Rebound Effect

The energy rebound effect can be divided into four categories [5]:

- The same consumer for the same goods or services;
- Different consumers for the same goods or services;
- The same consumer for different goods or services;
- Different consumers for different goods or services.

The first two categories correspond to the direct rebound effect; the latter two categories belong to the indirect rebound effect, and the macroeconomic rebound effect covers all of the types above [6]. In general, the estimation of direct rebound effect follows the "bottom-up" principle and examines the change of individual consumption patterns. However, the estimation of the macroeconomic rebound effect follows the "top-down" principle, which examines the change of total energy consumption without paying attention to the decomposition of the total energy consumption [7]. Some studies hold the view that if the direct and indirect rebound effects can be identified and calculated separately, and the macroeconomic effects are the sum of the two effects, but others have the opposite view that the macroeconomic rebound effect is different from the direct and indirect rebound effects [8–10]. The main economic mechanism of the macroeconomic rebound effect is composed of the economic growth effect [11] and the change effect [12]. The former refers to the technological progress in promoting economic growth, and in turn it results in increased energy consumption. The latter means that the technical progress can change consumer preferences and the industry, so energy consumption is also changed. In recent years, the study suggests that in addition to the secondary effects (indirect effect), the indirect effect also contains an implicit effect (a so called embedded effect). For example, although the consumer does not directly increase energy consumption with the increase of real income, they may increase their consumption of other goods or services. The process of production and transportation of these goods or services will consume energy, so the energy consumption increase is embedded in the non-energy goods and services [12,13].

In fact, the direct effect is the basis of the indirect effect and the macroeconomic rebound effect. The indirect effect is even considered to be a part of the direct effect in some studies [14,15]. For example, if the direct effect is 30%, the average direct and indirect rebound effect (DIRE) of the European Union's 27 countries is 73.6%. If the direct effect is 50%, the average DIRE is 81.16% [16], so restraining direct rebound effect is the foundation of restraining indirect and macroeconomic rebound effect.

1.2. Evidences of the Direct Rebound Effect

The existing empirical studies cover personal passenger transport [17–22], household heating [23,24] or other household services [25–30]. However, based on the data from different regions, or different energy services, the results are controversial. The direct rebound effect of developed countries is no more than 40%, meaning that improving energy efficiency will reduce energy consumption, and only a part of the expected savings is offset [31]. However, the direct rebound effect of developing countries is extremely serious, sometimes even exceeding 100% [27]. The income gap may be the main reason behind the difference between different regions [12]. Residents in developed countries have higher income and tend to demand saturation [32], so the energy consumption induced by the improvement of energy efficiency will decrease, and the magnitude of the direct rebound effect is smaller than that in developing countries. The energy demand in developing countries is far from saturated [6], so income growth may not inhibit the direct rebound effect in developing countries in the short term.

What's the magnitude of China's direct rebound effect? Taking residents' electricity consumption as an example, the direct rebound effect for urban residents' electricity consumption is less than 100% [18]. However, it may rise up to 165.22%, mainly due to "marginal consumer groups" [27]. If the heterogeneity of urban and rural direct rebound effect is ignored, the direct rebound effect for residents' electricity consumption would have a threshold effect based on per capita income [33]. With the steady growth of per capita income, the magnitude of direct rebound effect tends to decrease. To sum up, the magnitude and the change of direct rebound effect for China's residents' electricity consumption are still controversial.

There are three reasons that cause the difference mentioned above. First, the electricity consumption of urban residents in China is much larger than that of rural residents, which leads to heterogeneity between urban and rural residents. Taking the two groups as a whole to avoid differences will result in inaccurate results. Second, the effect of power price on residents' electricity consumption between price increase periods and price decline periods is not completely reversible [34]. The calculation result of the direct rebound effect for residential electricity consumption without price decomposition will be different between the two periods. Third, the definition of direct rebound effect given by Berkout et al. [3] and Greening et al. [35] which implies the assumption that energy consumption among regions is independent, is the basis of the empirical studies above. However, Tobler's First Law of Geography shows that everything is related to everything else, but near things are more related to each other. China' economic development and energy consumption have obvious clustering properties in geospatial space. Therefore, the spatial spillover effect cannot be ignored when the direct rebound effect is explored. In essence, economic activities cause widespread connections between regions [36]. The aggregation of users may improve the energy efficiency of users' communities; for instance, shared-use of common resources [37] and demand side management participation though an aggregator [38]. Users or local governments that actively cooperate for a common goal of reducing energy consumption may be one of the reasons for spatial aggregation. The improvement of electricity efficiency in a local area will affect not only the residents' electricity consumption in the local region, but also the residents' electricity consumption in neighboring areas, so the direct rebound effect will spill over between regions. Ignoring the spatial dependence will confuse the direct rebound effect and its spatial spillover effect, leading to incorrect results.

In view of this, the main contributions of this paper are in the following aspects. First, based on the perspective of spatial spillover, the measurement model of direct rebound effect is improved, so that the direct rebound effect can be measured more accurately and comprehensively. Second, considering the asymmetric influence of price on demand and the heterogeneity of the direct rebound effect between urban and rural areas, the spatial panel data of urban residents are used for empirical test, and multiple price decomposition models are introduced to ensure the robustness of the results. Finally, the trend of the direct rebound effect on urban residents' electricity consumption is examined. The research results have important reference to the realization of energy savings and emission reduction targets.

2. The Improved Method of Calculating Direct Rebound Effect

Improving energy efficiency will decrease energy consumption with the same level of energy service. For instance, if the rate of electricity use and cooling area decrease, residents will use less electricity cooling the same area. However, improving efficiency means a decrease in real power price, which will incentivize residents to use more electricity in turn. Calculating the direct rebound effect is to calculate the gap between the expected savings and the actual savings. Direct rebound effect is then defined as: Direct rebound = (expected savings − actual savings)/expected savings. The traditional calculation method of the direct rebound effect controls no other variables, so some studies recommend that the price elasticity could be an ideal proxy indicator of direct rebound effect with other variables controlled [39]. The definition and identification of direct rebound effect can be found in Appendix A.

The calculation method above only analyzes the energy consumed by the same consumers (consumers in the local region) and does not consider other consumers (consumers in the adjacent region). It implies the assumption that energy consumption in different regions is independent. However, if there is a spatial "convergence effect" in energy consumption, whereby the increased energy efficiency in a local region will have an influence on the energy consumption not only in the local region, but also in the adjacent regions, leading to a "spatial feedback effect". For the same reason, the improvement of energy efficiency in the adjacent region also induces a direct rebound effect in the local region. This paper views this as the spatial spillover effect of the direct rebound effect. It is impossible to distinguish whether the additional energy consumption in the local region is caused by the energy efficiency improvement in the local region or in the adjacent region without considering spatial spillover effect.

Based on spatial spillover effect, this paper improves the calculating model of direct rebound effect. The spatial lag of electricity consumption is introduced into the model, and the spatial lag model (SLM) controlling other variables is:

$$y_t = \lambda W y_t + X_t \beta + c + u_t \tag{1}$$

where both the explained variable and the explanatory variable are logarithmized. y_t is the urban residents' electricity consumption of n regions in year t. W is the space weight matrix, and $W y_t$ is the spatial lag of y_t. λ measures the effect of spatial lag $W y_t$ on y_t, reflecting spatial dependence. X_t is the explanatory variables matrix of n regions in year t. β is the coefficient of the explanatory variable. c is the individual effect of n regions. According to the individual effect, the model can be divided into fixed effect model and random effect model.

If the spatial correlation of urban residents' electricity consumption is not considered, Equation (1) is reduced to a standard static panel model.

Rewrite Equation (1) as a reduced form:

$$y_t = (I - \lambda W)^{-1}(X_t \beta + c + u_t) \tag{2}$$

$$E(y_t | X_t, W) = (I - \lambda W)^{-1}(X_t \beta) \tag{3}$$

Equation (3) shows that measuring the direct rebound effect should consider the spatial feedback effect. According to the research on direct and indirect effects in spatial econometric models by LeSage and Pace [40], the calculation of the direct rebound effect in the space lag fixed effect model is:

$$RE = -\frac{1}{nT} \sum_{t=1}^{T} \sum_{i=1}^{n} \frac{\partial E(\hat{y}_{it} | X_t, W)}{\partial \ln P_i} \tag{4}$$

where $\hat{y}_t = y_t - (I - \lambda W)^{-1} c$. The direct rebound effect calculated here is the average value of the direct rebound effect of n regions, so it can be called the average direct rebound effect (abbreviated as RE).

The calculation of the space spillover effect of direct rebound effect is defined as:

$$SRE = -\frac{1}{nT(n-1)} \sum_{t=1}^{T} \sum_{i=1}^{n} \sum_{j=1, j \neq i}^{n} \frac{\partial E(\hat{y}_{it} | X_t, W)}{\partial \ln P_j} \tag{5}$$

Equation (5) also calculates the average spatial spillover effects of n regions, so it can be called the average spatial spillover effect (abbreviated as SRE).

The spatial lag model only considers the endogenous interaction effects, ignoring the spatial correlation of unobservable random impacts. In the case of multiple spatial interactions, a more appropriate method is to use the spatial autoregressive model with spatial autoregressive disturbances (SARAR) model. Since urban residents' electricity consumption may have both the interaction of endogenous interaction and error terms, the SARAR model is introduced to measure the direct rebound effect and its spatial spillover effect:

$$y_t = \lambda W y_t + X_t \beta + c + \varepsilon_t, \varepsilon_t = \rho W \varepsilon_t + \varphi_t \tag{6}$$

where ρ is the coefficient of spatial lag $W \varepsilon_t$. The calculation of the average direct rebound effect and the average spatial spillover effect of the SARAR model is consistent with the SLM model.

The process of using spatial econometric models is as follows: before establishing a spatial econometric model, it is necessary to test the spatial autocorrelation and heterogeneity in the data, using two types of spatial autocorrelation test: local autocorrelation and global autocorrelation. Then multiple models are set up. For nested models, a likelihood ratio (LR) test can be used to choose the best model. Due to endogeneity problems, the ordinary least squares (OLS) estimators are inconsistent. This paper uses the maximum likelihood estimator method (MLE) to get the consistent estimator. For panel data, individual effects need to be tested, so the Hausman test is used to select the appropriate model between the fixed effect and random effect models. And we use the method proposed by Lee and Yu [41] to estimate the panel spatial econometric model. Firstly, the individual effects are eliminated, then the maximum likelihood estimator method is performed.

3. Variables and Data Description

3.1. Variables Selection

Electricity consumption (y). The electricity consumption is an endogenous variable, which is measured by the electricity consumption of urban residents.

Power price (P). The power price is the core explanatory variable, which is measured by the average selling price of electricity used by residents. The power price has both rising and falling periods, and the impact of rising and falling price on the demand for electricity is not completely reversible. However, the direct rebound effect is mainly related to the falling price. So the power price is decomposed into three parts [4]:

$$P_{it} = P_{max,it} \times P_{rec,it} \times P_{cut,it} \tag{7}$$

where P_{it}, $P_{max,it}$, $P_{rec,it}$ and $P_{cut,it}$ represent the actual price, maximum price, cumulative rising price and cumulative falling price in province i in year t, respectively. The decomposed price is calculated as follows:

$$P_{max,it} = \max\{P_{i1}, P_{i2}, \cdots, P_{it}\} \tag{8}$$

$$P_{rec,it} = \prod_{j=0}^{t} \max\left\{1, \frac{P_{max,ij-1}/P_{ij-1}}{P_{max,ij}/P_{ij}}\right\} \tag{9}$$

$$P_{cut,it} = \prod_{j=0}^{t} \min\left\{1, \frac{P_{max,ij-1}/P_{ij-1}}{P_{max,ij}/P_{ij}}\right\} \tag{10}$$

The power price is also decomposed into two parts [28]:

$$P_{inc,it} = P_{max,it} \times P_{rec,it} \tag{11}$$

$$P_{dec,it} = P_{max,it} \times P_{cut,it} \tag{12}$$

The two decomposition methods are both used for a robust test.

Degree day (DD). The degree day, referring to the deviation between the daily average temperature and the base temperature, is an environmental factor that should be controlled. It reflects the climate characteristics. Urban residents will use household appliances such as air conditioners more frequently with high degree days, so the electricity consumption is larger. Degree days are divided into heating degree days (HDD) and cooling degree days (CDD), and their calculation is as follows [28]:

$$HDD = \sum_{m=1}^{12} (1 - rd)(T_{b1} - T_m) \times M \tag{13}$$

$$CDD = \sum_{m=1}^{12} rd(T_m - T_{b2}) \times M \tag{14}$$

where *HDD* and *CDD* are the heating degree day value and the cooling degree day value. T_m is the monthly average temperature. T_{b1} and T_{b2} represent the base temperature of the heating degree day and the cooling degree day, respectively. *rd* is a dummy variable, and if the monthly average temperature is higher than the base temperature, it is 1. Then, $DD = HDD + CDD$.

Income (*I*). Income is an economic factor that should be controlled, which is measured by the per capita disposable income of urban residents. Income is an important factor affecting consumer spending. Since 2006, urban residents' income has been increasing with a high rate.

Population (*POP*). Population is measured by the number of permanent residents of urban residents. Obviously, the more people, the greater electricity consumption. In order to accurately measure the increase in electricity consumption induced by efficiency, it is necessary to control the population factor.

3.2. Data Sources

The data sample is from 2006 to 2016, and includes 29 provincial units. Data come from the China energy statistical yearbook (2007–2017), China electricity yearbook (2007–2017), China statistical yearbook (2007–2017), China national bureau of statistics website and the Wind database. The power price and the income are converted into constant prices based on 2006. Table 1 presents correlation coefficients between main variables as well as their means and standard deviations. We take logarithms of all variables to reduce heteroscedasticity.

Table 1. Statistical description and correlation table.

Variable	Mean	St.d	(1)	(2)	(3)	(4)	(5)
(1) *E*	105.4	74.88	1.000				
(2) *P*	432.2	63.41	0.545	1.000			
(3) *I*	170.5	60.85	0.579	0.252	1.000		
(4) *POP*	232.3	142.5	0.895	0.577	0.330	1.000	
(5) *DD*	105.4	33.01	−0.393	−0.314	−0.165	−0.346	1.000

Table 1 shows that the average of China's urban residents' electricity consumption and power price are 10.54 billion kilowatt hours and 0.432 yuan per kilowatt hours. It reflects that the cost of electricity consumption is very small, leading to rapid electricity demand growth. Thus, the magnitude of the direct rebound effect for urban residents' electricity consumption in China may be larger than in other countries whose power price is higher than in China. The standard deviation of all variables is small, meaning that all variables are distributed very evenly. The correlation coefficients show that there is no serious collinearity between variables.

4. Empirical Analysis

4.1. Analysis of Results of Static Panel Model

In order to compare with the calculation results of SARAR and SLM model, the static panel model is used. Table 2 shows the static panel model estimation results. Hausman test results reject the null hypothesis at the 1% level, meaning that the individual fixed effect model is superior to the individual random effect model. The fixed effect estimation results show that the direct rebound effect is 39.0%, which means that 39.0% of the electricity consumption of urban residents saved by improving electricity efficiency is offset by the direct rebound effect. Actually only 61.0% of the expected savings can be achieved. The fixed effect estimation results also show that population, income and degree day have significant effects on the electricity consumption of urban residents. Then, we use improved models to analyze the direct rebound effect and its spatial spillover effect.

Table 2. Estimation results of static panel model.

Variable	Fixed Effect	Random Effect
$\ln P_{inc}$	0.592 *** (0.007)	0.601 ** (0.005)
$\ln P_{dec}$	−0.390 * (0.066)	−0.455 ** (0.033)
$\ln DD$	0.344 *** (0.002)	0.025 (0.740)
$\ln POP$	0.917 *** (0.000)	0.911 *** (0.000)
$\ln I$	0.711 *** (0.000)	0.718 *** (0.000)
R^2	0.882	0.878
Hausman test	19.450 *** (0.004)	

Note: The number in parentheses is the level of significance. ***, **, and * indicate significance levels at 1%, 5%, and 10%, respectively.

4.2. Spatial Correlation Test

Before applying the spatial econometric model, it is necessary to analyze the local and global spatial correlation to test whether there is spatial dependence in the urban residents' electricity consumption between regions. First, the local correlation types are analyzed. Although the Moran scatter plot can infer spatial correlation to some extent, the Moran scatter plot cannot determine whether the local correlation type is statistically significant. So the local indicators of spatial association (LISA) map is used to analyze the local spatial autocorrelation. Figures 1 and 2 show the LISA maps of China's urban residents' electricity consumption in 2006 and 2016, respectively.

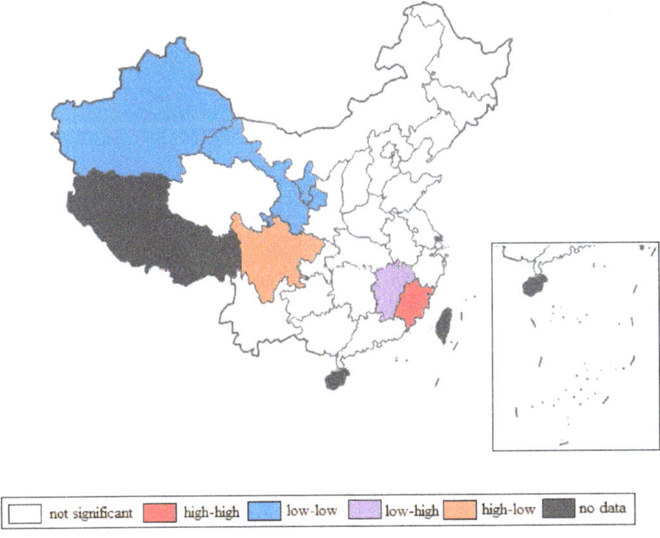

Figure 1. Spatial aggregation of urban residents' electricity consumption in 2006.

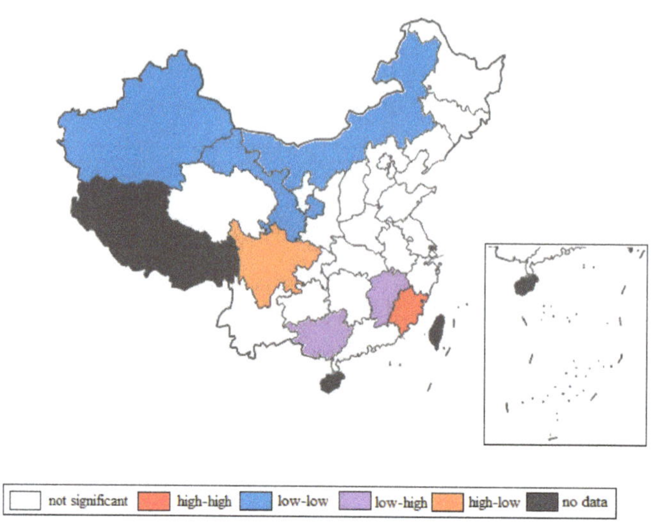

Figure 2. Spatial aggregation of urban residents' electricity consumption in 2016.

The LISA maps show that there are four types of spatial agglomeration in China's urban residents' electricity consumption, and there is little change in local correlation patterns over time. From the perspective of aggregation effect, various types of spatial aggregation reflect the spatial heterogeneity of urban residents' electricity consumption. In terms of time, although the provinces with low-low aggregation and low-high aggregation have a small increase, it does not show a significant leap (for example, high-high to low-low), indicating that the spatial aggregation in urban residents' electricity consumption is stable.

The local spatial aggregation in urban residents' electricity consumption indicates that the spatial dependence cannot be ignored when the direct rebound effect is examined. Table 3 lists the test results of Moran's *I* index of urban residents' electricity consumption, in order to judge the global correlation in urban residents' electricity consumption.

Table 3. Spatial autocorrelation test.

Year	Moran's *I*	Z	P
2007	0.210	2.186	0.020
2009	0.235	2.366	0.013
2011	0.201	2.058	0.019
2013	0.252	2.471	0.010
2015	0.187	1.999	0.028

Table 3 shows that there is a significant spatial autocorrelation in urban residents' electricity consumption, and the spatial correlation is positive, indicating that the urban residents' electricity consumption mainly reflects convergence effect.

4.3. Analysis of Estimation Results of SARAR and SLM Models

Table 4 displays the estimation results of the SLM and the SARAR model and the robust test results. Hausman test results of the SLM and the SARAR model reject the null hypothesis at the 1% level, meaning that the individual fixed effect model is superior to the individual random effect model. The SARAR fixed effect model has a larger log likelihood value than the SLE fixed effect model. The statistic of the LR test for the SLM model and the SARAR model is 7.448, rejecting the null hypothesis at the 1% level, showing that the SARAR fixed effect model is better than the SLM fixed effect model.

Then the direct rebound effect for residents' electricity consumption and its spatial spillover effect are calculated based on SARAR model estimation results.

Table 4. Estimation results of SLM and SARAR model and robust test.

Variable	SLM		SARAR		Robust Test
	Fixed Effect	**Random Effect**	**Fixed Effect**	**Random Effect**	
Wy	0.275 *** (0.000)	0.037 (0.156)	0.317 *** (0.000)	0.036 (0.233)	0.318 *** (0.000)
$W\varepsilon$	-	-	−0.363 *** (0.006)	0.017 (0.912)	−0.363 *** (0.006)
$\ln P_{inc}$	0.464 ** (0.023)	0.583 *** (0.006)	0.525 *** (0.003)	0.578 *** (0.008)	-
$\ln P_{dec}$	−0.355 * (0.074)	−0.422 ** (0.042)	−0.363 ** (0.034)	−0.422 ** (0.043)	-
$\ln P_{max}$	-	-	-	-	0.158 (0.321)
$\ln P_{rec}$	-	-	-	-	0.538 * (0.053)
$\ln P_{cut}$	-	-	-	-	−0.361 ** (0.036)
$\ln DD$	0.305 *** (0.003)	0.093 (0.331)	0.326 *** (0.000)	0.086 (0.468)	0.326 *** (0.000)
$\ln POP$	0.691 *** (0.000)	0.888 *** (0.000)	0.661 *** (0.000)	0.888 *** (0.000)	0.661 *** (0.000)
$\ln I$	0.541 *** (0.000)	0.692 *** (0.000)	0.508 *** (0.000)	0.693 *** (0.000)	0.508 *** (0.000)
Log likelihood	243.326	205.289	247.050	205.295	247.052
Hausman test	63.51 *** (0.000)		215.16 *** (0.000)		-

Note: The number in parentheses is the level of significance. ***, **, and * indicate significance levels at 1%, 5%, and 10%, respectively.

All the variable coefficients in the SARAR fixed effect models are significant. However, the absolute value of all variable coefficients in the SARAR fixed effect model is lower than that in static panel fixed effect model, indicating that ignoring the spatial correlation will overestimate the influence of these variables on electricity consumption.

This is because residents' electricity consumption in the local region is affected not only by power price, population and per capita income in the local region, but also by the positive impact of the spatial lag of residents' electricity consumption. The static panel model classifies the positive impact of spatial lag on residents' electricity consumption into other explanatory variables. So, the contribution of these explanatory variables is exaggerated.

4.4. Analysis of RE and SRE

In the SARAR fixed effect model, due to the existence of spatial lag, the spatial feedback effect should be considered to measure the direct rebound effect. Combined with Equation (4), the average direct rebound effect is 37.00%, indicating that improving the electricity efficiency does induce a direct rebound effect. However, the direct rebound effect for urban residents' electricity consumption is much lower than 100%, and is lower than that of the static panel model. This means that the direct rebound effect value is reduced after considering the spatial correlation. Increasing the efficiency of electricity consumption will ultimately reduce the urban residents' electricity consumption. 37% of the expected savings are offset, and 63% of the expected targets can be achieved actually. So, improving the efficiency plays an important role in reducing the urban residents' electricity consumption. Table 4 also shows that in addition to the decline in power price, the growth of population, per capita income and degree day value will also increase the urban residents' electricity consumption, especially when the inter-regional urban residents' electricity consumption has a mutual pulling effect. When the government measures the restraining effect of electricity efficiency on residents' electricity consumption, the factors above should be controlled. Otherwise, the direct rebound effect for residents' electricity consumption will be overestimated, and the inhibition effect of improving efficiency on electricity conservation will be underestimated.

The spatial spillover effect of direct rebound effect for urban residents' electricity consumption can be calculated and tested by using Equation (5). The test results confirm that the direct rebound effect for urban residents' electricity consumption has a significant spatial spillover effect at 1% level, and the spatial spillover effect is 13.30%. That is to say, per 1% decrease in power price due to the increased efficiency in adjacent areas will increase the urban residents' electricity consumption in the local region by 0.133%. Adding RE and SRE together, the total electricity consumption induced by the increased efficiency is 50.30%. The proportion of RE is 73.56%, and the proportion of SRE is 26.44%.

The calculating results above show that if the spatial dependence in urban residents' electricity consumption is not considered, the direct rebound effect and its spatial spillover effect will be confused. Due to the spatial spillover effect, the realization of energy-saving targets in local area depends on the implementation effect of energy efficiency in surrounding areas. Moreover, due to the low spatial spillover effect, direct rebound effect induced by efficiency improvement in the local region is still the main reason affecting the implementation effect of energy efficiency policies in the local region.

4.5. Robust Test

In addition to the two-part decomposition method adopted above, some studies also adopt a three-part decomposition method. Then the three-part decomposition method is used for a robustness test, shown in the last column of Table 4. The results of the robustness test are consistent with the empirical results above, indicating that the direct rebound effect measurement value for urban residents' electricity consumption is not sensitive to the price decomposition methods.

4.6. Analysis of the Temporal Change of Direct Rebound Effect

In order to investigate the change of direct rebound effect for urban residents' electricity consumption, the coefficient of $\ln P_{dec,it}$ is allowed to change with time. The estimated results are shown in Table 5.

Table 5. Estimation results of SARAR fixed-effect model with partial variable coefficients.

Variable	Coefficient	Variable	Coefficient
Wy	0.318 *** (0.000)	$\ln P_{dec}\_2009$	−0.336 * (0.098)
$W\varepsilon$	−0.419 *** (0.002)	$\ln P_{dec}\_2010$	−0.332 (0.105)
$\ln P_{inc}$	0.472 ** (0.033)	$\ln P_{dec}\_2011$	−0.328 (0.109)
$\ln DD$	0.330 *** (0.004)	$\ln P_{dec}\_2012$	−0.322 (0.115)
$\ln POP$	0.539 *** (0.001)	$\ln P_{dec}\_2013$	−0.315 (0.123)
$\ln I$	0.166 (0.340)	$\ln P_{dec}\_2014$	−0.312 (0.126)
$\ln P_{dec}\_2006$	−0.356 * (0.078)	$\ln P_{dec}\_2015$	−0.310 (0.130)
$\ln P_{dec}\_2007$	−0.347 * (0.086)	$\ln P_{dec}\_2016$	−0.309 (0.133)
$\ln P_{dec}\_2008$	−0.343 * (0.091)	Log Likelihood	250.522

Note: The number in parentheses is the level of significance. ***, **, and * indicate significance levels at 1%, 5%, and 10%, respectively.

According to Table 5, the calculating results of direct rebound effect for urban residents' electricity consumption in some years are shown in Table 6.

Table 6. Calculation results of direct rebound effect for urban residents' electricity consumption.

Year	2007	2009	2011	2013	2015
RE	35.4% * (0.086)	34.2% * (0.098)	33.4% (0.109)	32.1% (0.123)	31.6% (0.130)

Note: The number in parentheses is the level of significance. * indicates significance levels at 10%.

Table 6 shows that direct rebound effect for urban residents' electricity consumption declined from 2006 to 2009, but the decline is very small. The calculation results of direct rebound effect after 2009 are not significant, indicating that there is no obvious downward trend in direct rebound effect in the short term. The changes of RE, SRE and the total effect (abbreviated as TE) are displayed in Figure 3.

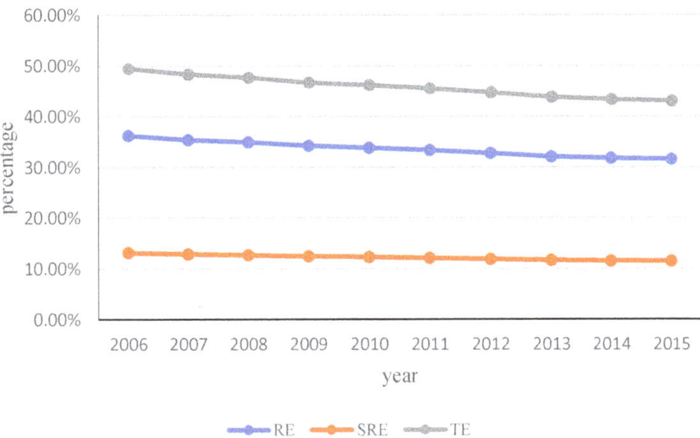

Figure 3. The changes of RE, SRE and TE from 2006 to 2015.

Figure 3 shows that the change characteristics of the three effects are similar, and the decrease range is small. In order to verify this conclusion, the significance of the power price and time interaction term are tested. The test results still cannot reject the null hypothesis at 10% significance level, meaning that the direct rebound effect is fixed over these years, so the direct rebound effect for urban residents' electricity consumption will not decrease currently.

According to Zhang et al. [33], consumers' energy demand tends to be saturated with income growth, and direct rebound effect will decline. However, the empirical test in this paper shows that direct rebound effect for urban residents' electricity consumption in China has not shown a significant downward trend although the urban residents' income has been increasing. The main reason is that China's urbanization rate increased by 1.31% annually from 2006 to 2016, indicating that China's urbanization is large and the process is relatively fast. It has caused the rigidity of electricity demand. In particular, the transfer of rural residents to urban areas will bring a large-scale marginal consumer group. Therefore, the rigidity of electricity demand and the large marginal consumer group will eventually offset the inhibition effect of income growth on the direct rebound effect.

5. Conclusions and Policy Implications

Based on price decomposition methods and spatial econometric models, the calculation method of the direct rebound effect is improved. The panel data of China's urban residents' electricity consumption are used for our empirical analysis. The conclusions are as follows:

First, spatial analysis indicates that there are four types of spatial aggregation in China's urban residents' electricity consumption, and the global spatial correlation has a significant positive value. Studies of the direct rebound effect for urban residents' electricity consumption should not ignore the spatial feedback effect and spatial spillover effect. The improved model can subdivide the calculation results into direct rebound effect and its spatial spillover effect, improving the accuracy and explanatory power of the results. In addition, due to the asymmetric influence of price on demand, the introduction of the price decomposition methods can avoid the upward bias of the calculation results to some extent.

Second, the direct rebound effect for urban residents' electricity consumption in China and its spatial spillover effect are 37.00% and 13.30%, respectively. This shows that although improving the electricity efficiency has induced a direct rebound effect, the direct rebound effect is not serious, and improving efficiency is still an important measure to curb the urban residents' electricity consumption. Moreover, compared with the spatial spillover effect of direct rebound effect, direct rebound effect induced by energy efficiency improvement in the local region is still the main factor affecting the implementation effect of energy efficiency policy in the same region.

Third, direct rebound effect for urban residents' electricity consumption without spatial spillover effects does not show a significant downward trend. The reason is that the rapid urbanization process at the current stage has caused rigid residents' electricity demand and large-scale marginal consumer groups, which offsets the inhibition effect of income growth on the direct rebound effect.

According to the analysis above, the main policy implications are as follows: first, the government must attach importance to the direct rebound effect, and establish a comprehensive, multi-sectoral monitoring system for direct rebound effect, so as to avoid failure of energy efficiency policy caused by serious direct rebound effect. Second, the direct rebound effect is mainly caused by the price effect. The government should promote the marketization of power prices through environmental regulations (such as resource taxes), and reduce the excessive consumption of electricity due to low cost. At the same time, in order to achieve the expected energy-saving goals of energy efficiency policies more effectively, local governments should focus on the synergy of policy formulation and implementation between the local region and adjacent areas.

Author Contributions: Data curation, Y.H.; Formal analysis, J.S.; Funding acquisition, Y.H. and J.S.; Investigation, Y.W.; Methodology, J.S.; Software, J.S. and Y.Y.; Writing—original draft, J.S.; Writing—Review & Editing, J.S.

Funding: This work was funded by the General Planning Project of the Social Science Fund of the Ministry of Education (18YJA790031), the Major projects of the National Social Science Fund (15ZDC034) and the Liaoning Natural Science Foundation (201602267).

Conflicts of Interest: The authors declare no conflict of interest.

Appendix A

There are some equivalent definitions of direct rebound effect, which allows identification of the rebound effect

Firstly, we define the energy efficiency. Energy efficiency at the household level can be expressed as the ratio of energy services to energy inputs:

$$\varepsilon = S/E \tag{A1}$$

where ε, S, and E denote the energy efficiency, energy services and energy inputs, respectively.

Definition A1.

$$\text{Direct rebound} = (\text{expected savings} - \text{actual savings})/\text{expected savings} \tag{A2}$$

- If the energy efficiency improvement does not lead to an increase in energy services, the actual savings are equal to the expected savings, so the direct rebound effect is equal to zero.
- However, energy efficiency improvement means that real energy service cost is reduced, and consumers will increase energy services. Therefore, the actual savings are less than the expected savings, and the direct rebound effect is greater than zero.
- If the increase of energy service caused by the decrease of real energy service cost is greater than the expected savings, the actual savings are less than zero, and the direct rebound effect is greater than 100%, which is called backfire effect.

Since it is difficult to distinguish between actual savings and expected savings, Definition A1 is rarely used for empirical research.

Definition A2.
$$\mathrm{DR} = 1 + \eta_\varepsilon^E \tag{A3}$$

where DR represents direct rebound and η_ε^E represents the elasticity of energy demand with respect to efficiency. Definitions A1 and A2 are equivalent, and we explain in detail below.

Equation (A1) can be rewritten as:
$$S = \varepsilon E \tag{A4}$$

Total differentiation of Equation (A4) after applying natural logarithms is:
$$\frac{\mathrm{d}S}{S} = \frac{\mathrm{d}E}{E} + \frac{\mathrm{d}\varepsilon}{\varepsilon} \tag{A5}$$

If energy efficiency improvement does not result in an increase in energy services, then $\mathrm{d}S = 0$. Equation (A5) is simplified to:
$$\frac{\mathrm{d}E}{E} = -\frac{\mathrm{d}\varepsilon}{\varepsilon} \tag{A6}$$

then:
$$\eta_\varepsilon^E = \frac{\mathrm{d}E}{\mathrm{d}\varepsilon}\frac{\varepsilon}{E} = -1 \tag{A7}$$

- The above analysis shows that if energy efficiency improvement does not lead to an increase in energy services, the proportion of energy demand reduction is the same as the proportion of energy efficiency improvement. That is to say, $\eta_\varepsilon^E = -1$. So, the direct rebound effect is equal to zero.
- However, energy efficiency improvement means that real energy service cost is reduced, and consumers will increase energy services. Therefore, $\mathrm{d}S > 0$ and $\eta_\varepsilon^E > -1$. So the direct rebound effect is greater than zero.
- If the increase of energy service caused by the decrease of real energy service cost is greater than the expected savings, $\eta_\varepsilon^E > 0$, and the direct rebound effect is greater than 100%, which is called backfire effect.

Energy efficiency at the household level is often unobservable, so Definition A2 is also rarely used for empirical research.

Definition A3.
$$\mathrm{DR} = -\eta_{P_E}^E \tag{A8}$$

where $\eta_{P_E}^E$ represents the elasticity of energy demand with respect to energy price.

Equation (A8) is equivalent to Equation (A3), and the derivation process of Equation (A8) will be described in detail below.

According to Equation (A1), the relationship between energy service price (or energy service cost) and energy price (or energy cost) is:

$$P_S = P_E / \varepsilon \tag{A9}$$

where P_S represents energy service price and P_E represents energy price.

Combined with Equations (A1), and (A9), Equation (A3) can be rewritten as:

$$
\begin{aligned}
\text{DR} \quad &= 1 + \eta_{\varepsilon}^{E} = 1 + \frac{\partial \ln E}{\partial \ln \varepsilon} = 1 + \frac{\partial \ln(S/\varepsilon)}{\partial \ln \varepsilon} = 1 + \left(\frac{\partial \ln S}{\partial \ln \varepsilon} - 1 \right) \\
&= \frac{\partial \ln S}{\partial \ln P_S} \frac{\partial \ln P_S}{\partial \ln \varepsilon} = \frac{\partial \ln S}{\partial \ln P_S} \frac{\partial \ln(P_E/\varepsilon)}{\partial \ln \varepsilon} \\
&= \frac{\partial \ln S}{\partial \ln P_S} \left(\frac{\partial \ln P_E}{\partial \ln \varepsilon} - 1 \right)
\end{aligned}
\tag{A10}
$$

Because nominal energy prices are not affected by energy efficiency, therefore $\partial \ln P_E / \partial \ln \varepsilon = 0$. Equation (A10) is simplified to:

$$\text{DR} = -\frac{\partial \ln S}{\partial \ln P_S} = -\frac{\partial S}{\partial P_S} \frac{P_S}{S} = -\eta_{P_S}^{S} \tag{A11}$$

where $\eta_{P_S}^{S}$ represents the elasticity of energy service demand with respect to energy service price.

Combined with Equations (A1), and (A9), $\eta_{P_S}^{S}$ can be rewritten as:

$$
\begin{aligned}
\eta_{P_S}^{S} \quad &= \frac{\partial \ln S}{\partial \ln P_S} = \frac{\partial \ln S}{\partial \ln P_E} \frac{\partial \ln P_E}{\partial \ln P_S} = \frac{\partial \ln(\varepsilon E)}{\partial \ln P_E} \frac{\partial \ln(\varepsilon P_S)}{\partial \ln P_S} \\
&= \left(\frac{\partial \ln \varepsilon}{\partial \ln P_E} + \frac{\partial \ln E}{\partial \ln P_E} \right) \left(\frac{\partial \ln \varepsilon}{\partial \ln P_S} + \frac{\partial \ln P_S}{\partial \ln P_S} \right)
\end{aligned}
\tag{A12}
$$

Assuming that energy efficiency is exogenous, then $\frac{\partial \ln \varepsilon}{\partial \ln P_E} = 0$ and $\frac{\partial \ln \varepsilon}{\partial \ln P_S} = 0$. Equation (A12) is simplified to:

$$\eta_{P_S}^{S} = \left(0 + \frac{\partial \ln E}{\partial \ln P_E} \right)(0 + 1) = \frac{\partial \ln E}{\partial \ln P_E} = \eta_{P_E}^{E} \tag{A13}$$

The electricity efficiency here mainly refers to energy efficiency ratio of household appliances, which is usually determined by the technical level of the manufacturer. Consumers can only improve the utilization efficiency of household appliances. In fact, some researches point out that higher efficiency may only be achieved by purchasing more expensive new equipment in China, so the electricity efficiency is exogenous.

Combined with Equations (A10), (A11) and (A13), Equation (A3) can be rewritten as:

$$\text{DR} = 1 + \eta_{\varepsilon}^{E} = -\eta_{P_E}^{E} \tag{A14}$$

Equation (A14) indicates that the price elasticity could be an ideal proxy indicator of direct rebound effect with other variables controlled.

Definition A3 allows identification of the rebound effect. Due to the ease of data acquisition, most empirical studies have adopted Definition A3.

The above analysis shows that the direct rebound effect is mainly related to the falling of real energy price induced by improvement in energy efficiency. However, the power price has both rising and falling periods in the real economy, and the impact of rising and falling price on the electricity demand is not completely reversible. Generally, price elasticity during price rise period is greater than during the price decline period. Direct use of Definition A3 will overestimate the direct rebound effect, and it is necessary to decompose the price, so the price decomposition method is not used to identify the rebound effect. However, it is introduced to improve the accuracy of measurement results.

References

1. Statistical Review of World Energy. 2017. Available online: https://www.bp.com/en/global/corporate/energy-economics.html (accessed on 28 April 2019).
2. Lin, M. The clean energy consumption, environment governance and the sustainable economic growth in China. *J. Quant. Tech. Econ.* **2017**, *12*, 4–22.
3. Berkhout, P.H.G.; Muskens, J.C.; Velthuijsen, J.W. Defining the rebound effect. *Energy Policy* **2000**, *28*, 425–432. [CrossRef]
4. Haas, R.; Biermayr, P. The rebound effect for space heating empirical evidence from Austria. *Energy Policy* **2000**, *28*, 403–410. [CrossRef]
5. Madlener, R.; Alcott, B. Energy rebound and economic growth: A review of the main issues and research needs. *Energy* **2009**, *34*, 370–376. [CrossRef]
6. Van den Bergh, J.C. Energy conservation more effective with rebound policy. *Environ. Resour. Econ.* **2011**, *48*, 43–58. [CrossRef]
7. Llorca, M.; Jamasb, T. Energy efficiency and rebound effect in European road freight transport. *Transp. Res. Part A Policy Pract.* **2017**, *101*, 98–110. [CrossRef]
8. Wei, T. Impact of energy efficiency gains on output and energy use with Cobb–Douglas production function. *Energy Policy* **2007**, *35*, 2023–2030. [CrossRef]
9. Wei, T.; Liu, Y. Estimation of resource-specific technological change. *Technol. Forecast. Soc. Chang.* **2019**, *138*, 29–33. [CrossRef]
10. Wei, T.; Zhou, J.; Zhang, H. Rebound effect of energy intensity reduction on energy consumption. *Resour. Conserv. Recycl.* **2019**, *144*, 233–239. [CrossRef]
11. Saunders, H.D. Fuel conserving (and using) production functions. *Energy Econ.* **2008**, *30*, 2184–2235. [CrossRef]
12. Saunders, H. Is what we think of as "rebound" really just income effects in disguise? *Energy Policy* **2013**, *57*, 308–317. [CrossRef]
13. Brockway, P.; Saunders, H.; Heun, M.; Foxon, T.; Steinberger, J.; Barrett, J.; Sorrell, S. Energy rebound as a potential threat to a low-carbon future: Findings from a new exergy-based national-level rebound approach. *Energies* **2017**, *10*, 51. [CrossRef]
14. Freire-González, J. Methods to empirically estimate direct and indirect rebound effect of energy-saving technological changes in households. *Ecol. Model.* **2011**, *223*, 32–40. [CrossRef]
15. Freire-González, J. A new way to estimate the direct and indirect rebound effect and other rebound indicators. *Energy* **2017**, *128*, 394–402. [CrossRef]
16. Freire-González, J. Evidence of direct and indirect rebound effect in households in eu-27 countries. *Energy Policy* **2017**, *102*, 270–276. [CrossRef]
17. Graham, D.J.; Glaister, S. Road traffic demand electricity estimates: A review. *Transp. Rev.* **2004**, *24*, 261–274. [CrossRef]
18. Wang, Z.; Lu, M. An empirical study of direct rebound effect for road freight transport in China. *Appl. Energy* **2014**, *133*, 274–281. [CrossRef]
19. Winebrake, J.J.; Green, E.H.; Comer, B.; Li, C.; Froman, S.; Shelby, M. Fuel price elasticities in the U.S. combination trucking sector. *Transp. Res. Part D* **2015**, *38*, 166–177. [CrossRef]
20. Stapleton, L.; Sorrell, S.; Schwanen, T. Estimating direct rebound effects for personal automotive travel in Great Britain. *Energy Econ.* **2016**, *54*, 313–325. [CrossRef]
21. Moshiri, S.; Aliyev, K. Rebound effect of efficiency improvement in passenger cars on gasoline consumption in Canada. *Ecol. Econ.* **2017**, *131*, 330–341. [CrossRef]
22. Sorrell, S.; Stapleton, L. Rebound effects in UK road freight transport. *Transp. Res. Part D Transp. Environ.* **2018**, *63*, 156–174. [CrossRef]
23. Nesbakken, R. Energy consumption for space heating: A discrete–continuous approach. *Scand. J. Econ.* **2001**, *103*, 165–184. [CrossRef]
24. Schwarz, P.M.; Taylor, T.N. Cold hands, warm hearth: Climate, net takeback, and household comfort. *Energy J.* **1995**, *16*, 41–54. [CrossRef]
25. Jin, S.H. The effectiveness of energy efficiency improvement in a developing country: Rebound effect of residential electricity use in South Korea. *Energy Policy* **2007**, *35*, 5622–5629. [CrossRef]

26. Davis, L.W. Durable goods and residential demand for energy and water: Evidence from a field trial. *Rand J. Econ.* **2008**, *39*, 530–546. [CrossRef]

27. Lin, B.; Liu, X. Electricity tariff reform and rebound effect of residential electricity consumption in China. *Energy* **2013**, *59*, 240–247. [CrossRef]

28. Wang, Z.; Lu, M.; Wang, J.C. Direct rebound effect on urban residential electricity use: An empirical study in China. *Renew. Sustain. Energy Rev.* **2014**, *30*, 124–132. [CrossRef]

29. Lin, B.; Liu, H. A study on the energy rebound effect of China's residential building energy efficiency. *Energy Build.* **2015**, *86*, 608–618. [CrossRef]

30. Labidi, E.; Abdessalem, T. An econometric analysis of the household direct rebound effects for electricity consumption in Tunisia. *Energy Strategy Rev.* **2018**, *19*, 7–18. [CrossRef]

31. Sorrell, S.; Dimitropoulos, J.; Sommerville, M. Empirical estimates of the direct rebound effect: A review. *Energy Policy* **2009**, *37*, 1356–1371. [CrossRef]

32. Ouyang, J.; Long, E.; Hokao, K. Rebound effect in Chinese household energy efficiency and solution for mitigating it. *Energy* **2010**, *35*, 5269–5276. [CrossRef]

33. Zhang, Y.J.; Peng, H.R. Exploring the direct rebound effect of residential electricity consumption: An empirical study in China. *Appl. Energy* **2017**, *196*, 132–141. [CrossRef]

34. Hymel, K.M.; Small, K.A. The rebound effect for automobile travel: Asymmetric response to price changes and novel features of the 2000s. *Energy Econ.* **2015**, *49*, 93–103. [CrossRef]

35. Greening, L.A.; Greene, D.L.; Difiglio, C. Energy efficiency and consumption-the rebound effect-a survey. *Energy Policy* **2007**, *28*, 389–401. [CrossRef]

36. Liu, M.; Zhao, Y.Y. Measurement and empirical study of spatial spillover effects of Chinese manufacturing industry based on input factors. *J. Appl. Stat. Manag.* **2018**, *37*, 122–134.

37. Carli, R.; Dotoli, M. Cooperative distributed control for the energy scheduling of smart homes with shared energy storage and renewable energy source. *IFAC-PapersOnLine* **2017**, *50*, 8867–8872. [CrossRef]

38. Chapman, A.C.; Verbic, G.; Hill, D.J. Algorithmic and strategic aspects to integrating demand-side aggregation and energy management methods. *IEEE Trans. Smart Grid* **2016**, *7*, 1–13. [CrossRef]

39. Sorrell, S.; Dimitropoulos, J. The rebound effect: Microeconomic definitions, limitations and extensions. *Ecol. Econ.* **2008**, *65*, 636–649. [CrossRef]

40. Lesage, J.P.; Pace, R.K. *Introduction to Spatial Econometrics*; CRC Press: Boca Raton, FL, USA, 2009.

41. Lee, L.F.; Yu, J.H. Estimation of spatial autoregressive panel data models with fixed effects. *J. Econom.* **2010**, *154*, 165–185. [CrossRef]

Article

An Analysis of Energy Use Efficiency in China by Applying Stochastic Frontier Panel Data Models

Xiaoyan Zheng [1] and Almas Heshmati [2,*]

[1] Department of Economics, Sogang University, 35 Baekbeom-ro (Sinsu-dong #1), Mapo-gu, Seoul 04107 Korea; hyaoyan6002@gmail.com

[2] Jönköping International Business School, Jönköping University, Room B5017, P.O. Box 1026, SE-551 11 Jonkoping, Sweden

* Correspondence: almas.heshmati@ju.se; Tel.: +46-36-101780

Received: 6 March 2020; Accepted: 1 April 2020; Published: 13 April 2020

Abstract: This paper investigates energy use efficiency at the province level in China using the stochastic frontier panel data model approach. The stochastic frontier model is a parametric model which allows for the modeling of the relationship between energy use and its determinants using different control variables. The main control variables in this paper are energy policy and environmental and regulatory variables. This paper uses province level data from all provinces in China for the period 2010–2017. Three different models are estimated accounting for the panel nature of the data; province-specific heterogeneity and province-specific energy inefficiency effects are separated. The models differ because of their underlying assumptions, but they also complement each other. The paper also explains the degree of inefficiency in energy use by its possible determinants, including those related to the public energy policy and environmental regulations. This research supplements existing research from the perspective of energy policy and regional heterogeneity. The paper identifies potential areas for improving energy efficiency in the western and northeastern regions of China. Its findings provide new empirical evidence for estimating and evaluating China's energy efficiency and a transition to cleaner energy sources and production.

Keywords: energy efficiency; time-variant efficiency; true fixed-effects model; four components stochastic frontier model; determinants of inefficiency; Chinese provinces

1. Introduction

After the 2008 economic crisis, the situation in the world stabilized in 2010. According to the yearly China Economic Report [1], China's annual GDP increased from 6.066 trillion US dollars in 2010 to 8.271 trillion US dollars in 2017, with a total growth rate of 36.35 percent showing the highest growth among the top 15 large global economies. But the China Energy Statistics Yearbook 2018 [2] shows that the GDP growth rate gradually decreased from 10.7 percent in 2010 to 6.9 percent in 2017. In 2017, the global primary energy consumption was 13.5 billion tons of oil equivalent. The annual consumption growth rate in 2010–2017 was 1.4 percent. The economic growth rate slowed down in China.

According to the BP World Energy Statistics Yearbook [3] during 2010–2017, the gap between China's energy demand and energy supply increased over time. As China continues to promote urbanization and industrialization and gradually upgrades its consumer energy consumption structure, inequalities between China's energy supply and energy demand will remain severe until 2020. This gap will play an increasingly important role in energy security. In the face of rigid growth in energy demand, China's energy supply is expected to face severe challenges with increased supply pressures.

The BP Energy Outlook [4] predicts a radical energy transition. The ongoing transition to a lower-carbon fuel mix is led by renewables and natural gas which account for 85 percent of the growth in energy and are gaining in importance relative to traditional primary sources of oil and coal. It is forecast that the consumption of liquid fuels will grow over the next decade, but it will plateau as efficiency improvements in the transport sector are realized. A reduced use of the abundant global oil resources is likely to lead to a more competitive market and lower oil prices that will boost oil demand. The use of natural gas has grown dramatically and this growth is driven by its use in industry and power generation. Europe and China are two of the largest importers of gas. The growth in renewable energy is faster than that in oil and dominated by the developing world with China, India, and other Asian countries accounting for almost half the growth in global renewables. China and India drive global economic growth and together with other developing countries account for over 80 percent of the expansion in world output. Improvements in living standards in developing countries lead to an increase in energy demand.

The BP Energy Outlook [4] further suggests that the pattern of energy used within industry is expected to shift as a result of China's changing economic role. The process leading to the growth in energy used in industry will shift from China to other developing countries. By 2040, renewables are expected to overtake coal as the largest source of power generation. Global coal demand flatlines, with the fall in China and the OECD, but will be offset by gains in India and other emerging Asian countries; however, the growth in coal consumption will still slow down. By the mid-2020s, India will be the world's largest economic growth market. China and India both started with relatively coal-intensive fuel mixes. In a scenario of energy transition, China's coal share will fall from 60 percent in 2017 to around 35 percent in 2040 and will be offset by increasing shares of renewables, natural gas, and nuclear energy to match the growth in Chinese energy demand over the Energy Outlook's period, which is 2017–2040.

Two transition scenarios are predicted—evolving and rapid transition. According to the evolving transition scenario, the energy consumption for 1995, 2017, and 2040 is estimated at 891, 3132 and 4017 Mtoe (million tons of oil equivalent). The transition (from 1995 to 2017 and from 2017 to 2040) will lead to changes in consumption estimated at 2241 and 885 Mtoe. This corresponds to a 252 and 28 percent change which, on an annual basis, is 5.9 and 1.1 percent, respectively. In a rapid transition scenario, the estimated energy consumption is 891, 3132, and 3700 Mtoe. The changes are estimated to reach 2241 and 568 Mtoe with 252 and 18 percent total changes or 5.9 and 0.7 percent changes annually (BP Energy Outlook [4] pp. 135–137).

China's energy consumption per unit of GDP is twice that of the world average and four times that of developed countries. In recent decades, industrialized countries have invested in and developed energy saving and alternative energy technologies. It is difficult to meet the fast-growing energy demand simply by increasing energy supply. Saving energy and improving energy efficiency are extremely important and effective ways for China to meet its energy related challenges and the challenges of climate change. In such a situation one can ask, what is the status of energy efficiency in China specifically at the province level?

The Chinese government's interventions in energy use and energy efficiency mainly include government investments in the energy industry and the enforcement of energy policies targeting the energy industry. However, from the perspective of energy utilization and environmental protection, government interventions should also consider such incentives as encouraging and punishing different energy consumption industries. These include various programs such as tax incentives and subsidies for the introduction of environmentally friendly energy-saving products.

In 2013, the State Council of China issued the 'Action Plan for Air Pollution Prevention and Control' called 'Atmosphere Ten' which clearly states that the overall improvement in air quality in the country in five years led to a reduction in heavy air pollution in Beijing-Tianjin-Hebei, Yangtze River Delta, and Pearl River Delta of 15–25 percent. In 2017, the government's work report proposed to win the 'blue sky defense war' and speed up the resolution on coal-fired air pollution. As a relatively efficient and clean energy source, natural gas is favored by the government and the market. The policy of 'coal to gas' is an important substitution measure for improving air quality and it has been widely promoted in the past. This requires Beijing, Tianjin, Hebei, Shanxi, Shandong, and Henan provinces and other cities to complete 3.55 million units of 'coal to gas' and 'coal to electricity' transformations in energy technology.

However, due to China's large regional heterogeneity as compared to other countries, the feedback on energy efficiency policies in its regions is different. Therefore, energy market reforms conductive with environmental policy must be actively promoted, and, in parallel, reduce government interventions in the energy market. Regional heterogeneity in energy consumption is evident in the demand for energy and its impact on economic growth. Giving full flexibility to the endowment of energy factors improving energy efficiency can effectively promote economic development. Given these conditions it will be interesting to know whether China's energy efficiency has improved with technological innovations, and what kind of typical regional heterogeneity exists in China's energy efficiency.

Based on existing research, the methods for measuring efficiency mainly include the data envelopment analysis (DEA) and the stochastic frontier analysis (SFA). The former does not need to estimate the specific production function form, thus avoiding the problems caused by the choice of a wrong functional form. DEA uses information on inputs and outputs, but it does not describe the production process fully. Conversely, SFA describes individual producers' production processes by estimating the production function, thus controlling efficiency estimates. In addition to inputs and outputs, SFA also uses production and market environmental factors. Thus, this approach assumes a functional form.

At present, most scholars adopt the DEA method for efficiency analyses, while the SFA method is less frequently used. Only a few scholars have used it for empirical research related to energy use efficiency. The simple Cobb-Douglas production function is also a commonly used functional form for describing a regional economy. Considering the heterogeneity of China's economic regions, it is appropriate to use the SFA approach for measuring regional energy efficiency. Unlike DEA, SFA is a parametric method which allows for modeling the relation between energy use and its determinants and in addition to the inputs and outputs that one can control for firm, industry, province, and other environmental and policy characteristics. Further, the importance of extra information can be tested statistically.

Literature and evidence on inefficiencies and differences in regional level energy use in China is vast. By analyzing panel data for 30 provinces in 2005–2014 using the DEA efficiency model to measure total factor energy efficiency in China, [5] showed that total factor energy efficiency was high in the East and low in the West of the country. The eastern region had higher total factor energy efficiency and characteristics of lack of energy resources. However, the western region had the characteristics of lower total factor energy efficiency while it was rich in energy resources. The allocation of production elements of 'more input and less output' also existed in the central region, leading to an enormous waste of energy resources in these areas.

Ref [6] points out that the government should and can solve the problems and inefficiencies of energy allocation in the market and enforce these using mandatory energy policies. By studying the relationship between government interventions, natural resources, and economic growth, [7] found that appropriate government interventions can reduce the negative impact of pollution of natural resources on economic growth. One can ask, what are the energy policies that China has adopted for improving energy efficiency during the development process?

It is evident that production in China is very energy intensive. Energy sources are mainly fossil fuel based with extremely negative health, environmental, and climate effects. This paper evaluates energy use efficiency as a tool for reducing energy consumption and air emissions. This research does a panel data analysis of energy use efficiency in China at the province level. The method is at the forefront of research and allows for accounting province heterogeneity and temporal changes in energy use efficiency, making the results informative and useful.

In analyzing energy use efficiency, this paper uses three different models—[8], the true fixed-effects model [9,10], and four components of the stochastic frontier model. The stochastic frontier panel model approach is parametric and allows for modeling the relationship between energy use and its determinants conditioned on different control variables. The main control variables are energy policy, and environmental and regulatory variables. The data is from the province level and covers all provinces in China (except Tibet due to lack of data availability) observed over the period 2010–2017. Three different models are estimated accounting for the panel nature of the data; province-specific heterogeneity and province-specific energy use inefficiency effects are separated. The models differ because of their underlying assumptions but also complement each other considering the directions that literature has developed in, namely assumptions about the distribution of inefficiency effects, estimation methods, and time-variance of inefficiency and its separation from province heterogeneity. The degree of inefficiency in the use of energy is also explained by its possible determinants including those related to public energy policy and environmental regulations. This research supplements existing research from the perspective of energy policy and regional heterogeneity. It shows that there is enormous potential for improving energy efficiency in the western and northeastern regions of China. These findings provide new empirical evidence for estimating and evaluating China's energy use efficiency and transition to cleaner energy sources.

The rest of the paper is organized as follows. After this brief introduction, Section 2 presents a literature review on energy efficiency. The evolution of methods for estimating energy efficiency and the approaches used are also discussed in this section. Section 3 outlines the methodologies of the three different models used. Section 4 describes the data and the specifications of the empirical model. Section 5 discusses the results both by comparing the models and by distinguishing between regional heterogeneity in China. Section 6 gives the conclusion and implications of the findings of the study.

2. Literature Review

This section is divided into two sub-sections elaborating on the significance, concept, meaning, and evolution of the methods of measuring and estimating energy efficiency.

2.1. Significance and Concepts of Energy Efficiency

Energy experts and research scholars in China and elsewhere have reached a general consensus about the important role that energy plays in an economy and in society. It is believed that improvements in energy efficiency can significantly reduce energy consumption and environmental pollution and help in gradually achieving sustained and steady economic growth. In 1995, the World Energy Commission defined energy efficiency as reducing energy inputs to provide equal energy services interpreted as producing the same amount of goods and services. However, this definition is broad and does not accurately define the concept of energy efficiency.

Ref [11] defines energy efficiency on the basis of its traditional meaning, that is, the production of the same amount of services or desirable outputs but with less energy inputs and undesirable outputs. [12] define and separate energy efficiency through economic and technical perspectives. By summarizing and analyzing existing energy efficiency measurement indicators, [13] divide energy efficiency into a number of categories—energy macro efficiency, energy physical efficiency, energy factor utilization efficiency, energy element allocation efficiency, energy value efficiency, and energy economic efficiency. Similarly, [14] point out that energy efficiency means producing the same amount of effective outputs or services with less energy. They believe that the key to defining energy efficiency is scientifically identifying effective outputs and inputs.

Based on different research fields, energy efficiency uses various quantitative indicators. Based on an analysis of the theoretical framework, energy efficiency in this paper is defined as the overall efficacy of energy economic efficiency and energy environmental efficiency.

2.2. Evolution of Methods for Estimating Energy Efficiency

Looking at relevant literature on energy use efficiency, we see that the research methods used for analyzing energy use efficiency are mainly divided into two types: 'single factor efficiency' without considering other factors and 'all-factor energy efficiency' with multiple inputs and multiple outputs.

The former's results only consider the proportional relationship between energy input and production output, while the latter adds the results of all other input factors including energy in the calculation.

Because the method of measuring single factor energy efficiency is simple and intuitive and it has strong operability, it has been favored by many scholars both in China and elsewhere, and it has been the main method for studying energy efficiency problems over time. However, with the continuous progress in research in the area of energy efficiency, the traditional single factor energy efficiency measurement method has been questioned and replaced with multi-factors energy efficiency measurement methods.

Ref [15] evaluated various indicators of traditional energy efficiency and maintained that traditional indicators did not describe the essence of 'energy efficiency' because they have many defects. Using a single-factor approach and three full-factor methods, [16] compared the energy efficiency of various regions in China based on data for 2005. They found that the total factor approach was promising, as it revealed the impact of a regional factor endowment structure on energy efficiency.

Because of the shortcomings and limitations of single factor efficiency research, scholars started investigating more systematic and scientific methods for evaluating and studying energy efficiency. [17] proposed the concept of total factor energy efficiency based on the total factor productivity framework and measured the total factor energy efficiency of 29 provinces in China. Their results showed that total factor energy efficiency was a more realistic measure of energy use efficiency. The approach used in this research differs from the single factor efficiency approach, by conditioning the model on other factors such as GDP, exports, education investments, R&D investments, environmental protection, population, and urbanization, all of which influence energy use. Thus, the derived demand for energy is conditional on other factors accounting for the multiple factor nature of energy use.

Researchers agree that two papers by [18] and [19] mark the birth of the stochastic frontier methodology. Subsequently, [20] proposed a new method for effectively dividing the error terms of the production and cost functions into technical inefficiency terms and random error terms and using these for measuring enterprises' technical efficiency. However, these methods are based on cross-sectional data and cannot be technically efficient for multiple production unit observations. In short, the measure of energy efficiency is time-invariant and restrictive. [21] applied the fixed-effects model and the random-effects model for estimating enterprises' technical efficiency. However, their model assumed that the technical efficiency of each enterprise was fixed or time-invariant. To make up for this shortcoming, [22] and [8] developed different models for estimating the time-varying technical efficiency of enterprises.

3. Methodology

The stochastic frontier (SF) approach for estimating technical efficiency is based on the idea that an economic unit may operate below its production potential or frontier due to low performance, errors, and some uncontrollable factors. A study of the frontier approach started with Farrell [23] who suggested that efficiency could be measured by comparing realized or actual output with the maximum or potential attainable output. Other than comparing output, we can also compare the actual input use with the minimum required input use. The two methods are called output oriented and input oriented approaches. Their aim is maximizing output with available inputs and technology or minimizing costs for given outputs and technology. The former is more adaptable for industry/firm data and the latter for services data. The empirical part of this study is based on three different models— [8], the true fixed-effects model [9,10], and four error components of the SF model with determinants of inefficiency (following [24] and [25]).

Most theoretical stochastic frontier production functions have not explicitly formulated a model for technical inefficiency effects in terms of appropriate determinants. By using panel data, one can remove the limitations of depending on the distributional assumption of noise and inefficiency components and observing each unit at several different points of time. However, the extended dimension in time adds to the complexity, as it requires the modeler to take into account some heterogeneity effects that may exist beyond what is possible to control using a cross-sectional approach, which lumps individual effects with random errors. This can be achieved by introducing an 'individual (unobservable) effect,' say, α, that is time-invariant but individual-specific. The limitation of such a model is eliminated when using panel data methods.

We can examine whether inefficiency has been persistent over time or whether a unit's inefficiency is time-varying since we have information about units over time. One component of inefficiency may have been persistent over time while another may have varied over time. Regarding time-invariant individual effects, we also need to consider whether an individual effect represents persistent inefficiency or persistent unobserved heterogeneity, as well as whether individual effects are fixed parameters or are they realizations of a random variable [26]. Thus, it is important that policies promote an efficient use of resources that are scarce, and it can serve as an effective policy tool by separating unobserved heterogeneity and inefficiency components.

This study outlines three panel data models which differ in terms of the underlying assumptions made for the temporal behavior of the inefficiency components. All the models treat inefficiency as being individual-specific. This is consistent with the notion of measuring the efficiency of decision-making units. Model 1 allows for inefficiency to be both individual-specific and time-varying and explains the determinants of inefficiency. Model 2 separates inefficiency effects from unobserved individual non-inefficiency heterogeneity effects. Model 3 separates persistent inefficiency and time-varying inefficiency from unobservable individual heterogeneity effects. Thus, the three models are complementary and jointly provide information on province heterogeneity, province inefficiency, the random error term, and the variations in inefficiency in energy use. The three models are now outlined.

3.1. Model 1: The Time-Variant Efficiency Model

Ref [8] considered a production model wherein technical inefficiency effects were modeled in a stochastic frontier function for panel data. In this paper, we specify a factor demand version of the model. The objective is to minimize the use of a factor in the production of a given output, factor price, and technology. This is similar to [27] who analyzed labor use efficiency in the banking industry. Here we use the same approach but in the context of energy use. Separability between energy and other inputs is assumed. The assumption is supported by the fact that we use aggregate output and aggregate individual inputs. A cost function is appropriate for the current case as energy use is cost for producing a given output, which is desirable to be minimized. Provided the inefficiency effects are stochastic, the model permits the estimation of both technical change or a shift in function over time and time-varying technical inefficiencies. The model is estimated using the maximum likelihood method which allows for estimating the effects of inefficiency's determinants. In this case inefficiency is a function of time.

In Model 1 we use the following generic formulation to discuss the various components in a unifying network:

$$ENE_{it} = f(x_{it}, \beta) + \epsilon_{it}, \ \epsilon_{it} = v_{it} + u_{it},$$
$$u_{it} = G(t)u_i, \ v_{it} \sim N\!\left(0, \sigma_v^2\right), \ u_{it} \sim N^+\!\left(\mu, \sigma_u^2\right), \ G(t) = \left[1 + \exp\!\left(\gamma_1 t + \gamma_2 t^2\right)\right]^{-1}$$

where *ENE* is energy use and $G(t) > 0$ is a function of time (t); in this model, inefficiency (u_{it}) is not fixed for a given individual, instead it both changes over time and across individuals. Inefficiency is composed of two distinct components: the nonstochastic time component, G(t) and a stochastic individual component, u_i. The stochastic component, u_{it}, uses the panel structure of the data in this model. The u_i component is individual-specific and the *G(t)* component is time-varying and is common for all the individuals. We consider some specific forms of *G(t)* used in [28] model which assumes $G(t) > 0$, given that $u_i > 0$, and thus $u_{it} \geq 0$ is ensured by having a non-negative G(t). G(t) can be monotonically increasing (decreasing) or concave (convex) depending on the signs and magnitude of γ_1 and γ_2. Inefficiency changes in this model are time driven and a nonlinear exponential function of time. However, the trend pattern is similar for all individuals; the differences in performance among individuals are due to the u_i component. The random and nonlinear nature of the model requires iterative estimation by the maximum likelihood (ML) estimation method. Cost efficiency is estimated assuming truncated normal distribution using the product of the individual specific u_i and the time variant $G(t)$. The product of the two is in the interval between 0 and 100 where 100 represents a full cost-efficient unit.

3.2. Model 2: The True Fixed-Effects Model

Model 1 is a standard panel data model where α_i is an unobservable individual effect. The model can be estimated using the standard panel data fixed and random-effects estimators to estimate the model's parameters to obtain the estimated value of u_i. The highest estimated value of $\hat{\alpha}_i$, namely $\hat{u}_i$, is used as a reference for the frontier.

However, there is a notable drawback in Model 1's approach as it does not allow individual heterogeneity to be distinguished from inefficiency. In other words, all time-invariant heterogeneity such as enterprise infrastructure that is not necessarily inefficient is included as inefficiency [9,29]. Also, the time-invariant assumption of inefficiency is a potential issue with Model 1. If T is large, it seems implausible that the inefficiency in energy use will stay constant for an extended period of time, since the technological progress will eventually replace less efficient technologies. So, should one view the time-invariant component as persistent inefficiency or as individual heterogeneity? The optimal choice lies somewhere in between, that is, a part of the inefficiency might be persistent, while another part may be transitory.

To solve the problem that the two parts cannot be separated from time-invariant individual heterogeneity effects, we have to choose either a model wherein α_i represents persistent inefficiency, or a model wherein α_i represents an individual-specific heterogeneity effect.

Following Kumbhakar and Heshmati [29] we consider both specifications in this paper. Thus, the models we examine can be written as:

$$ENE_{it} = \alpha_i + x'_{it}\beta + \epsilon_{it}, \ \epsilon_{it} = v_{it} + u_{it},$$
$$v_{it} \sim N(0, \sigma_v^2), \ u_{it} = h_{it}u_i, \ h_{it} = f(z'_{it}, \delta), \ u_i \sim N^+(\mu, \sigma_u^2),$$

The key feature that allows for the model's transformation is the multiplicative form of inefficiency effects, u_{it}, in which individual-specific effects, u_i, appear in multiplicative forms with individual and time-specific effects, h_{it}. As u_i^* does not change with time, the within and first-difference transformations leave this stochastic term intact. Thus, the difference between Model 1 and Model 2 is that inefficiency in Model 2 is explained by its observable determinants (z), while in the former, the time patterns of inefficiency are explained by a trend, but inefficiency is not explained by any determinants. Thus, cost efficiency is obtained based on the separated u_{it} components of the residual.

3.3. Model 3: Four Components of the Model with Determinants of Inefficiency

To fully satisfy the assumptions made in the model, we introduce a final model by [24] and [25] that overcomes some of the limitations of the earlier models. In this model, the error term is split into four components. The four components in this paper's context capture:

- Provinces' latent heterogeneity [9], which has to be disentangled from provinces' persistent inefficiency effects;
- Short-run time-varying transitory inefficiency;
- Persistent or time-invariant inefficiency as in [30,31] and [29]; and Random shocks.

Then, our final model based on these characteristics is the Kumbhakar et al. [25] model which is specified as:

$$ENE_{it} = \alpha_0 + f(x_{it}; \beta) + \mu_i + v_{it} + \eta_i + u_{it}$$

where μ_i is two-sided individual province heterogeneity, v_{it} is a two-sided random error term, η_i is one-sided time-invariant individual inefficiency, and u_{it} is one-sided time-variant inefficiency. In production models, the signs on the front of the inefficiency components are negative, reflecting production below the frontier output, while in cost or energy use models they are positive, suggesting higher cost or energy use above the minimum or frontier.

Instead of using a single stage ML estimation method based on the distributional assumption of the four components ([32], a simpler multi-step procedure is considered and we write the model as:

$$ENE_{it} = \alpha_0^* + f(x_{it};\beta) + \alpha_i + \epsilon_{it}$$

where $\alpha_0^* = \alpha_0 - E(\eta_i) - E(u_{it})$; and $\alpha_i = \mu - \eta_i + E(\eta_i)$.

This model can be estimated in three steps. In the first step, we use the standard random-effects panel regression to estimate $\hat{\beta}$. This procedure also gives predicted values of α_i and ϵ_{it}, which we denote by $\hat{\alpha}_i$ and $\hat{\epsilon}_{it}$. In the second step, we estimate the time-varying technical inefficiency, u_{it}, and in the final step, we estimate η_i following a procedure similar to that in Step 2. Lastly, we estimate the persistent efficiency, PE, as $PE = -\exp(\eta_i)$. The residual efficiency, RE, is obtained as in Models 1 and 2, assuming a half normal distribution or truncated normal distribution u_{it}. The overall efficiency, OE, following Kumbhakar et al. [28], is obtained from the product of PE and RE, that is, OE = PE × RE.

Table 1 gives the main characteristics of the three different efficiency models. The characteristics are related to the underlying assumptions of the different models, decomposition of the error components, time variation patterns of inefficiency, and the estimation procedure.

Table 1. Main characteristics of the different models.

	Model 1	Model 2	Model 3
General firm effects are treated as:	Fixed	Fixed	Random
Energy use inefficiency components:			
Persistent inefficiency	No	No	Yes
Residual inefficiency	No	No	Yes
Overall energy use inefficiency:			
Mean	Time-inv.	Zero trunc.	Zero trunc.
Variance	Homosc.	Homosc.	Homosc.
Symmetric random error term:			
Variance	Homosc.	Homosc.	Homosc.
Estimation method:	ML	ML	Multi-step

Notes: Fixed-effects (Fixed), random-effects (Random), homoscedastic variance (Homosc.), time invariant efficiency (Time-inv.), zero truncated error term (Zero trunc.), and maximum likelihood (ML).

4. Data

The data used in this study are from the province level observed for the period 2010–2017. It is obtained from the National Bureau of Statistics of China [1]. The dataset is the best available and frequently used in research and planning. This section describes the data source and provides a list of key and control variables; it also gives a descriptive analysis of the data.

4.1. Main Variables

In this research, energy use is defined as the economic value of total energy used per capita. It covers all economic sectors. It is reflected in both the price and quantity of energy. It is also reflected in the value of production. The definition of energy used here is close to the one used by [11], who defined energy efficiency as the production of the same amount of services or desirable outputs but with less energy inputs and undesirable outputs. In the current study, the undesirable output is controlled for by environmental stringency, carbon dioxide, and fine particulate matter. Provinces' per capita GDP is used as the main explanatory variable. It reflects labor productivity, size or scale in the economy as well as opportunity for energy use or consumption.

It should be noted that one may consider income to be endogenously determined and, as such, it can induce biased estimation results. One way of endogenizing income is by using predicted income or lag income as the explanatory variable. However, the two approaches may in turn lead to a bias. Here, we ignore the issue of endogeneity with the argument that we use province level data which is average per capita income and not endogenous to private and public users. Variations in income levels within the province that could be a source of endogeneity are not observed. At the level of aggregate income there is one-to-one correspondence between income and expenditure, and work is part of social life and most people, regardless of their income per hour, work 40 h per week.

4.2. Control Variables

A review of the factors affecting energy efficiency in existing literature shows that these are mainly focused on three aspects: technological progress; structural factors including industrial, economic, and energy consumption structures; and system factors including energy prices, the degree of opening up to the outside world, and the government's environmental regulations.

Technological changes: It is generally believed that improvements in energy efficiency are mainly through structural adjustments and technological progress. In the process of economic development, technological progress accelerates the process of eliminating backward industrial sectors, transforming the original industrial sectors, and improving the industry which also promotes establishing new industrial sectors. Progress directly improves energy efficiency through the transformation of traditional technologies, development of new technologies, and adoption of new processes. This paper uses R&D internal expenditure (in 10,000 yuan) of industrial enterprises in each province as a proxy for technological progress. Changes in product mix and manufacturing mix are partially controlled for over time through investments in R&D and education, as well as time variance efficiency.

The government's environmental regulations or environmental protection investments: [33] targeted 14 prefectures in Xinjiang and used three indicators of the government's environmental pollution treatment investments to characterize the government's environmental regulations. Their results showed that the policy on pollution treatment investments and resource tax both generated energy inefficiencies. [34] used Xinjiang as their research subject for measuring the intensity of environmental regulations using the entropy method. Their results showed that the government's environmental regulations had an inhibitory effect on energy efficiency, which was not only reflected in the current period, but also in three periods lagged. [35] showed that environmental protection investments had a negative impact on energy efficiency probably because pollution treatment was not effective and investments in treatment were often passive.

Openness defined as ((export + import)/GDP) characterizes foreign trade. Foreign trade is an important component of economic development. The structure of foreign trade products and the structure of foreign trade itself can affect energy efficiency. [36–39] show that the degree of openness is positively related to energy efficiency. Some scholars have come to different conclusions though. [40] shows that at the national level, economic openness is significantly positively correlated with the development of electrical equipment. At the regional level, economic openness is only significantly positively related to energy efficiency in the middle Yellow River. [14] show that for every 1 percent increase in the value of imports and exports in GDP, energy efficiency will decrease by 0.18 percent, but due to its dual effect, performance will vary in different regions. [41] research on single factor energy efficiency shows that the relationship between openness and energy efficiency in typical provinces is inconsistent, and he believes that the impact of openness on energy efficiency is a sufficient condition and not a necessary condition.

Population and urbanization: [42] show that both endogenous innovations and human development have a positive impact on single factor energy efficiency. [43] examined the impact of urban morphology and transportation modes on national and regional energy efficiency. His results showed that the former had a significant negative impact on regional energy efficiency while the latter had no significant impact. In some previous research the impact of urban agglomeration scale density on urban energy efficiency is examined. The former can improve the latter, but impact on energy efficiency can be heterogeneous. It should be noted that in this research all explanatory variables and determinants of energy use inefficiency are province-specific and some, such as education and R&D investments, have spillover effects. In this research, we do not account for spatial effects of investments across provinces.

4.3. A descriptive Analysis of the Data

The energy consumption structure in China by sectors is very skewed (transport 8.2 percent, industry 29.0 percent, building 16.7 percent, electricity 40.1 percent, and others 6.1 percent) [44]. Concerning primary energy consumption, the problems facing China's energy use include a very high proportion of coal use, low thermal efficiency, high unit energy consumption, high growth rate of consumption, and trade disputes with the US which influence energy efficiency with an impact on industry. From a spatial perspective, the level of economic development in different regions of China is very different. While there are differences in climate, geographical environment, and resources, there are also differences in energy structures in different regions.

The model used in this study is parametric and it allows for modeling the relationship between energy use and its determinants conditioned on different control variables. The main control variables are energy policy (investments in environment protection) (x_{env}); the degree of trade openness (x_{exp}); and environmental and regulatory variables including education investments (x_{edu}), R&D investments ($x_{R\&D}$), population (x_{pop}), and urbanization (x_{urb}). The variables which may influence energy use efficiency are z_1 (PM2.5), z_2 (CO$_2$), and (municipal solid waste treated). PM2.5 refers to atmospheric fine particulate matter (PM) that has a diameter of less than 2.5 micro-meters. We also use the log of GDP per capita (x_{gdp}) as a main indicator. To see the variations in energy use, we use the cost function approach and the log of energy use per capita (ENE$_{cost}$) as the dependent variable. The series used in this analysis is at the province level and contains all provinces in China (except Tibet due to lack of data) observed yearly from 2010 until 2017.

Table 2 shows that all the indicators are logarithmically transformed, except for investments in environment protection, which are defined as a percentage of regional GDP or gross regional product GRP (x_{env}) and urbanization (x_{urb}) in the production function variables. Energy use cost per capita ranges between 427.638 and 5665.779 CNY among the sample provinces, with a mean of 1556.498 and dispersion of 1039.165 CNY. The GRP per capita varies in the interval of 1350.430 and 89,705.230 CNY in the provinces. The mean value is 21,652.784 with a dispersion of 16,997.766 CNY.

Table 2. Summary statistics of input and output data (2010–2017) (30 × 8 = 240 observations).

Variable	Definition	Mean	Std. Dev.	Minimum	Maximum
A. Energy cost function variables:					
ENE$_{cost}$	Energy use per capita	1556.498	1039.185	427.638	5665.779
x_{gdp}	GRP per capita	21,652.784	16,997.766	1350.430	89,705.230
x_{exp}	Value of export	69,598,470.492	123,875,644.740	424,174.000	646,000,000.000
x_{edu}	Education investments (in 10,000)	9,596,653.987	5,952,569.714	994,671.000	36,587,681.000
$x_{R\&D}$	R&D investments	2,872,909.717	3,713,827.748	57,760.000	18,650,313.000
x_{env}	Investments in environmental protection, as % of GRP	2.956	0.935	1.200	6.700
x_{pop}	Population (10,000 people)	4522.296	2705.794	563.000	11,169.720
x_{urb}	Urbanization (%)	0.560	0.127	0.338	0.896

Table 2. *Cont.*

Variable	Definition	Mean	Std. Dev.	Minimum	Maximum
B.	Determinants of energy use inefficiency:				
z_1	PM2.5 $(\mu g/m^3)$	39.903	15.703	10.487	82.379
z_2	CO_2 intensity (tons/billion yuan)	19.750	12.089	3.129	69.052
z_3	Municipal solid waste treated (tons/day)	17,269.510	13,961.950	931.000	78,185.000

Note: Monetary variables are in fixed Chinese yuan, CNY. Source: Based on data from the National Bureau of Statistics of China (2018).

5. An Analysis of the Results

The three stochastic frontier models are specified and estimated using the data described earlier, and the estimation results are given in Table 3.

Table 3. Stochastic frontier models' estimation results (NT = 240 observations).

Variable	Description	Model 1	Model 2	Model 3
x_{gdp}	Log GDP per capita	−0.563 **	−0.461 **	−0.583 **
x_{exp}	Log Exportation	0.008	−0.024	0.013
x_{edu}	Log Education Investments	0.016	0.053	0.016
$x_{R\&D}$	Log R&D Investments	0.144 *	0.124 *	0.143 *
x_{env}	Environment Protection	−0.010	0.006	−0.012
x_{pop}	Log Population (10,000 people)	−0.412 *	−0.954 *	−0.436 **
x_{urb}	Urbanization (%)	2.380 **	3.390 **	2.260 **

Note: significant at less than the 0.05 (*) and less than the 0.01 (**) percent level of significance.

In Table 3 we present the estimation results of the three energy efficiency models. In Model 1, GDP, R&D investments, and environment protection are all statistically significant predictors of energy use. In Model 2, GDP and R&D investments are predictors of energy use. However, environment protection is a statistically insignificant predictor of energy use. In Model 3, GDP and R&D investments are significant variables that predict variations in energy use. However, environment protection is not found to be a significant predictor of energy use.

Another result that can be attained from Table 3 is attributed to the use of time as a driver of efficiency, which reduces the inefficiency component of the overall residual.

The Wald test is a joint test for multiple regressors. It mainly tests how much the model changes if the variables added are removed. In other words, the distance from the coefficient of each variable to zero is measured. The test results (see Table 4) show that the independent variable contributes significantly to the model and cannot be eliminated. The p-values of the fit of the three models are all less than 0.01, indicating that the models fit the data well.

Table 4. Model fit test's results.

Model Fitted	Model 1	Model 2	Model 3
Wald test statistics	92.49	7684.93	109.70
Wald test p-value	<0.001	<0.0001	<0.001

The rest of this section analyzes the results. The analysis is in the form of a comparison of the different model's estimation results and an analysis of time-variance patterns as well as regional differences in energy use efficiency.

5.1. A Comparative Analysis of the Models' Estimation Results

Table 5 gives the descriptive statistics for mean energy use efficiency according to the three models. Model 1 shows that province level energy use efficiency ranged from 0.091 to 0.937 with large dispersions. The energy use efficiency in Model 2 ranged from 0.387 to 1.000. In Model 3 the residual efficiency ranged from 0.013 to 0.990, the persistent efficiency ranged from 0.179 to 0.897, and the overall efficiency ranged from 0.176 to 0.876. A number of 0.80 for Province A in a given year indicates that province A is 80 percent efficient in energy use compared to the frontier reference Province B with the best energy use technology. Province A has the potential of improving its efficiency by 20 percent.

Table 5. Descriptive Statistics for Energy Efficiency Measures by Different Models.

	Energy Efficiency	Mean	Std. Dev.	Minimum	Maximum
	Model 1	0.371	0.184	0.091	0.937
	Model 2	0.968	0.092	0.387	1.000
Model 3	Residual efficiency	0.973	0.013	0.092	0.990
	Persistent efficiency	0.625	0.179	0.209	0.897
	Overall efficiency	0.609	0.176	0.202	0.876

Notes: Model 1: The time-variant efficiency model. Model 2: The true fixed-effects model (Greene, 2005a). Model 3: Four components of the SF model with determinants of inefficiency.

5.2. An Analysis of Trends in Energy Use Efficiency

Table 6 gives the yearly mean of provincial energy use efficiency for the three models. The results show that, according to the time-variant Model 1, energy use efficiency decreased during the study period. But Models 2 and 3 show increasing energy use efficiency. However, the changes over time are extremely small.

Table 6. Development of mean energy efficiency over time (2010–2017).

Year	Model 1	Model 2	Model 3		
			Residual Efficiency	Persistent Efficiency	Overall Efficiency
2010	0.374	0.954	0.970	0.621	0.602
2011	0.373	0.971	0.973	0.622	0.606
2012	0.372	0.963	0.971	0.623	0.606
2013	0.372	0.970	0.976	0.626	0.611
2014	0.371	0.973	0.977	0.627	0.613
2015	0.370	0.971	0.974	0.627	0.612
2016	0.370	0.970	0.972	0.627	0.610
2017	0.369	0.974	0.976	0.628	0.613

Table 6 shows that the trends of national mean energy use efficiency over 2010–2017 were practically constant over time. Although energy demand increased constantly, there was a technological revolution and policies for improving energy efficiency were introduced continuously, there were no significant improvements in energy use efficiency throughout the country. The possible small improvements in energy use efficiency are eliminated by increased consumption of energy due to economic growth in energy intensive industries.

5.3. Regional Heterogeneity in Energy Efficiency

For investigating the performance of different provinces and their positions as compared to the best performing province, energy use efficiency was compared across provinces and major regions in China. In the latter case, the provinces were divided into East (Beijing, Fujian, Guangdong, Henan, Hebei, Jiangsu, Shandong, Shanghai, Tianjin, and Zhejiang), Center (Anhui, Hubei, Henan, Hunan,

Jiangxi, and Shanxi), West (Chongqing, Gansu, Guangxi, Guizhou, Inner Mongolia, Ningxia, Qinghai, Shaanxi, Sichuan, Xinjiang, and Yunnan), and Northeast (Heilongjiang, Jilin, and Liaoning).

Table 7 gives the summary of average energy use efficiency values by provinces for the period 2010–2017. Different models' estimated measures of efficiency show that there were differences between provinces in terms of energy use efficiency.

Table 7. Average energy use efficiency by provinces (2010–2017).

Provinces	Model 1	Model 2	Model 3		
			Residual Efficiency	Persistent Efficiency	Overall Efficiency
East					
Beijing	0.724	1.000	0.986	0.857	0.845
Fujian	0.365	0.998	0.972	0.684	0.665
Guangdong	0.332	1.000	0.980	0.634	0.621
Hainan	0.136	0.928	0.966	0.343	0.332
Hebei	0.313	0.998	0.975	0.599	0.583
Jiangsu	0.437	1.000	0.987	0.738	0.728
Shandong	0.145	1.000	0.988	0.341	0.337
Shanghai	0.437	0.999	0.984	0.743	0.731
Tianjing	0.520	1.000	0.989	0.795	0.786
Zhejiang	0.347	1.000	0.980	0.658	0.645
Center					
Anhui	0.478	0.999	0.982	0.772	0.758
Hubei	0.352	1.000	0.983	0.668	0.656
Henan	0.402	1.000	0.986	0.717	0.707
Hunan	0.417	1.000	0.982	0.731	0.718
Jiangxi	0.437	0.994	0.977	0.750	0.733
Shanxi	0.699	0.974	0.962	0.849	0.816
West					
Chongqing	0.937	0.997	0.980	0.892	0.875
Gansu	0.128	0.962	0.973	0.301	0.293
Guangxi	0.232	0.994	0.975	0.509	0.497
Guizhou	0.443	0.952	0.963	0.748	0.720
Inner Mongolia	0.516	0.893	0.945	0.777	0.734
Ningxia	0.281	0.537	0.940	0.555	0.522
Qinghai	0.401	0.946	0.974	0.728	0.709
Shaanxi	0.215	0.998	0.975	0.469	0.457
Sichuan	0.309	0.999	0.971	0.617	0.599
Xinjiang	0.093	0.987	0.972	0.216	0.210
Yunnan	0.383	0.932	0.959	0.702	0.674
Northeast					
Heilongjiang	0.199	0.984	0.964	0.429	0.413
Jilin	0.320	0.980	0.969	0.624	0.605
Liaoning	0.138	0.999	0.972	0.311	0.302

According to the models' results reported in Table 7, most of provinces in East China had relatively higher energy use efficiency as compared to provinces in the Center, West, and Northeast of the country. Provinces in the East such as Beijing, Chongqing, and Shanxi had high efficiency above 80 percent. Conversely, an energy use efficiency of less than 40 percent was observed in Gansu, Xinjiang, Shandong, and Liaoning provinces.

It can be seen in Table 7 that there is very obvious regional heterogeneity of energy use efficiency. Beijing, as the main energy efficiency policy implementation region, has always maintained high energy efficiency. Because of hosting a large proportion of secondary and tertiary industries, Changsha and Chongqing have also maintained high values in terms of energy efficiency.

Industrial cities such as Gansu, Shandong, and Liaoning have a very high proportion of production using coal. It can be speculated that the use of nonclean energy and the level of technology are the reasons for the low energy efficiency in these cities. As Xinjiang is a minority autonomous region that lacks resources, it has low technological levels, and slow implementation of energy efficiency policies which could have contributed to its low energy efficiency levels.

In looking at average energy use efficiency by provinces it is noted that Models 1 and 3 have similar trend calculation results, while Model 2 shows higher results that are similar to the results of residual efficiency in Model 3, which cannot reflect regional heterogeneity well. What we are concerned with is why the cities/provinces of Fujian, Guangdong, Shandong, Zhejiang, Hubei, Gansu, Shaanxi, Heilongjiang, and Liaoning have different efficiency results across different models. The reason could be that the energy structures in these provinces are basically dominated by energy-intensive secondary industries and there is congestion in resource inputs for achieving economic growth.

Figure 1 shows the average value of energy use efficiency by regions in the three models. It can be seen in the figure that the central region has higher energy efficiency, which has much to do with the good implementation of energy efficiency policies and human resource allocation structures in this region.

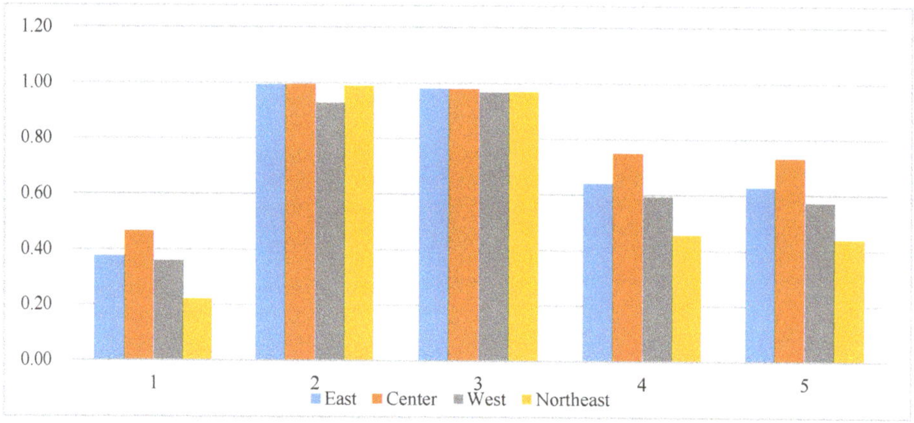

Figure 1. Estimated energy use efficiency by regions (2010–2017).

Notes:

1. Model 1: The time-variant efficiency model.
2. Model 2: The true fixed-effects model.
3. Model 3: Four components residual efficiency.
4. Model 3: Four components persistent efficiency.
5. Model 3: Four components overall efficiency.

A table giving the full results (not reported here but available on request) shows all 30 provinces' yearly energy use efficiency for the three models. From this table, we can compare the trends of energy efficiency between provinces and regions more comprehensively, and we can also see that energy efficiency showed slow and steady growth.

Energy efficiency in the central region before 2010 was low, and its energy efficiency in 2005–2010 was lower than that in the eastern and western regions, indicating that the central region had a weak capacity to absorb production capacity, and the industrial market had not been fully developed. After 2010, as the country's 'Central Rise' policy entered the implementation phase, the central region's industrial structure was adjusted, its capacity to absorb production was continuously enhanced, and energy resource utilization technology was improved, leading to continuous improvements in energy efficiency year by year.

Energy efficiency in the western region declined steadily. The reason for this declining pattern is that the western region has abundant energy endowments and the gradual implementation of the western development policy enhanced its economic development, expanded its market capacity, and helped achieve improved energy efficiency. However, with the country's excessive dependence on the western region's policies, this region's market could not absorb too much capacity, and energy productivity and energy consumption capacity did not match, resulting in serious overcapacity which led to energy efficiency falling for several years.

Affected by the world financial crisis in 2008, China's economic development, in particular the development of energy intensive secondary industries, was hit hard. Therefore, after experiencing a decline in energy efficiency, the Chinese government adopted a large-scale investment stimulus package to protect its high rate of economic growth. Vigorous development of infrastructural investments and construction drove the development of the secondary industries. As a result, from 2010 to 2017, energy efficiency in the eastern and central regions increased significantly and steadily. However, the improvements were far below the optimal level required by health and environmental standards.

6. Conclusions and Implications

This study estimated three different models accounting for the panel nature of the data and determined separate province-specific energy use inefficiency effects. It also explained the degree of inefficiency in the use of energy using its possible determinants including those related to the public energy policy and environmental regulations. This research supplements existing research from the perspective of energy policy and regional heterogeneity. We observed a large potential for improving energy use efficiency, particularly in the western and northeastern regions. This study provides new empirical evidence for evaluating China's energy efficiency and transitioning to cleaner energy sources.

Energy use efficiency in most provinces of China improved slowly after 2010 as did the trend of steady regional economic growth, but the magnitude of energy efficiency improvements was small compared to investments in technological innovations. A comparison of the results of the three stochastic frontier models shows that there was provincial and regional heterogeneity in energy use and its efficiency. The models complement each other and being based on different distributional assumptions and estimation methods together provide a picture of energy consumption in China at the province level for the period 2010–2017.

We can also see that the impact of the government's policies on energy efficiency were significant. As the country's 'Central Rise' policy entered the formal implementation phase, the central region showed improvements in energy efficiency. This also means that there is potential for improving energy efficiency in the western and northeastern regions. With the 'coal to gas' and 'coal to electricity' policy, energy efficiency in Beijing-Tianjin-Hebei, Yangtze River Delta, and Pearl River Delta showed relatively high levels of progress.

With the country's excessive dependence on policies for the western region, this region's market could not absorb as much capacity and energy productivity and energy consumption capacity did not result in production capacity, which led to decreased energy efficiency. The results of the western region's policy imply that the government's energy policy should be adjusted considering regional heterogeneity. But the low level of energy efficiency in the northeastern region still needs more empirical analysis to find out why this is the case. The 'Central Rise' policy could be modified to account for specific characteristics of the western and northeastern regions, such as resource endowments, production capacity adjustments, and infrastructure to increase their energy use efficiency. Further, the determinants of energy use (in) efficiency can be identified and the models be specified such that each model can explain possible outcomes of energy use and environmental protection.

A possible and interesting extension of this study is expanding the data period to include the period before the 2008 global economic crisis and disaggregating the province level data to the industry level. This will help control for energy intensity and targeted energy saving policies and an evaluation of their impact.

Author Contributions: Conceptualization, X.Z. and A.H.; Methodology, A.H.; Data collection, formal analysis, X.Z.; Writing—original draft preparation, X.Z., Writing—Review and Editing, X.Z. and A.H. All authors have read and agreed to the published version of the manuscript.

Funding: This research received no external funding.

Acknowledgments: The authors are grateful to an editor of the journal and three anonymous referees for their comments and suggestions on an earlier version of this paper.

Conflicts of Interest: All authors declare that they have no conflict of interest.

References

1. National Bureau of Statistics. 2018. Available online: http://data.stats.gov.cn/ (accessed on 1 March 2020).
2. China Energy Statistical Yearbook. China Statistics Press, 2019. Available online: https://www.chinayearbooks.com/tags/china-energy-statistical-yearbook (accessed on 8 April 2020).
3. BP World Energy Statistics Yearbook. 2018. Available online: http://www.bp.com/zh_cn/china/reports-and-publications/_bp_2018.html (accessed on 8 April 2020).
4. BP Energy Outlook (2019) edition. Available online: https://www.bp.com/content/dam/bp/business-sites/en/global/corporate/pdfs/energy-economics/energy-outlook/bp-energy-outlook-2019.pdf. (accessed on 8 April 2020).
5. Chen, Z. Energy Resource Endowment, Government Intervention and Analysis on China Energy Efficiency. Master's Thesis, Zhongnan University of Economics and Law, Wuhan, China, 2017.
6. Hirst, E. Improving energy efficiency in the USA: The Federal Role. *Energy Policy* **1991**, *19*, 567–577. [CrossRef]
7. Ding, J.; Deng, K. Government Intervention, Natural Resources and Economic Growth: Based on Regional Panel Data Study of China. *China Ind. Econ.* **2007**, *7*, 55–64.
8. Battese, G.E.; Coelli, T.J. Frontier production functions, technical efficiency and panel data: With application to paddy farmers in India. *J. Product. Anal.* **1992**, *3*, 153–169. [CrossRef]
9. Greene, W. Fixed and Random Effects in Stochastic Frontier Models. *J. Product. Anal.* **2005**, *23*, 7–32. [CrossRef]
10. Greene, W. Reconsidering heterogeneity in panel data estimators of the stochastic frontier model. *J. Econom.* **2005**, *126*, 269–303. [CrossRef]
11. Patterson, M.G.; Makarov, V.I.; Khmelinskii, I.V. What is energy efficiency? Concepts, indicators, and methodological issues. *Energy Policy* **1996**, *5*, 377–390. [CrossRef]
12. Bosseboeuf, D.; Chateau, B.; Lapillonne, B. Cross-country comparison on energy efficiency indicators: The on-going European effort towards a common methodology. *Energy Policy* **1997**, *25*, 673–682. [CrossRef]
13. Wei, Y.; Liao, H. The Seven Types of Measurement Indicators of Energy Efficiency and Their Measurement Methods. *China Soft Sci.* **2010**, *1*, 128–137.
14. Wei, C.; Shen, M. Energy Efficiency and Energy Productivity: A Comparison of based on the Panel Data by Province. *J. Quant. Tech. Econ.* **2007**, *9*, 110–121.
15. Wei, C. Energy Efficiency and Energy Productivity: Comparison of Provincial Data Based on DEA Method. *Econ. Tech. Econ. Res.* **2007**, *24*, 110–121.
16. Yang, H.; Shi, D. Energy Efficiency Research Methods and Comparison of Energy Efficiency in Different Regions of China. *Econ. Theory Econ. Manag.* **2008**, *3*, 12–120.
17. Hu, J.-L.; Wang, S.-C. Total-factor energy efficiency of regions in China. *Energy Policy* **2006**, *34*, 3206–3217. [CrossRef]
18. Aigner, D.; Lovell, C.; Schmidt, P. Formulation and estimation of stochastic frontier production functions models. *J. Econom.* **1977**, *6*, 21–37. [CrossRef]
19. Meeusen, W.; Broeck, J.V.D. Efficiency Estimation from Cobb-Douglas Production Functions with Composed Error. *Int. Econ. Rev.* **1977**, *18*, 435–444. [CrossRef]
20. Jondrow, J.; Lovell, C.K.; Materov, I.S.; Schmidt, P. On the estimation of technical inefficiency in the stochastic frontier production function model. *J. Econom.* **1982**, *19*, 233–238. [CrossRef]
21. Schmidt, P.; Sickles, R.C. Production frontier and panel data. *J. Bus. Econ. Stat.* **1984**, *2*, 367–374.

22. Cornwell, C.; Schmidt, P.; Sickles, R.C. Production frontiers with cross-sectional and time-series variation in efficiency levels. *J. Econom.* **1990**, *46*, 185–200. [CrossRef]
23. Farrell, M.J. The Measurement of Productive Efficiency. *J. R. Stat. Soc. Ser. A (General)* **1957**, *120*, 253. [CrossRef]
24. Colombi, R.; Kumbhakar, S.C.; Martini, G.; Vittadini, G. Closed-skew normality in stochastic frontiers with individual effects and long/short-run efficiency. *J. Product. Anal.* **2014**, *42*, 123–136. [CrossRef]
25. Kumbhakar, S.C.; Lien, G.; Hardaker, B. Technical efficiency in competing panel data models: A study of Norwegian grain farming. *J. Product. Anal.* **2014**, *41*, 321–337. [CrossRef]
26. Kumbhakar, S.C.; Wang, H.J.; Horncastle, A.P. *A Practitioner's Guide to Stochastic Frontier Analysis Using Stata*; Cambridge University Press: New York, NY, USA, 2015.
27. Battese, G.; Heshmati, A.; Hjalmarsson, L. Efficiency of labour use in the Swedish banking industry: A stochastic frontier approach. *Empir. Econ.* **2000**, *25*, 623–640. [CrossRef]
28. Kumbhakar, S.C. Production frontiers and panel data and time varying technical efficiency. *J. Econom.* **1990**, *46*, 201–211. [CrossRef]
29. Kumbhakar, S.C.; Heshmati, A. Efficiency Measurement in Swedish Dairy Farms: An Application of Rotating Panel Data, 1976–88. *Am. J. Agric. Econ.* **1995**, *77*, 660–674. [CrossRef]
30. Kumbhakar, S.C.; Hjalmarsson, L. Technical efficiency and technical progress in Swedish dairy farms. In *The Measurement of Productive Efficiency: Techniques and Applications*; Fried, H.O., Schmidt, S., Lovell, C.A.K., Eds.; Oxford University Press: Oxford, UK, 1993; pp. 256–270.
31. Kumbhakar, S.C.; Hjalmarsson, L. Labour-use efficiency in Swedish social insurance offices. *J. Appl. Econom.* **1995**, *10*, 33–47. [CrossRef]
32. Colombi, R.; Martini, G.; Vittadini, G. *A Stochastic Frontier Model with Short-run and Long-run Inefficiency Random Effects*; Department of Economics and Technology Management, Universita di Bergamo: Bergamo, Italy, 2011.
33. Chen, L.; Zhao, G.C. Impact of Local Government Environmental Regulations on Total Factor Energy Efficiency based on the Panel data of Xinjiang. *J. Arid Land Resour. Environ.* 2014. *8*, 7–13.
34. You, J.H.; Gao, Z.G. An Empirical Study of the Impact of Government Environmental Regulations on Energy Efficiency: A Case Study of Xinjiang. *Resour. Sci.* **2013**, *35*, 1211–1219.
35. Zeng, G.X. China's Energy Efficiency, CO2 Analysis of emission reduction potential and influencing factors. *China Environ. Sci.* **2010**, *30*, 1432–1440.
36. Guo, W.; Sun, T.; Zhou, P. Evaluation of Regional Total Factor Energy Efficiency in China and Its Spatial Convergence-Based on Improved Unexpected SBM model. *Syst. Eng.* **2015**, *33*, 70–80.
37. Wang, Q.; Feng, B. Empirical Study on Energy Efficiency in Beijing-Tianjin-Wing Area Considering the Haze Effect. *Arid Land Resour. Environ.* **2015**, *29*, 1–7.
38. Wang, Q.; Feng, B. Analysis of Total Factor Energy Efficiency in China and Its Influencing Factors: Based on the Provincial Level from 2003 to 2010 Board data. *Syst. Eng.-Theory Pract.* **2015**, *35*, 1361–1372.
39. Wu, Q.; Li, X. Study on Industrial Energy Efficiency and Its Influencing Factors in Shandong Province. *China Popul. Resour. Environ.* **2015**, *25*, 114–120.
40. Shen, N. Study on Energy Input, Pollution Output and Spatial Distribution of Energy Efficiency in China. *Financ. Trade Econ.* **2010**, *1*, 107–113.
41. Shi, D. Analysis of Regional Differences in Energy Efficiency and Energy Saving Potential in China. *China Ind. Econ.* **2006**, *10*, 49–58.
42. Chen, J.; Cheng, J.H. Endogenous Innovation, Humanistic Development and China's Energy Efficiency. *China Popul. Resour. Environ.* **2010**, *20*, 57–62.
43. Wang, Q. Energy Efficiency and Its Policies and Technologies (Part 1). *Energy Conserv. Environ. Prot.* **2010**, *6*, 11–14.
44. Dong, K.; Sun, R.-J.; Li, H.; Jiang, H.-D. A review of China's energy consumption structure and outlook based on a long-range energy alternatives modeling tool. *Pet. Sci.* **2017**, *14*, 214–227. [CrossRef]

MDPI
St. Alban-Anlage 66
4052 Basel
Switzerland
Tel. +41 61 683 77 34
Fax +41 61 302 89 18
www.mdpi.com

Energies Editorial Office
E-mail: energies@mdpi.com
www.mdpi.com/journal/energies